KEY THINKERS ON THE ENVIRONMENT

Key Thinkers on the Environment is a unique guide to environmental thinking through the ages. Joy A. Palmer Cooper and David E. Cooper, themselves distinguished authors on environmental matters, have assembled a team of expert contributors to summarize and analyze the thinking of diverse and stimulating figures from around the world and from ancient times to the present day. Among those included are:

- philosophers such as Rousseau, Kant, Spinoza and Heidegger
- activists such as Chico Mendes and Wangari Maathai
- literary giants such as Virgil, Goethe and Wordsworth
- major religious and spiritual figures such as Buddha and St Francis of Assisi
- eminent scientists such as Darwin, Lovelock and E.O. Wilson.

Lucid, scholarly and informative, the essays contained within this volume offer a fascinating overview of humankind's view and understanding of the natural world.

Joy A. Palmer Cooper was formerly Professor of Education and Pro-Vice-Chancellor at Durham University. She is a Vice-President and former National Chair of the National Association for Environmental Education, Chairman of Trustees of the UK Charity Project Sri Lanka and a Trustee of the British and Foreign School Society. She is the author and editor of numerous books and articles on environmental issues and environmental education.

David E. Cooper is Emeritus Professor of Philosophy at Durham University and Secretary of the UK Charity Project Sri Lanka. He has been a Visiting Professor at universities in the USA, Canada, Malta, South Africa, Sri Lanka and Germany. He is the author of many books, including ones in the field of environmental philosophy.

KEY THINKERS ON THE ENVIRONMENT

*Edited by Joy A. Palmer Cooper
and David E. Cooper*

Routledge
Taylor & Francis Group

LONDON AND NEW YORK

First published 2018
by Routledge
2 Park Square, Milton Park, Abingdon, Oxon OX14 4RN

and by Routledge
711 Third Avenue, New York, NY 10017

Routledge is an imprint of the Taylor & Francis Group, an informa business

British Library Cataloguing-in-Publication Data
A catalogue record for this book is available from the British Library

Library of Congress Cataloging-in-Publication Data
A catalog record for this book has been requested

ISBN: 978-1-138-68472-0 (hbk)
ISBN: 978-1-138-68473-7 (pbk)
ISBN: 978-1-315-54365-9 (ebk)

Typeset in Bembo
by Saxon Graphics Ltd, Derby

MIX
Paper from
responsible sources
FSC
www.fsc.org
FSC™ C013985

Printed in the United Kingdom
by Henry Ling Limited

CONTENTS

CONTENTS

ALPHABETICAL LIST
OF CONTENTS

CONTRIBUTORS

Adams, Paul C. is Associate Professor in the Department of Geography and the Environment at the University of Texas at Austin, USA.

Allaby, Michael is an author based in Argyll, Scotland.

Bannon, Bryan E. is Associate Professor of Philosophy and Director of the Environmental Studies and Sustainability Programme at Merrimack College, North Andover, MA, USA.

Barry, John is Professor of Green Political Economy in the School of Politics, International Studies and Philosophy, the Queen's University of Belfast, Northern Ireland.

Barsam, Ara received his Doctorate in Theology at the University of Oxford and is currently leading a large international education project in Thailand.

Bekoff, Marc is Professor Emeritus of Ecology and Evolutionary Biology at the University of Colorado, Boulder, USA.

Bilimoria, Purushottama is Senior Lecturer at Graduate Theological Union at the University of California, Berkeley; Honorary Research Professor of Philosophy and Comparative Studies at Deakin University, Australia; and Senior Fellow at the University of Melbourne, Australia.

Birns, Nicholas is Associate Professor in the School of Professional Studies, New York University, USA.

Brady, Emily is Professor of Environment and Philosophy at the University of Edinburgh, UK.

Brentari, Carlo is Research Fellow at the Department of Humanities, the University of Trento, Italy.

Callicott, J. Baird is retired University Distinguished Research Professor and Regent's Professor of Philosophy at the University of North Texas, USA.

Casal, Paula is Senior Professor at ICREA (Catalan Institution for Research and Advanced Studies) and also teaches at Pompeu Fabra University, Barcelona, Spain.

Chryssavgis, John taught in Sydney and Boston; he is now theological advisor to Ecumenical Patriarch Bartholomew on environmental issues.

Coletta, W. John is Professor of English at the University of Wisconsin–Stevens Point, USA.

Cooper, David E. is Emeritus Professor of Philosophy at Durham University, UK.

Cooper, Joy A. Palmer is former Professor of Education and Pro-Vice-Chancellor of Durham University, UK.

Corcoran, Peter Blaze is Professor of Environmental Studies and Environmental Education and Director of the Center for Environmental and Sustainability Education at the Florida Gulf Coast University, USA.

Dreier, Peter is Dr E.P. Clapp Distinguished Professor of Politics and Chair of the Urban and Environmental Policy Department at Occidental College, Los Angeles, USA.

Dumble, Lynette J. is Founder and International Director of the Global Sisterhood Network and Associate of the Gramya Resource Centre for Women, Hyderabad, India.

Gates, Phillip J. is an author and naturalist and Visitor in the Department of Biosciences at Durham University, UK.

Gill, Erin is Business Development, Marketing and Communications Manager with Arup in the UK and an independent researcher.

Glotfelty, Cheryll is Professor of Literature and Environment in the Department of English at the University of Nevada, Reno, USA.

Graves, Gregory is Lecturer in Environmental History at the University of California, Santa Barbara, USA.

Griffin, Nicholas is Canada Research Professor in Philosophy at McMaster University, Ontario, Canada.

Hardie, Philip R. is Senior Research Fellow and Honorary Professor of Latin at Trinity College, the University of Cambridge, UK.

Hart, Thomas E. is Adjunct Assistant Professor of Philosophy at Ryerson University, Toronto, Canada.

Holland, Alan is Emeritus Professor of Philosophy at Lancaster University, UK.

Jaeger, Peter is Professor of Poetics, Department of English and Creative Writing, Roehampton University, London, UK.

James, Simon P. is Reader in the Department of Philosophy at Durham University, UK.

Knights, Paul is British Academy Postdoctoral Fellow at Manchester University, UK.

Krueger, Frederick W. is Executive Coordinator for the National Religious Coalition on Creation Care (NRCCC) while facilitating the ministry of Orthodox Fellowship of the Transfiguration (OFT).

Kumar, Satish is Visiting Fellow at Schumacher College, Dartington, Devon, UK, and Editor of *Resurgence* magazine.

Linzey, Andrew is Director of the Oxford Centre for Animal Ethics and a Member of the Faculty of Theology at the University of Oxford, UK.

MacDonald, Mia is the founder of Brighter Green, a New York based public policy NGO, working at the intersection of environmental, animal and global sustainability issues.

MacDonald, Paul S. was formerly a senior academic in the Department of Philosophy at Murdoch University, Australia.

McCarter, Robert is an author and Ruth and Norman Moore Professor of Architecture in the Graduate School of Architecture and Urban Design at Washington University in St Louis, USA.

McDowell, Michael is co-editor of *Windfall: A Journal of Poetry and Place* and former Instructor in English at Portland Community College, Oregon, USA.

McKinnell, Elizabeth is Teaching Fellow in Philosophy, Durham University, UK.

Moss, Ann is Emeritus Professor of French at Durham University, UK, and a Fellow of the British Academy.

Mossley, David J. is an independent philosopher, educator, author and consultant based in the UK.

Nelson, Michael Paul is Ruth H. Spaniol Chair of Renewable Resources and Professor of Environmental Philosophy and Ethics, Department of Forest Ecosystems and Society, Oregon State University, Corvallis, Oregon, USA.

Neuteleers, Stijn is a postdoctoral researcher in the Faculty of Philosophy, University of Groningen, Netherlands.

Peters, Jason is Professor of English at Augustana College, Illinois, USA.

Riordan, Colin is President and Vice-Chancellor of Cardiff University, UK.

Robinson-Bertoni, Sarah E. is Lecturer in Religious Studies at Santa Clara University, California, USA.

Rolston III, Holmes is Professor of Philosophy Emeritus and University Distinguished Professor at Colorado State University, USA.

Rupke, Nicolaas A. is Emeritus Professor of the History of Science at Göttingen University, Germany, and Rupert H. Johnson Jr Professor of History at Washington and Lee University, Virginia, USA.

Schnadelbach, R. Terry is Emeritus Professor of Landscape Architecture at the University of Florida, USA.

Sen Gupta, Kalyan is Emeritus Professor of Philosophy at Jadavpur University, Calcutta, India.

Simmons, Ian G. is Emeritus Professor of Geography at Durham University, UK, and a Fellow of the British Academy.

Smith, Richard is Professor of Education at Durham University, UK.

Walls, Laura Dassow is William P. and Hazel B. White Professor of English at the University of Notre Dame, Indiana, USA.

Warner, Diane is Librarian for the James Sowell Family Collection in Literature, Community and the Natural World, at Texas Tech University, Lubbock, Texas, USA.

Weir, Jack is Professor of Philosophy at Morehead State University, Morehead, Kentucky, USA.

Weisbaum, Elli is an Instructor in the Applied Mindfulness and Meditation Certificate Programme at the University of Toronto, Canada, and is a member of the Order of Interbeing, Thich Nhat Hanh's core community.

Yamauchi, T. is Professor Emeritus of Osaka University of Education and Professor of East Osaka Junior College, Japan.

Zhuang, Yue is Senior Lecturer in Chinese at the University of Exeter, UK.

PREFACE

This book is intended to be a valuable resource for readers with an interest in the lives and work of men and women who have decisively contributed to environmental thinking and practice. The entries follow a common format. An opening quotation sets the scene and readers are then provided with an overview of the subject's work and basic biographical information. Each contributor then illuminates the influence and importance of a subject's thinking and, in some cases, actions. At the end of each essay, the author provides information that will lead interested readers into further and more detailed study. First, there are references for the notes to which the numbers in the text refer; second, there is a cross-referencing with other subjects in the book whose thought or influence significantly relates to that of the subject of the entry; third, there is a list of the subject's major writings (where applicable); and finally, there is a list of further reading.

In 2001 Routledge published a volume entitled *Fifty Key Thinkers on the Environment*. The new volume contains most of the entries in that work. These have been amended and updated wherever appropriate. The new work also contains more than twenty further entries. The volume does not pretend to be exhaustive and the selection of a relatively small number of people among the many who have powerfully influenced environmental thought and practice has, of course, been a difficult task. There are seminal figures, such as Jean-Jacques Rousseau, Rachel Carson and Aldo Leopold, whom no volume on traditions of environmental thought could possibly omit. But in the case of figures whose place in these traditions is less well-known, the Editors have had to exercise their judgement.

We have been guided, in part, by the aim of maintaining a number of balances – between thinkers from different historical periods, from different regions of the world, and from different backgrounds and approaches. The figures included range from Confucius and the Buddha to contemporary authors and activists. They come from Europe, the

Americas, Africa, India, East Asia and Australia. There are entries on poets, philosophers, geographers, nature writers, economists, historians and religious teachers, as well as environmental scientists and activists.

Some readers might legitimately have made a different selection of figures. However, they will agree, we hope, that all the figures discussed in the volume are people who have made distinctive and influential contributions – through their teachings, writings or actions – to the ways in which, today, we understand the inter-relationships among human beings, other species, and the natural world.

DAVID E. COOPER and
JOY A. PALMER COOPER

CONFUCIUS 551–479 BCE

[Zen Dian says:] 'In late spring, after the spring clothes have been newly made, I should like, together with five or six adults and six or seven boys, to go bathing in the River Yi and enjoy the breeze on the Rain Altar, and then go home chanting poetry.' Upon hearing this, the Master sighed and said, 'I am all in favour of Dian.'[1]

Confucius was in many regards a surprising person to have so influenced the course of civilization that the cultures of China, Japan and Korea are, even today, characterized as 'Confucian'. Yet his own life was undramatic. Though of noble descent, Confucius – the name is a Latinization of Kong Fuzi (Grand Master Kong) – was born in rather humble circumstances in the state of Lu, today's Shandong. He held a few minor public offices in his early years and travelled from state to state in his late life, hoping to achieve political influence, but in vain. He left no great works of literature, although his name is associated with the editorship of Chinese classics such as the *Book of Changes* and the *Book of Songs*. Yet Confucius did teach, and after his death, he left a small group of devoted disciples. The *Analects* (*Lunyu*, lit. meaning 'selected words'), upon which the present text relies, were written by those disciples and their own followers.

Living in a time when the imperial rule of the West Zhou dynasty was breaking down, Confucius offered a new basis for restoring order, that is, cultivating a moral philosophy of *ren* (humaneness, benevolence, or virtue in its inclusiveness). In more concrete terms, *ren* means the application of love (1.6; 12.22). What has not been stressed, however, is that for Confucius, *ren* was not exclusively among human beings, but was inclusive of myriad things – birds, beasts, mountains and water, or what is known today as nature. *Ren* as love embracing both nature and humanity laid a cornerstone for a Confucian ecology that regards creatures as humans' companions; as emotional engagement, *ren* paved the way for the development of an aesthetic and poetic landscape tradition; carrying social-economic concerns, *ren* stressed the balance between the wellbeing of both nature and humanity. Confucius' view of *ren* as inclusive love for all things is suggested as follows, 'Does Heaven speak? The four seasons pursue their courses, and all things are continually being produced, but does Heaven say anything?' (17.19).

Heaven (or the universe), for Confucius, is the model of greatest virtue as exemplified in its power of giving and maintaining life: 'The great virtue of Heaven and Earth is giving and maintaining life.'[2] Like all early Chinese philosophers, Confucius holds the view that the universe is a spontaneously self-generating life process, or great transformation: 'Production and reproduction is what is called (the process of) change'.[3] But unlike the Daoist philosophers, who maintained a non-interference approach, Confucius considered it a responsibility for humans to assist the organismic process of life-giving and -maintaining, a point that is made clear in *Doctrine of the Mean* (chapter 22), by Zisi (481–402 BCE). To cultivate the virtue of *ren* (or perfect one's nature) is the foundation for such assistance.

When Confucius says, 'Heaven is author of the virtue that is in me' (7.23), he is talking about the virtue of *ren* – humaneness or love derived from the life-giving power of Heaven.[4] Since the Dao of Heaven is life-giving and -maintaining, humaneness or love, as heavenly derived virtue, thus has its core as a sensitivity towards life. In other words, loving all life is the basic content of *ren*. This love or sensitivity is universal – it does not only include men, but all creatures endowed with life by Heaven.

The *Analects* abounds with examples. *Ji* (horses which can run a thousand *li* in a day), Confucius says, should not be praised for their strength, but for their inner qualities (14.33). This shows that Confucius does not treat horses as instruments, but as companions of humans that deserve respect and to be cared for. Female pheasants in the mountains, which rise up (when disturbed by the sight of men) and circle around before alighting (assuming men would not harm them), are praised by Confucius for their timely actions and earned Zilu, Confucius' disciple's cupped hands in respect (10.18). Confucius also admired pine trees and cypresses, for they are the last to lose leaves when the cold season comes (9.28). In Confucius and his disciples' world, we see an origin of the idea of all creatures as companions of human beings, a concept often attributed to the Neo-Confucians in the eleventh century. As companions, they possess admirable inner qualities as well as emotions which correspond to human feelings. For instance, another disciple of Confucius, Zengzi, notes that 'when a bird is about to die its call is mournful, when a man is about to die his words are good' (8.4). Rather than a mere literary instrument, these examples reveal an emotional and affective approach that Confucius and his disciples ascribe to all creatures, or nature: empathy, compassion and admiration were the bonds between humanity and nature, the self and the myriad things.

Previously scholars have often noted this kind of all-embracing approach to nature as something distinctive of the Neo-Confucians influenced by Buddhism (Zhang Zai's [1020–77] notion of 'forming one body with the universe', for example).[5] It is also with Neo-Confucianism that today's scholars address the issue of Confucian ecology.[6] Yet the above reading reveals that *ren* as love towards all things as companions was clearly embodied in Confucius' thought. Moreover, with Confucius, this approach was distinctively aesthetic – an approach that forms an affective attachment between man and nature, thus laying a cornerstone for a distinguished tradition of Chinese landscape poetics: 'The Master said, "The wise delight in water; the benevolent delight in mountains. The wise are active; the benevolent are still. The wise are joyful; the benevolent are long-lived"' (6.23).

A typical reading of the above has invoked Confucius' belief that nature exemplifies and provides us with models of ethical virtues. This is sensible, but nonetheless incomplete, because it fails to appreciate the fundamental fact that there is emotional engagement with nature. The aesthetic delight is not a private sensation of the individual, or an egoistic expanding of the self to include things (because it does not allow for a human-thing dichotomy). But rather, the desire to realize humaneness (*qiu ren*) already assumes that the person is an integral part of nature; as receptivity and true resonance, the person is both responsive and responsible to the cosmos – the active (yang) and the tranquil (yin) – and accordingly perfects his nature. In such emotional engagement, the person cultivates his humaneness through the harmonious blending of inner feelings with the cosmic principle of Dao as exemplified by nature – thus an individual merges into the great Dao of Heaven.

The idea of the self merging into nature was more explicitly demonstrated in the story of Zeng Dian (quoted initially). Dian's aspiration of chanting and dancing on the Rain Altar in accord with the rhythm of the cosmic life-giving force of spring budding resonated strongly with Confucius' thoughts. Such stories initiated the cultivation of an emotional and aesthetic 'feel' for nature – an approach differing from Laozi's metaphysical approach to Dao – which became an eternal inspiration for later poetic and artistic conceptions of Chinese landscapes. Compatible with Zhuangzi and Buddhist intuitive approaches,[7] this Confucian moral and aesthetic attitude towards nature provides the basis for later Confucian scholars being sensitive to the detail of nature's presence (as an embodiment of Dao of Heaven) and their intimate interaction with their natural surroundings (as their engagement with Dao).

What distinguishes Confucius' approach to nature from those of Daoism and Buddhism most strongly is Confucius' insistence on human beings' sociability as well as their responsibility towards society and nature. The Confucian love of creatures (nature) is inseparable from love of people.[8] This essence was articulated by Confucius' eminent follower, Mencius (372–289 BCE), who states that we should be 'affectionate to relatives, benevolent to the people, and loving to creatures [*ren min ai wu*]' (*Mencius*, 7A45). King Xuan of Qi, for example, was sufficiently kind to animals, yet no benefits were extended from this kindness to the people (*M.* 1A7). Such behavior, Mencius criticized, was against the principle of *ren min ai wu.* What Mencius stressed is that love of creatures must be in unity with being benevolent to the people. They are interdependent.

'Loving creatures' (*ai wu*) in social terms means that people obtain necessary resources from nature through agriculture, hunting, as appropriate, so as to maintain a sustainable relationship between man and nature. As examples of 'appropriateness', Confucius earlier 'used a fishing line but not a cable; used a corded arrow but not to shoot at roosting birds' (7.26). Mencius further advised, 'do not interfere with the seasons of husbandry', 'do not use close nets in the pools and ponds', 'axes and bills enter the hills and forests only at the proper time' (*M.* 1A3).

The ultimate aim of this Confucian notion of 'benevolent to the people, and loving to creatures' is to form an ideal society, in which both people and creatures (nature) can flourish, a vision that is in accord with the Dao of Heaven. To ensure this vision, both government and people should take responsibility rather than leave it alone. Mencius insists on the Confucian idea of the cultivation of humaneness being applied to both the ruler and the people. As the ultimate exemplary person in society, the king should 'dispense benevolent government to the people, being sparing in the use of punishments and fines, and making the taxes and levies light, thus ensuring that the fields shall be ploughed deeply, and the weeding of them be carefully attended to' (*M.* 1A5). This way the king sets up a model of being responsible – for his actions in and with nature – that his people can emulate. Similarly, the people ought to cultivate their own human nature in order to be able to assist Heaven: 'He who has perfected his *xin* (mind-heart) knows his nature. Knowing his nature, he knows Heaven. To preserve one's mind-heart, and nourish one's nature, is the way to serve Heaven' (*M.* 7A1).

For Confucius and Mencius, cultivating humaneness then is not only about ensuring social harmony, but it is also fundamental for attending to the wellbeing of both people and nature: 'If it [nature/human nature] receives its proper nourishment, there is nothing which will not grow. If it loses its proper nourishment, there is nothing which will not decay away' (*M.* 6A8).

The emotional engagement and aesthetic approach to nature, discussed above, is by no means separate from the socio-economic discourse of being 'benevolent to the people, and loving to creatures'. As Confucian thought was adopted as official ideology since the Han dynasty, *ren*, the Confucian criterion of love for people as well as for creatures (nature), was imbued into China's cultural mentality and its socio-politics. Generations of scholar-officials promoted Confucian cultivation of *ren* through both their literary accomplishment and their socio-political actions. Their landscape poems, prose works and inscriptions nurtured affective attachment with nature among both elites and the masses. Their dedicated engagement with agriculture, water management, flood control as well as in cities, villages and infrastructure construction, aspiring to follow Dao of Heaven, maintained a surprising balance between cultural prosperity and natural resources in China for the long span of over two thousand years.

Notes

1   11.26, Lau, *The Analects*, p. 111. References to the *Analects* in the text are to the chapters and sections into which translators standardly divide the text. My citations are primarily from Lau.
2   Xi Ci xia (Commentary II), *Book of Changes*, trans. James Legge, http://ctext.org/book-of-changes/xi-ci-shang
3   Xi Ci shang (Commentary I), Ibid.
4   A number of Chinese scholars made similar observations on this; see, for example, Meng Peiyuan, 'Cong kongzi sixiang kan zhongguo de shengtai wenhua' (Chinese ecological culture from the perspective of Confucius' thought), in *Zhongguo wenhua yanjiu*, winter (2005), p. 7.
5   Tu Weiming, 'The Continuity of Being', in Tucker and Berthrong, *Confucianism and Ecology*, p. 113.
6   See Rodney L. Taylor, 'Companionship with the World: Roots and Branches of a Confucian Ecology', in Tucker and Berthrong, *Confucianism and Ecology*, p. 43.
7   See David Edward Shaner, 'The Japanese Experience of Nature', in Callicott and Ames, *Nature in Asian Traditions of Thought*, esp. 175–6.
8   *Mencius*, trans. James Legge, http://ctext.org/mengzi

See also in this book
Buddha, Wang Yang-ming, Zhuangzi

Confucius' major writings
Lau, D.C. *The Analects*, London: Penguin Books, 1979.
Waley, Arthur. *The Analects*, London: Everyman's Library, 2000 (Orig. Publ. George Allen & Unwin, 1938).

Further reading
Callicott, J. Baird and Ames, R.T. (eds), *Nature in Asian Traditions of Thought: Essays in Environmental Philosophy*, Albany, NY: SUNY, 1989.
Tucker, Mary Evelyn and Berthrong, J. (eds), *Confucianism and Ecology: The Interrelation of Heaven, Earth, and Humans*, Harvard, MA: Harvard University Press, 1998.

YUE ZHUANG

BUDDHA fifth century BCE

> How astonishing it is, that a man should be so evil as to break a
> branch off the tree, after eating his fill.[1]

Born Siddharta Gotama into a royal family in northern India, *c.* fifth
century BCE, the young prince was overwhelmed by the universality of
suffering, old age, illness and death that he witnessed whenever he was
allowed outside the palace gates. He took early to a life of contemplation,
meditation, austerity and simple living so as to fathom the riddle of life
and death, and to resolve his insufferable despair over the endless,
meaningless cycle of re-death and continual rebirth, until he attained
enlightenment (*nirvana*). The natural settings surrounding Buddha's
whole life appeared to have inspired, if not Buddha's own thinking
directly, the imagery attributed to the sequence of events leading to his
enlightenment. It has been remarked that 'the Buddha Gotama was born,
attained enlightenment, and died under trees'. What textual records we
have, furthermore, testify to 'the importance of forests, not only as an
environment preferred for spiritual practices such as meditation but also
as a place where laity sought instruction'.[2]

> [So said the Buddha] ... Seeking the supreme state of sublime peace,
> I wandered ... until ... I saw a delightful forest, so I sat down
> thinking, 'Indeed, this is an appropriate place to strive for the
> ultimate realization of ... Nirvana.'[3]

Gotama was likely reacting to rapid commercial urbanization and the rise
of merchant and artisan classes in his region, and a concomitant agrarian
economy responsible for the deforestation of the Ganges region and
consequent vanishing of animal life from its natural habitat.

In Buddha's collected sermons there are compassionate calls to show
due care and loving kindness towards all sentient creatures. Birds and
animals bear witness to the Buddha's testimony, and they also become
dialogic partners in the ensuing discourses. 'The Buddha Among the
Birds' is only one of the 550 stories from the Jataka tradition that narrates
Buddha's life among animals, and there are stories that recall Buddha's
experiences *as* an animal in his former births. It would seem that the
Buddha was reevaluating the human–cosmos relationship prevalent in the
Indic civilization since the arrival of the Vedic Aryans with their proclivity
towards sacrifice, exploitation of animals for agriculture and warfare, and
subservience to a Brahmanic pan-naturalism, with its ingrained fear of

nature. Buddha succeeded in shifting perception from one of fearful warring nature-forces to that of the benign disposition of nature.

The Buddha interacted in deep empathy with people from all stratas of life, including the settled merchant classes and trading groups travelling to the region, and from his reflections developed a form of social ethics which he practised and preached. These teachings were handed down and later recorded in the Pali canons, brought together into 'three baskets'. The coded teachings of the enlightened one (or 'buddha') on a broad ethical paradigm that connected with the path of liberation from suffering, despite their heavy emphasis on ascetic life (i.e. renunciation or withdrawal from society), contain innovative and vital knowledge about Buddha's thinking on the environment. One insight that is nowadays seen as holding a key to the growth of Buddhist ecological consciousness over the course of two millennia and across Asia is that of 'dependent arising' (*pratitya-samutpada*): 'on the arising of this, that arises'. The causal principle of interdependence registers an ecological vision that, as a recent scholar aptly put it, 'integrates all aspects of the ecosphere – particular individuals and general species – in terms of the principle of mutual codependence'.[4] The relational model undermines the sovereignty and presumed autonomy of the self over other beings and creatures (animals or plants). The ideals of *dharma* and virtues developed in accordance with this insight have been topics of intense reflection and debate among Buddhist schools, and have also been implemented at different historical junctures, such as by Emperor Ashoka after his conversion to Buddhism. He institutionalized care and welfare towards animals, as the following edict poignantly records for us:

> Here no animal is to be killed for sacrifice ... the Beloved of the Gods has provided medicines for man and beast ... medicinal plants ... [R]oots and fruits have also been sent where they did not grow and have been planted along the roads for use of man and beast.[5]

Another side of the causal principle of interdependence is the consequent or karmic continuum, which suggests that every action conditions a being's personal history of suffering, the cessation thereof and subsequent liberation from the karmic continuum: 'on the cessation of this, that ceases'. From the particularity of individual suffering (karmic action-effect), the Buddha was able to generalize to humankind, the animal world and natural environment themselves as distinctive manifestations of the cumulative effect of karmic conditioning. He eschewed any hierarchical dominance of one order of being over the other. A social and ecological ethic (*dharma*) based on undoing the cyclical and all-devouring chain of karmic effects or conditionings was the primary goal of Buddha. His followers applied the teachings in several different directions. This is borne out in the Buddha's expectation that monks and lay Buddhists alike ought to strive always for the 'welfare of the many', 'the happiness of the many', 'compassion for the world'.[6]

Buddha's teachings, however, did not separate out a unidimensional emphasis on environmental ethics from an ontology and ethics of spiritual transformation or sacred-making *dharma* of the human and natural worlds alike. It has been argued that ontological notions such as Buddha-nature or Dharma-nature provide a basis for unifying all existent entities in a common sacred universe, even though the tradition has come to privilege human life vis-à-vis spiritual realization.[7] In other words, Buddhism underscores the inherent moral worth and 'considerability', in principle at least, of all beings towards which there are certain mutual and reciprocal obligations. We might not ordinarily consider the humble gurgling stream as having any particular obligation towards human beings, but the small schools of fish might be very appreciative of the sustenance and safe ecosystem that the cool water provides for them. For contemporary Buddhists this qualification has become even more urgent, given that some of the kinds and patterns of disjuncture of human–earth relationships that we face nowadays did not exist and might not even have been foreseen by Gotama in his despondent wanderings through the comparatively less disruptive urban environment of his day.

While Buddha may have realized the diversity and interconnectedness of the biocommunity, his worldview was neither entirely naturalistic nor as biocentric as Buddhism in its different forms has sometimes become. In this context, while the relevance and role of the environment is recognized in the ecology of individual movement towards Nirvana, the blurring of individual autonomy and particularity necessary for an ethic of duties, rights and the legal protection of minorities and endangered species, weakens the empowering strength needed for a balanced ethic of poly-ecoism. The Buddha's refusal to prescribe unqualified vegetarianism, it is often argued, is indicative of such a weak link in the Buddha's otherwise noble and promising prolegomenon for all future environmental ethics.[8] Nevertheless, the Buddha's plea for compassion for all life forms in their mutual interdependency and the aestheticization of nature that undergirds his wisdom-teachings, paved the way for a radical transformation of attitude towards nature in regions to which Buddhism travelled. For instance, the Dalai Lama is an ardent advocate of environmental compassion and an ethic of universal responsibility, which he sees very much lacking in the present, modern-day hectic world. Again, the Vietnamese monk Thich Nhat Hanh has evolved another strand of wisdom-concentration as a necessary ingredient in the development of a sustainable natural habitat for humans and natural beings alike.

More generally, despite the predominance of non-vegetarianism in Buddhist communities, the rights and ethical protection of certain liberties of animals have been recognized in Buddhism. Many Buddhist monasteries across east Asia have banned the cooking of animal flesh as this involves the killing of animals, whether or not the direct intention is the act of consumption at the dinner table. Buddhist environmentalists are active in modern-day Sri Lanka in their efforts to preserve the lush beauty of the island state from despoilment through extensive technological development and the ravages of an ethnic war. They too can be said to be continuing a practical environmental ethics fostered centuries ago when Buddhism was brought to Sri Lanka.

Likewise, the arrival of Buddhism in Tibet in the seventh century engendered a nationwide programme for the preservation of the heavenly-natural oasis that remained a mysterious land for much of the outside world. The ruling Lamas proscribed injuring and killing of animals, big and small. The moral practice of showing respect for and responsibly using nature became a way of life for the Tibetans. Even though Tibetan Buddhist metaphysics continued the influential Indian Buddhist doctrine of the absence of self-nature or intrinsic existence of properties and substances alike, proclaiming this 'emptiness' of all things has strengthened its moral framework on three counts.

1   Moral properties such as those of the good, compassion, and loving kindness or respect, though by no means absolute, have a solid presence (contingently supervenient on 'emptiness'), in as much as human interaction or ethical life generally presupposes these properties.
2   A pluralistic ontology that has fair regard for members within it, without privileging any particular species, easily gets translated into a non-anthropocentric respect for biodiversity.
3   The religious-soteriological 'end' requires certain self-motivated ethical practices and norms, including restraint on desires, meditation on the limits of the ego-self, altruism based on the moral properties of reverence and deep (but not condescending) compassion for all living and non-sentient beings. In other words, the normative constructs for monks, nuns, lay people, farmers and nomads too, underscore concern for the environment.

The Buddhist ethic of living in harmony with the earth accordingly pervaded all aspects of Tibetan culture. Perched on the 'roof-top' of the world, Tibet's environment was recognized as being crucial to the stability of ecological environs and crop cycles in much of neighbouring Asia. For instance, the ten or so major rivers that wind through Asia feed off the river valleys and smooth glacial icescapes of Tibet, and the monsoon relies on Tibet's abundant natural vegetation and dense forests. Its wildlife and natural animal sanctuaries maintained an equilibrium and contributed in different ways to the enrichment of the environment, providing manure for controlled husbandry and organic re-vegetation, as well as fuel (from yak dung), and so on. However, after the Chinese occupation of Tibet, the situation has dramatically altered: massive deforestation, land erosion, pollution of rivers, depletion of resources, excessive killing of animals and general degradation of the environment have had an adverse environmental impact on South East Asia, which has been subjected to uncontrollable flooding from the monsoon deluge each year. Buddhists are also concerned that the construction of the world's largest dam on the Yangtze river in China will add to the eco-disequilibrium being visited upon much of Asia by China's modernist ambitions and imperiousness.

Transcending the human-centric (ego-bounded) perspective is one of the great strengths of Buddha's interdependent or interconnected vision of all things within the natural–human–social matrix. As de Silva puts it: 'The Buddhist environmental philosophy may be described as a shift from an egocentric stance towards an ecocentric orientation'.[9] The key ontological and moral concepts that help ground Buddha's ecological thinking comprise:

- *pratitya-samutpada* – interdependent conditioning;
- *karma (Pali kamma)* – the law of moral causation;
- *duhkha (Pali dukka)* – unsatisfactoriness;
- *dharma (Pali dhamma)* – reciprocity of obligations or rights within the bounds of duties;
- *sila* – cultivation of virtues, disciplines, and the overcoming of vices. Among the virtues highlighted are: restraint, simplicity, loving-kindness, compassion, equanimity, patience, wisdom, non-injury, and generosity.

These concepts rest on: (1) a general principle of consequentialism (the gravity and impact of one's actions judged by their consequences), moderated by (2) a teleology (a larger purpose or particularity of ends towards which each species strives even in the apparent absence of agency – hence the mountain having its own silent *telos*), and (3) a deontology (*dharma* for *dharma*'s sake), intended as a check against excessive altruism, unmitigated utilitarianism, and ritualized narcissism or a 'grand narrative' teleology.

Notes

1 *Anguttara Nikaya, vol.* III, p. 262.
2 Lewis Lancaster, 'Buddhism and Ecology: Collective Cultural Perception', in M.E. Tucker and D. Williams (eds), *Buddhism and Ecology*, p. 11.
3 *Ariyapariyesana Sutra, Majjhima Nikaya*, cited in Donald K. Swearer, 'Buddhism and Ecology: Challenge and Promise', *Earth Ethics*, Fall, pp. 19–22, p. 21, 1998.
4 Ibid.
5 *Sources of Indian Tradition, vol.* I, rev. edn by A.T. Embree, New York: Columbia University Press, pp. 144–5, 1988.
6 *Middle Length Sayings*, cited in Padmasiri de Silva, *Environmental Philosophy and Ethics in Buddhism*, New York: St Martin's Press; London: Macmillan Press, p. 31, 1998.
7 Swearer, op. cit., p. 20.
8 P. Bilimoria, 'Of Suffering and Sentience: The Case of Animals (revised)', in H. Odera Oruka (ed.), *Philosophy, Humanity and Ecology: Philosophy of Nature and Environmental Ethics*, Kenya: African Centre for Technology Studies, pp. 329–44, 1994; for a humorous counter-argument on 'me-eat', see, Arindam Chakrabarti, 'Meat and Morality in the Mahabharata', *Studies in Humanities and Social Sciences*, III (2), pp. 259–68, 1996.
9 De Silva, op. cit., p. 31.

See also in this book
Bashō, Cage, Gandhi, Nhat Hanh, Schumacher

Buddha's major writing
Samyutta Nikaya, ed. L. Freer, London: Pali Text Society, 1884–1904.
Anguttara Nikaya, eds H. Morris and H. Hardy, vols I–V, London: Pali Text Society, 1885–1900.
Digha Nikaya, ed. T.W. Rhys Davids, London: Pali Text Society, 1920–1.
Gradual Sayings, vols I, II, V, trans. F.L. Woodward, London: Pali Text Society, 1932–6.
Suttanipatta, ed. T.W. Rhys Davids, London: Pali Text Society, 1948.
Dialogues of the Buddha, parts I, II, trans. T.W. Rhys Davids and C.A.F. Rhys Davids, London: Pali Text Society, 1956–7.

Further reading
Bilimoria, P., 'Duhkha & Karma: The Problem of Evil and God's Omnipotence', in *Sophia, International Journal for Philosophical Theology and Cross-cultural Philosophy of Religion*, 34 (1), 1995.
Bilimoria, P., Prabhu, J., and Sharma, R. (eds), *Indian Ethics Classical and Contemporary*, Volume I. Aldershot, UK: Ashgate Publishers (2007); Delhi: Oxford University Press, 2008.

Callicott, J.B. and Ames, R. (eds), *Nature in Asian Traditions of Thought: Essays in Environmental Philosophy*, Albany, NY: State University of New York Press, 1989.

Chapple, C.K., *Nonviolence to Animals, Earth, and Self in Asian Traditions*. Albany, NY: State University of New York Press, 1993.

Gross, Rita, 'Toward a Buddhist Environmental Ethic', *Journal of the American Academy of Religion*, 65 (2), 1997.

Tucker, M.E. and Williams, D. (eds), *Buddhism and Ecology: The Interconnection of Dharma and Deeds*, Cambridge, MA: Harvard Center for the Study of World Religions, 1997.

PURUSHOTTAMA BILIMORIA

ZHUANGZI *c.*370–*c.*286 BCE

All the fish needs is to get lost in water. All man needs is to get lost in Dao.[1]

The two most famous and enduring works of Daoism (or Taoism), both composed during the classical period of Chinese thought (*c.*500–200 BCE), are the *Daodejing* (or *Tao Te Ching*) and the Book of *Zhuangzi* (or *Chuang Tzu*). They are works, moreover, in which subsequent generations, right down to the present, have claimed to find an enlightened attitude towards the natural world, a 'doctrine of harmony with the natural environment'.[2] Traditionally the *Daodejing* was attributed to one Laozi (Lao Tzu) or Lao Tan, supposedly a contemporary of Confucius (sixth–fifth century BCE), and the *Zhuangzi* to a later Daoist thinker. Modern scholars, however, favour the view that the latter work was the earlier, with the *Daodejing* being a third-century BCE compilation by unknown authors who, in a manner then familiar, annexed their thoughts to the name of an ancient, and perhaps mythical, sage.[3]

Unlike Laozi, the actual existence of the reputed and eponymous author of the *Zhuangzi* is reasonably well attested. A later chronicle asserts that Zhuang Zhou or Zhuangzi (Master Zhuang) was an official in a lacquer garden in present-day Honan and that he refused higher royal office on the grounds that he would prefer to live like an ordinary tortoise, free to 'drag its tail in the mud', than to live artificially like the ones pampered at court (17). Other anecdotes in the Book suggest that he was an engaging, genial and ironic individualist, with scant respect for artifice and convention, especially for the Confucian rites of burial. We find him, from his deathbed, chiding his disciples for preparing a 'sumptuous burial' for him (32).

Although the Book is traditionally attributed to Zhuangzi, it is now accepted that he wrote only some of its thirty-three chapters, no more, perhaps, than the so-called 'inner chapters' (1–7), and that other chapters were later assembled by thinkers of his 'school'.[4] Like one of his translators, then, 'when I speak of Zhuangzi, I am referring … to the mind, or group of minds, revealed in the text called *Chuang Tzu*', rather than to a specific, historical individual.[5]

The *Zhuangzi* is a classic of what is often labelled 'philosophical Daoism'. The qualifier is important, since the form of Daoism the Book represents needs distinguishing, despite certain affinities, from the organized religion of Daoism that developed after the second century CE. The distance between the two may be gauged from reflecting that whereas Zhuangzi taught calm acceptance of, even indifference to, death, a main ambition of later religious Daoists was the discovery of the elixir of eternal life.

Before trying to characterize philosophical – or Lao-Zhuang – Daoism, three points should be borne in mind. First, the division of classical Chinese philosophers into 'schools' – Daoist, Confucian, Legalist, etc. – was the work of a later taxonomist and encourages exaggeration of the differences between, and the similarities within, these 'schools'. Thus Confucius, though he is often a critical target in the *Zhuangzi*, also figures in the Book as an admired sage. Second, Daoism is not to be distinguished from other 'schools' having a concern for the nature of the Dao (Way, Path), since it was the common and primary concern of Chinese philosophers to determine the proper Way for human beings to follow. Third, while there is significant affinity between the two Daoist classics, their emphases are different. For example, while many of the chapters of the *Daodejing* address the problem of how rulers should govern in turbulent, warring times, the *Zhuangzi* is occupied with how the individual who, like Zhuangzi himself, steers clear of political affairs should live in these, or any other, times.

Distinctive of Daoism is the very general and abstract notion of the Dao which it invokes. For Confucius, the Dao was the proper Way for human beings to live, but for Laozi and Zhuangzi the central notion is that of a 'Great Dao', the 'complete, universal, all-inclusive' Way of the universe itself, a Way with which human lives should harmoniously accord (22). This Way cannot be precisely articulated. In keeping with the famous opening lines of the *Daodejing* – 'The Way that can be told is not the constant Way' – Zhuangzi says that 'The Great Way is not named … If the Way is made clear, it is not the Way' (2). Nevertheless, some things can be said about it, and some lessons for human conduct and the relationship of human beings to the natural world may be drawn.

The Dao is the Way of nature as a whole, so that 'the true man', who is 'lost in Dao', is one who, like a fish, lives 'naturally'. The ills of social and individual life, for Zhuangzi, stem from the fact that, uniquely among living creatures, human beings are able to, and for the most part *do*, live unnaturally. This means, above all, that most people think and act on the basis of artificial distinctions – between good and bad, beautiful and ugly, men and animals, culture and nature, and so on – whose dependence on partial and pragmatic perspectives they fail to recognize, treating them instead as rigid and as mirrors of reality itself. 'Those who discriminate fail to see' (2), for the Dao itself, as the *source* of all differences and distinctions in the world, is itself seamless and fluid. 'That which makes beings beings is not separated from them by borders' (22).

Moreover, as the Way of nature *as a whole*, the Way is 'spontaneous' (*ziran*) and 'free'. It is not constrained by any aim or purpose and faces no obstacles which it must endeavour to overcome and by which it is limited. Likewise, therefore, 'the true man' will behave spontaneously: indeed, his life, like that of the Dao itself, will be one of 'non-action' (*wu wei*), in the sense of being non-deliberative and non-striving. Many of the most attractive stories in the *Zhuangzi* are in praise of skilled craftsmen – like the butcher whose carving effortlessly responds to the natural grain and joints of the ox (3) – who dispense with rules and verbal instructions in favour of a spontaneous, wordless 'know-how'. What is true of the butcher - or the swimmer who simply 'follows the course of the water itself' (19) – is also true of the Daoist sage. Ignoring the doctrines and principles taught by philosophers – 'the dregs of the men of old' – the sage lives in intuitive appreciation of the Way, recognizing that 'he who knows does not say; he who says, does not know' (13).

To live 'naturally', 'lost in Dao', is to act spontaneously, flexibly and intuitively, without rigid attachment to conventional rules and distinctions, linguistic, moral or other. The 'true man', as it were, 'hangs loose'. This does not mean that he acts on whims or caprices, but that his actions are appropriate responses to calm, mindful attention to how things themselves are. Nor does it mean people should try to go 'back to nature', to live like cavemen or 'wild men'. In the so-called 'primitivist' chapters (8–12) of the Book – though they are not written by Zhuangzi himself – there are passages which advocate a life of extreme simplicity. These have encouraged the idea that, in Lin Yutang's words, the person 'with a hidden desire to go about with bare feet goes to Taoism'.[6] But this 'primitivism' is contradicted by other passages, and is anyway of too extreme a kind to be relevant to modern environmental discussion.

Its main relevance to this discussion is found the *Zhuangzi*'s rejection of attitudes which have played pernicious roles in the comportment of human beings towards their natural environments, and in its promotion of attitudes conducive to greater harmony with these environments. To begin with, Zhuangzi is critical of anthropomorphic perspectives on which the lives and the goods of animals and other living beings are judged by criteria applicable only to humans. Several anecdotes highlight the difference between our and an animal's own conception of its good: for example, the one about the marsh pheasant which, though it must struggle to get food and drink, does not therefore desire to live trapped inside a well-provided cage (3). Second, his rejection of rigid distinctions incorporates a criticism of those between humans and other creatures, between 'great' and 'small', which encourage an anthropocentric elevation of humans above the rest of nature. It was said of Zhuangzi that he never ignored other creatures' 'views of right or wrong', and never 'arrogantly separated himself off' from these creatures (33). Looked at 'in the light of the Dao, nothing is best, nothing is worst ... seen in terms of the whole, no one thing stands out as "better"' (17).

Finally, the sage's life of *wu wei* is incompatible with precisely those types of desire and ambition – for profit, esteem, control – which have led many people to exploit nature. These men and women are 'driven', 'penned in by things ... Pitiful, are they not?' (24). For Zhuangzi, the person who has put aside such pernicious attitudes, and in whom, therefore, 'the Dao acts without impediment', will 'harm no other being' (17) – not because he now adheres to some moral principle, but because he now lacks any motivation to cause harm. Indeed, it belongs to the paramount virtue (*de*) of spontaneity that the spontaneous person will respect and foster the spontaneity of all beings, rather as the good gardener will protect her plants so that they can grow naturally and flourish. As the *Daodejing* (Ch 64) put it, people should nourish and 'support the myriad things in their natural condition'.

The two Daoist classics have been enormously influential. For more than two millennia, classical Daoism – despite the entrenchment of Confucian precepts in imperial education and government and, later, Mao Zedong's brutal animosity – has powerfully shaped aspects of Chinese life, not least a sense of intimacy with nature attested to in Daoist landscape painting and poetry. No less important was its impact upon religious developments in China and Japan, not simply on the various sects of institutionalized 'religious' Daoism, but on that intriguing blend of Daoist and Buddhist thought known as Chan – or, in Japan, Zen – Buddhism. The tone of the *haiku* verses of Bashō, discussed elsewhere in this volume, owe as much to the Daoist tradition as to the Buddha.

Over the last century, Daoist philosophy has attracted several Western thinkers, not least on account of its view of human beings' relation to nature – one which, in many Western eyes, favourably contrasts with that of their own societies. Martin Heidegger, arguably the twentieth century's most penetrating critic of technology, once began a translation of the *Daodejing*, and the Daoist influence on his thinking was much larger than his occasional acknowledgements suggest. During the last few decades, moreover, many environmental ethicists, both Chinese and Western, have enthusiastically invoked Daoist ideas.[7]

It can, however, be misleading to speak of Zhuangzi's 'environmental ethic'. The point is not simply that he did not envisage, let alone discuss, the global problems of pollution, species conservation and climate change that preoccupy today's environmental ethicists. More importantly, he would have been without sympathy for the vocabulary and concepts that dominate current moral debate about the environment. We find in the *Zhuangzi* no mention of, say, the 'rights' of animals, our 'obligations' to natural environments, or the 'intrinsic value' of nature. In his view, the need for talk of morality, justice, righteousness, benevolence, or value is a sure sign that people have 'lost the Way' and, therefore, are no longer 'lost *in* the Way' (22). The virtuous Daoist is someone who 'follows along with the way each thing (spontaneously) is of itself, going by whatever *it* affirms, without trying to add' or 'impose' anything (5). The person who, in this way, naturally 'let things be' does not stand in need of moral principles and reminders of rights and duties, for these come into play only when people have forgotten how to be mindful and spontaneous.

Notes
1   Ch. 6. References to the Book of *Zhuangzi* in the text are to the chapters of the work. I have used the various translations listed under 'Major writings'.
2   Fritjof Capra, *The Tao of Physics*, London: Fontana, p. 340, 1983.
3   See A.C. Graham, *Disputers of the Tao*, pp. 215ff.
4   Ibid., pp. 170ff.
5   Watson, *The Complete Works of Chuang Tzu*, p. 3.
6   *My Country and its People*, London: Heinemann, pp. 109–10, 1936.
7   See, eg., The Alliance of Religions and Conservation (ARC) website, www.arcworld.org/faiths.asp?pageID=11, for examples.

See also in this book
Bashō, Confucius, Heidegger

Zhuangzi's major writings
Among the many translations of the *Zhuangzi*, the following are especially reliable and well-annotated:

Graham, A.C., *Chuang-Tzu: The Inner Chapters*, Indianapolis, IN: Hackett, 2001.

Watson, Burton, *The Complete Works of Chuang Tzu*, New York: Columbia University Press, 1968.

Ziporyn, Brook, *Zhuangzi: The Essential Writings with Selections from Traditional Commentaries*, Indianapolis, IN: Hackett, 2009.

Further reading
Cooper, D.E., 'Daoism', in J.B. Callicott and R. Foreman (eds), *Encyclopedia of Environmental Ethics and Philosophy*, Belmont, CA: Wadsworth, 2008.

Cooper, D.E., *Convergence with Nature: A Daoist Perspective*, Totnes, UK: Green Books, 2012.

Girardot, N., Miller, J., and Xiaogan, Liu (eds), *Daoism and Ecology: Ways within a Cosmic Landscape*, Cambridge, MA: Harvard University Press, 2001.

Graham, A.C., *Disputers of the Tao: Philosophical Argument in Ancient China*, La Salle, IL: Open Court, 1989.

Hansen, C., 'Laozi' and 'Zhuangzi', in R. Arrington (ed.), *Blackwell Companion to the Philosophers*, Oxford: Blackwell, 1998.

Ivanhoe, P. (tr.), *The Daodejing of Laozi*, Indianapolis, IN: Hackett, 2002.

DAVID E. COOPER

ARISTOTLE 384–322 BCE

In all natural things there is something wonderful.

Parts of Animals 645[a1]

The Greek philosopher and scientist, Aristotle, was born in Macedon, where his father was physician to the king. The son was himself to enter royal service as tutor of a future and more famous king, Alexander the Great – a grateful pupil, if we credit the story that he instructed his far-flung subjects to provide Aristotle with specimens for his biological research. Much of his life, from 367 to 347 and again from 335 to 322, was spent in Athens, first as pupil and teacher at Plato's Academy and later at the school he himself founded, the Lyceum. Both sojourns were ended by outbursts of anti-Macedonian sentiment in Athens. After the first he lived in Lesbos, where his most important scientific work was done; after the second he moved to Chalcis, where he died a few months later.

Aristotle's life was one of unremitting study, the voluminous writings bequeathed to us forming only some 20 per cent, perhaps, of his original output. He wrote and lectured on an extraordinary range of subjects – including biology, astronomy, logic, metaphysics, ethics, poetics, and politics – as well as compiling massive records of, inter alia, the Pythian and Olympic Games. To say only this, however, underestimates his unique achievement: for, in the case of many subjects, he did not so much contribute to them as *invent* them. In some areas, moreover, such as logic and zoology, the taxonomies and general principles proposed by Aristotle remained almost unchallenged for more than 2,000 years. It is no exaggeration, therefore, to hold that 'an account of Aristotle's intellectual afterlife would be little less than a history of European thought'.[2] To equal his achievement would require someone totally to redraw the map of intellectual enquiry.

Aristotle was not, of course, an environmental scientist or philosopher in the contemporary sense. The 'eco-crises' which have stimulated recent environmental concern were happily unknown in ancient Greece. Indeed, the very concept of 'the environment' was not one available to Aristotle. Nor did he address such issues as our moral obligations to non-human life. It is clear, however – from my opening citation, for example – that Aristotle experienced and urged a profound regard for the living world and, as we shall see, several elements in his thinking prove attractive to contemporary environmental thought.

Some of those elements were integral to his general conception of the natural world, one which remains alive in a way that some of his pioneering studies in zoology and biology no longer do. (One should recall, though, that Aristotle was responsible for the modern notion of species, since it was he who proposed classifying animal kinds by reproductive criteria, rather than on the basis of less explanatory similarities.)

Aristotle divided the domain of enquiry into the theoretical, practical and productive sciences. The first are concerned with obtaining truth for its own sake, something of the first importance given that 'all men by nature desire to know' (*Metaphysics* 980ᵃ); the other two – ethics and poetics, for example – with how people should, respectively, behave and produce things. The theoretical sciences, Aristotle divides into 'theology', 'physics' and mathematics. The former terms are misleading: 'theology', for Aristotle includes logic and metaphysics, while 'physics' is the study of the natural world in general. The distinction between 'physics' and the other theoretical sciences is that it deals with things subject to movement and change, where 'change' includes coming into and passing out of existence.

'Physics', therefore, directly addresses one of the two main questions Aristotle raises in his *Metaphysics*: What are the most basic entities in reality, the 'primary substances', upon which everything else depends?

and What explains regular processes and changes, such as the growth and decay of organisms? The questions are related for Aristotle, since not only is it, in the final analysis, substances which 'become' and change, but we have knowledge of a substance only 'when we have found its primary causes' (*Physics* 184ª).

Aristotle rejected two views of substance or basic reality prevalent in his times: the doctrine that substance was some stuff, 'matter', out of which things are composed, and Plato's theory that what is truly real are immaterial Forms or Ideas of which ordinary things are both products and pale copies. For Aristotle, we should not confuse what something is with what it is made of, while Plato's view, by placing the Forms outside the ordinary world, is therefore 'useless' for explaining how, in that world, there are 'comings-to-be' of things (*Metaphysics* 1033ᵇ). Aristotle's own proposal is that a primary substance is a *unity* of form and matter. It is only through having a certain form that a region of matter constitutes a man, say, and it is this form which provides the essence – the 'what-it-is-to-be-a-...' – of something.

Aristotle connects his concept of form with the question of the causes of change and movement, since a being's form is also 'an end', a final 'cause as that for [the sake of] which' it begins and develops (*Physics* 199a). Achievement of its fully developed form, that is, is part of the explanation of why a plant, say, grows from a seed (which contains the form 'potentially'). Indeed, it is the main part of the explanation, the factor on which the biologist must focus to understand the process of development. Aristotle's notion of 'final causes', his teleology, has been much misunderstood. He did not mean, absurdly, that all living beings intentionally strive to attain their forms; nor, despite a notorious passage in *Politics* where he writes that 'since nature makes nothing ... in vain, ... she has made all animals for the sake of man' (1256ᵇ), is it his considered view that nature is a divine, purposeful intelligence. (That one-off passage is inconsistent with: (1) Aristotle's usual view of 'ends' as internal to, or 'immanent' in, natural beings; (2) his general attitude of admiration for nature in its own right; and (3) his theology, in which God, absorbed in self-contemplation, is unconcerned with creaturely life.) Aristotle's notion is, rather, a 'functionalist' one. Unless we know what something is 'for', what its normal developed state should be – the tree to grow fruit, the duck to live aquatically – we cannot fully understand the changes which we observe it undergoing – the growth of roots or of webbed feet, say.

Although only humans can intentionally aim at their *telos* or end, all living beings, according to Aristotle, have 'soul'. But 'soul', with its connotation of an immaterial homunculus 'inside' the body and surviving it, is a poor translation of the Greek term *psuche* (lit. 'breath'). The 'soul', he writes, is 'the substance *qua* form of a natural body' (*On the Soul* 412a), that which, so to speak, 'holds together' the body in a cohering whole, the 'principle' of its organization. That Aristotle intends something very different from the Christian concept is clear from his view that nutrition and sensation are faculties of the soul. Only at the level of human life is the faculty for rational thought present, and even there it is only the operations of an obscurely described 'active reason' which could intelligibly occur in the absence of body.

Aristotle's picture of the natural world, then, is a graduated one of beings ranging from 'mere things' through plants, the lower and higher animals, to human beings, each with a specific form which both constitutes its essence and plays the crucial role in explaining its behaviour. Despite its variety and complexity, therefore, the natural order is precisely that, an *order* – an interconnected and intelligible whole whose 'excellence', as Aristotle puts it, resembles that of an 'orderly' army, whose individual members, its soldiers, perform their appropriate functions (*Metaphysics* 1075ª).

For many Arab thinkers from the tenth to the thirteenth century and, from the thirteenth century on, for many Christian ones too, Aristotle was 'The Philosopher', 'the master of those who know' (as Dante put it). It would be rash to assume, however, that their 'Aristotelian' conception of the natural world was the Greek's own, for it incorporated Stoic and theological elements foreign to Aristotle's thinking. In particular, the doctrine of 'final causes' was given an 'anthropocentric' twist to make it accord with the conviction that the divinely ordained purpose of each being was to serve the ends of man – a twist encouraged, admittedly, by the *Politics* passage. Later champions of science, such as Francis Bacon and Galileo, who took themselves to be refuting Aristotle, were more often than not only refuting a bowdlerized 'Aristotelianism'. The same is true of nineteenth-century Darwinian critics, although Darwin himself fully recognized Aristotle's contribution, declaring that even the greatest of recent naturalists, such as Linnaeus, were 'mere schoolboys to old Aristotle'.[3] (Incidentally, Aristotle was well aware of the 'Darwinism' of his day. Although he rejected Empedocles' theory of, in effect, random mutation and natural selection, he did so on grounds that many contemporary naturalists take very seriously.)

Aristotle was not, to repeat, an 'environmentalist', but it is not difficult to see why some environmentalists, especially perhaps 'deep ecologists', are attracted to him. To begin with, they share his perception of the natural world as an integrated, continuous whole, without sharp 'breaks', especially between human and non-human life. Second, they can applaud his concept, once properly understood, of the 'end', the 'for-the-sake-of-which', each creature develops: for this favourably contrasts with a purely 'mechanistic' picture of nature which is, in their view, both impoverished and dangerous. Third, they can welcome Aristotle's view that a being's 'end' is 'a good' for it. A tree or a duck whose 'end' is actualized 'flourishes' *qua* the being it is, and to prevent anything from flourishing, Aristotle strongly implies, is to do a wrong.

Finally, some environmental ethicists have looked for inspiration to Aristotle's general approach to ethics, which focuses not on questions about rights and duties, but on ones about the 'excellences' appropriate to a flourishing human life, one which realizes the human *telos*. Such a life is one of 'virtue', and some of the practical virtues Aristotle prescribes, like self-restraint in material pursuits, are ones which surely have application to the treatment of our environments. More important, perhaps, are the implications of that 'intellectual virtue', 'theoretical wisdom', which, for Aristotle, is 'the highest form of activity' (*Nicomachean Ethics* 1177a). The 'contemplation' of 'the loftiest things' in which that wisdom consists is not, for Aristotle, a passionless investigation into truth, but something imbued with a sense of wonder at, and admiration for, the cosmos and its ingredients. The passage from which my opening citation comes – one which 'expresses some of the best in Aristotelian man'[4] – continues with an endorsement of Heraclitus' rebuke to some visitors disappointed to find the sage doing something as mundane as warming himself by a stove. 'There are gods here too', Heraclitus told them. The lesson Aristotle draws is that even the humblest creatures should be contemplated 'without aversion', for in them too 'there is something natural and beautiful' (*Parts of Animals* 645a).

Notes
1 References are to the standard pagination of Aristotle's works, which is retained in all good translations.
2 Jonathan Barnes, *Aristotle*, p. 86.
3 Quoted by W.D. Ross, *Aristotle*, p. 112.
4 Barnes, op. cit., p. 87.

See also in this book
Bacon, Darwin, Humboldt

Aristotle's major writings
The Complete Works of Aristotle, 2 vols, ed. J. Barnes, various translators, Princeton, NJ: Princeton University Press, 1984.

Aristotle: Introductory Readings, ed. T.H. Irwin, Indianapolis, IN: Hackett, 1996. Contains generous excerpts from works cited in this entry.

Further reading

Barnes, J., *Aristotle: A Very Short Introduction*, Oxford: Oxford University Press, 2000.

The Cambridge Companion to Aristotle, ed. J. Barnes, Cambridge: Cambridge University Press, 1995.

Ross, W.D., *Aristotle*, 5th edn, London: Methuen, 1949.

Westra, L. (ed.), *The Greeks and the Environment*, Lanham, MD: Rowman & Littlefield, 1997.

DAVID E. COOPER

VIRGIL 70–19 BCE

> Then there are all the famous cities, laboriously built, all the towns piled up by human hand on sheer rocks, with rivers gliding beneath their ancient walls. Shall I mention the Adriatic and the Tuscan seas? Or the great lakes? – you, Como the greatest lake, and you, lake Garda, whose rising waves imitate the roar of the sea? Or shall I mention the harbours, and the dykes imposed on the Lucrine lake, and the sea crashing out its indignation? ... This same land brings forth from her veins streams of silver and mines of bronze, and pours out floods of gold. This was the land that produced a fierce breed of men, the Marsians and the Sabine race, the Ligurians used to hardship and the Volscians with their javelins, this the land that produced the Decii, the Marii, the great Camilli, and the Scipios toughened in war, and you, greatest Caesar, who now victorious on the furthest shores of Asia drive off the unwarlike Indians from the hills of Rome.
>
> *'Praise of Italy'*, *Georgics*, 2.155–76

The Roman poet Virgil is said to have been born on 15 October 70 BCE in Andes, a village near Mantua; the late-antique writer Macrobius said that he was 'born in the Veneto of country parents and brought up amongst the woods and shrubs'.[1] He was educated in Cremona and Milan before going to Rome; he then for a while became part of an Epicurean community in Naples, a sect which advocated philosophical retreat from urban society and politics. In the late 40s he was writing his first major work, the *Eclogues* (probably published in 39–38 BCE), a book of ten pastoral poems one of whose subjects is the confiscations of land in 42 BCE for the settlement of veterans by Octavian (the future Augustus) after the civil war against the murderers of Julius Caesar. According to the ancient biographical tradition, Virgil's father's farm was one of those confiscated.

Virgil now came into the circle of the literary patron and intimate of Octavian, Maecenas, to whom he dedicated his four-book *Georgics*, in form a didactic poem on farming, probably published in 29 BCE, the year of Octavian's triple triumph two years after the decisive battle of Actium in which Octavian had defeated Antony and Cleopatra and so brought to an end two decades of civil war. During the last ten years of his life, Virgil was working on the *Aeneid*, an epic in the Homeric manner on the wanderings and wars of the Trojan hero Aeneas, archetypal city-founder and ancestor of Augustus (the name taken by Octavian in 27 BCE, when he consolidated his rule in Rome). On returning from a journey to Greece, Virgil died of a fever at Brindisi on 20 September 19 BCE, with the *Aeneid* lacking its final touches. His dying wish that the poem be burned was overruled by Augustus.

Virgil was immediately canonized as the national poet of Rome, and his works, above all the *Aeneid*, became the central classics of the later Western tradition. Indeed the history of the classical tradition could largely be written in terms of a history of the reception of Virgil. In all three of his major works he reveals a deep interest in the individual's relationship to his wider environment and to the natural world. Virgil's complex and sympathetic sensibility has left its mark on the ways in which succeeding generations of Europeans and North Americans have conceptualized and visualized the place of their culture and society in the world. It is difficult to generalize about the nature of this influence, since Virgil was a poet, not a systematic thinker. One of the marks of his greatness as a poet is his openness to the whole range of ancient traditions and attitudes, popular and philosophical, concerning the natural world and man's place in it. Furthermore, attitudes to man and his environment are to an extent determined by the different genres in which Virgil worked.

The three major works form a seemingly inevitable sequence, sometimes viewed in antiquity as reflecting the history of human civilization, from a pastoral to an agricultural to an urban way of life. The *Eclogues* stage a simple form of human society: individual herdsmen bonded to each other by friendship, ideally enjoying a close and unproblematic relationship with the animals they tend and the landscape they inhabit. The *Georgics* deal with the expertise and technology required to farm the land; the relationship of man to nature – animal, mineral, and vegetable – is now as much one of an imperialist and militaristic domination, as of a more collaborative coexistence. The *Aeneid*'s ultimate subject is the foundation of the great city of Rome and of a people whose military machine will conquer the world. But it is also a poem about Italy, and the agricultural societies and landscapes of Italy (like many famous Romans born outside the capital, Virgil felt a loyalty both to Rome and to his home town, Mantua); echoes of the earlier *Eclogues* and *Georgics* are not out of place in a poem written for an urban ruling elite many of whom had working country estates and a real interest in agriculture,[2] and the walls of whose houses were painted with romantic landscapes.[3]

For much of the last two thousand years, the pastoral idea has been a mainly Virgilian tradition. The inaccessibility to Greekless centuries of Virgil's own model, Theocritus' *Bucolics*, obscured the origins of pastoral as a semi-realist, earthy genre. Virgil is often credited with the invention in his *Eclogues* of 'Arcadia', a dream landscape which men attempt to enter either through art or direct manipulation of their physical surroundings.[4] While the world of the *Eclogues* is a more stylized and artificial creation than its Theocritean predecessor, Arcadia is but one of the Virgilian pastoral landscapes, and arguably marginal. The modern notion of an idyllic Arcadia is the product of a Renaissance elaboration of Virgilian hints, above all in the Italian poet Jacopo Sannazaro's *Arcadia*.

Central to Virgil's pastoral vision is a sense of life in harmony with nature, but under threat from disruptions both external, in the shape of civil war and land confiscations, and internal, above all in the form of erotic passion. Perfection may wear either a private face, in the shepherd happy with his girlfriend and whose love songs are echoed back by a sympathetic nature (*Eclogue* 1), or a public face, in the apotheosed hero at whose ascension all nature rejoices (*Eclogue* 5), alluding to the deification of Julius Caesar. The idea that the natural world flourishes or fails in sympathy with the justice or injustice of the city and its rulers is an ancient one, and deeply embedded in the political imagery of all Virgil's works. Virgil is also chiefly responsible for the widespread currency in the Western tradition of the Hesiodic idea of the Golden Age, both as a primitive Eden, but also as a paradise to be regained through the intervention of a salvific ruler (*Eclogue* 4; *Aeneid* 6.791–4).[5] In the *Eclogues*, philosophical (Epicurean and Stoic) notions of a life lived according to nature intersect with a popular moralizing tradition that opposes the simple and contented life of the countryside to the discontented luxury of the city; this complex of ideas also plays an important role in the *Georgics'* advocacy of a virtuous life on the farm, programmatically in the epilogue of the second book, and, in more nuanced forms, in the various pictures of life in primitive Italy in the *Aeneid*.

The last two books of the *Georgics* deal with animals, viewed from two very different perspectives. On the one hand animals are to be exploited without mercy for their utility to mankind. The old horse is to be put away without pity. On the other hand, there is a sustained and often sentimental anthropomorphism in the description of animal behaviour and feeling, which reaches a climax with the instructions on bee-keeping in book four, where the hive is at times a miniature replica of an idealized Roman society. The view that the bees are a uniquely advanced species goes back at least to Aristotle, but in general pagan antiquity, in contrast to the speciesism prevalent in Christian cultures, accommodated a wide range of views favourable, at least in theory, to the claims to respect of the animal kingdom.

History and an acute sense of time enter the landscape in the *Georgics* and the *Aeneid*. The environment bears the traces of the lives of past generations, evoking a patriotic and antiquarian nostalgia, as in the description in the opening quotation of the rivers flowing at the foot of the ancient walls of hill-towns perched on their rocky prominences. The landscape may bring a remote past before the eyes of the present day, but it may also be changed beyond recognition. In *Aeneid* 8 Aeneas visits the site of Rome, hundreds of years before the birth of Romulus and Remus, and is guided by the virtuous Arcadian king who then lived at the place round a settlement of primitive huts set amid scrub and cattle. But the narrator Virgil constantly reminds his contemporary reader of the marble and gilt buildings that now dazzle the eye. There is a typical Roman pride in the staggering growth of their civilization from humble beginnings, but also a nostalgia for a simpler and more virtuous past when luxury held no temptations, and also a half-conscious anxiety that the process might be reversed, and Rome return to the semi-pastoral landscape over which Edward Gibbon would have gazed as he sat on the Capitol while the barefooted friars sang Vespers, and came to the idea of writing *The Decline and Fall of the Roman Empire*.

Virgil's sense of the past merges with a post-classical nostalgia for antiquity in seventeenth- and eighteenth-century landscape painters such as Poussin and Claude Lorrain, both profoundly influenced by Virgil. Claude's paintings were particularly popular in England, where their look and classical associations were imitated in the art of landscape gardening. Some landscape gardens, such as that at Stourhead, were designed to an explicitly Virgilian programme.[6] A Virgilian vision of the world was also transmitted through eighteenth-century descriptive poetry modelled on the *Georgics*, particularly in James Thomson's very popular *The Seasons*.[7]

Virgil's view of man's relationship to his environment is multifaceted. The Roman empire is now the historical realization of a Stoically coloured cosmic sympathy between man and the natural world, now the violent and morally questionable subjugation of indignant peoples and landscapes. Man cuts down the forest to bring the blessings of agriculture, but trees are also sacred living objects that should not be violated. The urban landscape is proof of man's cultural and political progress, but the city is also the scene of the luxurious corruption of a virtuous primitivism. Human science and technology are objects of wonder and admiration, but there is a place for mystery and awe in approaching the secrets of the natural world. One could accuse Virgil of inconsistency, or one could see in him a supremely sensitive commentator on the complexities and dilemmas of an advanced urban civilization. For all the great differences between Rome of the late first century BCE and a post-Christian and high-technology global society of the twenty-first century, some of these complexities are still familiar.

Notes

1 *Saturnalia*, 5.2.1.
2 On the upper-class Roman's relationship to the natural world see G.B. Miles, *Virgil's Georgics: A New Interpretation*, chap. 1.
3 On Roman landscape painting and its cultural contexts see E.W. Leach, *The Rhetoric of Space. Literary and Artistic Representations of Landscape in Republican and Augustan Rome*, Princeton, NJ: Princeton University Press, 1988.
4 B. Snell, 'Arcadia: The Discovery of a Spiritual Landscape', in *The Discovery of the Mind*, Oxford: Blackwell, chap. 13, 1953.
5 P.A. Johnston, *Vergil's Agricultural Golden Age: A Study of the Georgics*, Leiden: Brill, 1980; H. Levin, *The Myth of the Golden Age in the Renaissance*, London: Faber & Faber, 1970.
6 M.J.H. Liversidge, 'Virgil in Art', in C. Martindale (ed.), *The Cambridge Companion to Virgil*, Cambridge: Cambridge University Press, pp. 99–101, 1997.
7 L.P. Wilkinson, *The Georgics of Virgil. A Critical Survey*, Cambridge: Cambridge University Press, pp. 299–305, 1969.

Virgil's major writings
Georgics, trans. L.P. Wilkinson, Harmondsworth: Penguin, 1982.
Eclogues, trans. G. Lee, Harmondsworth: Penguin, 1984.
Aeneid, trans. D. West, Harmondsworth: Penguin, 1991.

Further reading

Jenkyns, R., *Virgil's Experience. Nature and History; Times, Names, and Places*, Oxford: Oxford University Press, 1998.

Miles, G.B., *Virgil's Georgics: A New Interpretation*, Berkeley and Los Angeles, CA, and London: University of California Press, 1980.

Putnam, M.C.J., *Virgil's Pastoral Art: Studies in the Eclogues*, Princeton, NJ: Princeton University Press, 1970.

PHILIP R. HARDIE

SAINT FRANCIS OF ASSISI 1181/2–1226

> When [St Francis] considered the primordial source of all things, he was filled with even more abundant piety, calling creatures no matter how small, by the name of brother and sister because he knew they had the same source as himself.[1]

At first sight, the life of Saint Francis of Assisi presents us with a paradox. On the one hand, Francis is one of the most popular and venerated saints within Christendom. His love and care for creation have become legendary. When Pope John Paul II in 1980 declared Francis Patron Saint of Ecology, he was doing nothing less than acknowledging the universal appeal of his powerful creation-friendly example. Yet, on the other hand, the Christian tradition which canonized him, and which now venerates, lauds and champions him, is the same tradition which – not without justification – has itself been charged with a distinct lack of care for creation, even to the point of being directly responsible for current environmental crises. Understanding this paradox may provide the key both to the life of St Francis and its contemporary eco-relevance.

Although soon swallowed up in legend, basic details of Francis' life are still recoverable. He was born in 1181 or 1182 in Assisi, the son of a wealthy cloth merchant, Peter Bernardone. As a young man, Francis obtained a reputation as a profligate and a squanderer. In 1204, he was ill for a prolonged period, which put an end to his military career. A series of encounters and experiences then drastically changed his life. At the end of 1204 or early 1205, Francis apparently received his first visionary experience. During that same year, he was brought face to face with poverty and suffering through chance encounters with paupers. But it was his meeting with a leper, the most despised and feared of all medieval outcasts, which apparently changed his life.

Much to his father's chagrin, he renounced his early military and commercial ambitions, sold his possessions and embraced a life of poverty. Charged with having brought humiliation on his father's house, he was brought before the episcopal tribunal in 1206, but Bishop Guido II of Assisi befriended him. At San Damiano in about 1206, Francis experienced his famous vision in which a voice called upon him to rebuild the Church. From 1206 to 1208, he restored the chapels of San Pietro and Santa Maria degli Angeli at the Portiuncular while living as a hermit. Around 1209/10 Francis compiled his Rule and sought papal approval. Eager to secure reform of the Church, Pope Innocent III granted Francis an audience and subsequently authorized Francis and his followers as an itinerant preaching order within the Catholic Church. 'The friars' zeal for the proclamation of the Gospel, their highly acclaimed ministry of preaching, their rejection of material possession in imitation of Jesus Christ and their itinerant lifestyle recommended them to Innocent 111.'[2] The community grew and expanded over the following ten years and became an instrument of papal reform of the Church culminating in the decrees of the Fourth Lateran Council in 1215.

From the start, we can see that Francis' work was a licensed reform experiment within the Catholic Church. Although Francis was impeccably loyal to the Church, and especially to the papacy which endorsed him, his unusual status granted him free rein to preach the Gospel in all its radical simplicity as he saw it. It is said that Francis' life was decisively transformed when he attended Mass at the Portiuncular in February 1208 and heard 'the Gospel passage in which the apostles were commissioned to preach'.[3] From the standpoint of ecological theology, there are four aspects of his ministry which deserve particular attention.

The first concerns *simplicity*. As we have seen, Francis caused scandal by his rejection of his father's wealth and by dressing in a threadbare tunic and sandals. This was not affectation. It was an attempt to imitate Jesus in his identification with the poor and outcast. In doing so, Francis lived the notion deeply rooted in the Gospels that material wealth is a handicap to spiritual progress. Unlike most other Christians of his day – including it must be said bishops and priests – who saw no difficulty in the accumulation of riches, Francis saw simplicity of life as a moral requirement of the Gospel. Accordingly, his Rule forbade his friars from eating luxurious food,[4] wearing expensive garments or accumulating money. Simplicity required living as the poorest of the poor and sharing all things in common.

The second concerns *kinship*. Francis took literally the claim that the Gospel should be preached to 'all creation'. As the above lines from his biographer, St Bonaventure, show, Francis celebrated the kinship of all creatures created by the same God and whose Gospel of love extended to the smallest thing, both animate and inanimate, within creation. Fellow creatures are our 'brothers' and 'sisters'. Although such a notion of kinship or cosmic fellowship is implicit in the Gospels, and arguably required by a doctrine of God the Creator, Francis' high regard for creation was – in terms of conventional theology – highly eccentric. Medieval theology saw sharp distinctions between humans and animals and was deeply dualistic in its thinking, making contrasts (as most of the tradition has done) between things earthly and things spiritual. Francis' sense of friendship and kinship with other creatures, while wholly orthodox, was nevertheless deeply counter-cultural.

The third concerns *generosity*. Francis did not just perceive an ontological bond between all creatures by virtue of their common Creator, he sought to manifest that unity through acts of moral generosity. 'He overflowed with the spirit of charity', writes early biographer Thomas of Celano, 'pitying not only men who were suffering need, but even the dumb brutes, reptiles, birds, and other creatures without sensation.'[5] The key to understanding Francis at this point is to be found in his profound sense that humans were called to imitate Christ, hence they were to reflect a Christlike generosity even and especially to the least of all. Innumerable stories of Francis testify to his filial relations with other creatures. He loved even the worm not solely because it reminded him of the saying that 'I am a worm and no man', but primarily because – as Celano put it – 'he glowed with exceeding love ... wherefore he used to pick them up in the way and put them in a safe place, that they might not be crushed by the feet of the passers-by'.[6]

In order to appreciate the radicality of this approach, one has only to contrast it with the thought of Francis' near contemporary, St Thomas Aquinas. For St Thomas, there was an absolute distinction between animals and humans, and humans could have 'no fellowship' with animals because they were non-rational. Although both men were canonized saints and celebrated figures within the Catholic Church, the difference between them is almost total. While Francis accepted that humans had dominion over animals, he interpreted this power Christologically, that is, in terms of service. As Paul Santmire notes, the saint displayed 'a concrete Christocentric devotion [to others] of radical proportions ... He became the Christlike servant of nature'.[7]

The fourth concerns *celebration*. Again in contrast to wholly instrumentalist views of creation as simply here for our use, Francis saw the world of creation as a place of celebration. He took seriously those verses in the Psalms which speak of creatures praising their Creator and saw in all things, even inanimate ones, a response to the love of God. His famous 'Canticle to Brother Sun' is a tremendous theophany of creation in praise of its Creator. Normally viewed as unconscious matter, he sees the sun, moon, wind, water and fire as part of the divine cosmic consciousness. As one commentator observes, '[F]or Francis, what we refer to as "dumb nature" is far from dumb; it is eloquent in singing and testifying to the beauty of its creator.'[8]

The theological significance of Francis' life may be understood as a prefiguring of that state of peaceableness within creation which will finally be accomplished at the end of time. Such eschatological consciousness was prevalent in Francis' time and, as several writers suggest, the saint's anticipation of the immanent consummation of the Kingdom of God led him to live those laws of the coming kingdom – poverty, humility, selfless love, obedience – in this world. As Roger Sorrell explains, '[T]here is no doubt that Francis shared his hagiographers' conceptions [that] ... creatures' responses to him demonstrated their respect for God's servant and the beginning of the restoration of harmony between God, humanity, and the rest of creation.'[9] The accounts of Celano and Bonaventure lend strong support to this view.

For example, Celano believed that when Francis was submitted to Brother Fire and was not injured, 'he had returned [the fire] to primitive innocence [*ad innocentiam primam*], for whom, when he wished it, cruel things were made gentle'.[10] Bonaventure similarly reports, 'so it was that by God's divine power the brute beasts felt drawn towards him and inanimate creation obeyed his will. It seemed as if he had returned to the state of primeval innocence, he was so good, so holy.'[11] If such an eschatological motivation is accepted, Francis' writing and ministry, far from being romantic rhetoric or eccentric practice, is a manifestation in time and space of God's eternal purpose.

Perhaps inevitably, Francis' example has been eclipsed by the centuries of Christian thought and practice which followed. The sharply contrasting approach of St Thomas – in many ways the founding father of modern Roman Catholicism – has been vastly more influential and has ushered in centuries of neglect of, and even callousness towards, the non-human world. Francis is remembered and honoured, and even lip service is paid to his example, and yet he has had little effect on the development of scholastic theology. It must be said that still many Christians, even and especially Franciscans, play down the eco- and animal-friendly dimensions to his ministry.

But there are some signs that increasing dissatisfaction with the instrumentalist and utilitarian attitudes to creation embodied in historical theology are encouraging churchpeople and theologians to re-examine the tradition and rediscover genuine but neglected creation-friendly elements within it – and not least of all, Francis himself. 'St Francis is before us as an example of unalterable meekness and sincere love with regard to irrational beings who make up part of creation', maintained Pope John Paul II in his sermon at Assisi on 12 March 1982. 'We too are called to a similar attitude', he continued. 'Created in the image of God, we must make him present among creatures "as intelligent and noble masters and guardian of nature and not as heedless exploiters and destroyers".'[12]

Notes

1 St Bonaventure, in *The Life of St Francis*, ed. Ewert Cousins, New York: Paulist Press, pp. 254–5, 1978.
2 Michael Robson, *St Francis of Assisi: The Legend and the Life*, p. 90.
3 Ibid., p. xxxi.
4 Francis' vegetarianism is disputed but it is clear that his community followed an ascetical, frugal diet, which very rarely made use of flesh foods.
5 Thomas of Celano, *Vita Prima* 59, in H. Paul Santmire, *The Travail of Nature*, p. 108.
6 Thomas of Celano, *Vita Prima* 59, in Roger Sorrell, *St Francis of Assisi and Nature*, p. 46.
7 Santmire, op. cit., p. 109.
8 David Kinsley, 'Christianity as Ecologically Responsible', in *This Sacred Earth: Religion, Nature, Environment*, ed. Roger Gottlieb, London: Routledge, p. 123, 1996.
9 Sorrell, op. cit., p. 54.
10 Thomas of Celano, *Vita Secunda* 166, in Sorrell, ibid., p. 52.
11 St Bonaventure, *Legenda Minor* 3: 6, in Sorrell, ibid.
12 Pope John Paul II, Message on 'Reconciliation', *L'Osservatore Romano*, 29 March 1982, pp. 8–9. The final two lines are a quote from the previous encyclical, *Redemptor Hominis*, 15.

See also in this book
Bartholomew

St Francis' major writings
The Writings of St Francis, in *The Omnibus of Sources*, ed. Marion Habig, O.F.M., Chicago, IL: The Franciscan Herald Press, 1973. (This volume contains early biographies of St Francis by St Bonaventure and Thomas of Celano.)
The Life of St Francis, ed. Ewert Cousins, *The Classics of Western Spirituality*, New York: Paulist Press, 1978.

Further reading
Armstrong, Edward, *St Francis: Nature Mystic, The Derivation and Significance of the Nature Stories in the Franciscan Legend*, Berkeley, CA: University of California Press, 1973.

Cunningham, Lawrence, *Saint Francis of Assisi*, New York: Harper & Row, 1976.

Cunningham, Lawrence, *Brother Francis: An Anthology of Writings by and about Saint Francis of Assisi*, New York: Harper & Row, 1972.

Linzey, Andrew and Cohn-Sherbok, Dan, *After Noah: Animals and the Liberation of Theology*, London: Mowbray, 1997.

Robson, Michael, *St Francis of Assisi: The Legend and the Life*, London: Geoffrey Chapman, 1997.

Santmire, H. Paul, *The Travail of Nature: The Ambiguous Ecological Promise of Christian Theology*, Minneapolis, MN: Fortress Press, 1985.

Sorrell, Roger, *St Francis of Assisi and Nature: Tradition and Innovation in Western Christian Attitudes toward the Environment*, Oxford: Oxford University Press, 1988.

ANDREW LINZEY AND ARA BARSAM

WANG YANG-MING 1472–1528

Man is the mind of the universe: at bottom Heaven and Earth and all things are my body. Is there any suffering or bitterness of the masses that is not disease and pain in my own body? Those who are not aware of disease and pain in their body are people without the sense of right and wrong. The sense of right and wrong is knowledge possessed by men without deliberation and ability possessed by them without their having acquired it by learning. It is what we call innate knowledge (liang-chih).[1]

Wang Yang-ming (or Wang Shou-jen), the most influential Confucian thinker in the Ming dynasty (1368–1644) in China, was a critical inheritor of the two main tendencies of Neo-Confucianism: that is, the philosophies of Ch'eng I-ch'uan (1033–1107) and Chu Hsi (1130–1200) on the one hand and those of Ch'eng Ming-tao (1032–85) and Lu Shiang-shan (1138–92) on the other. Thus he was thought to be a kind of synthesizer who had perfected the philosophy of Neo-Confucianism. Wang was, in his youth, much influenced by Zen Buddhism and Taoism, though he rejected them later on the grounds that they represent a sort of quietism which escapes from social relationships. Like a Zen master, he was suddenly enlightened at the age of 37, after long years of concentration in thinking in very harsh situations. Once, when he was young, Wang had concentrated his thoughts on the things outside of his mind, because his forerunner Chu Hsi interpreted the important thesis of 'Great Learning' as meaning that 'the investigation of things' will lead to 'the extension of the knowledge', thus rectifying the will. So, Wang concentrated his thought by watching the bamboos in his garden for seven days and became sick. Later he changed his course to concentrate on the inside of the mind. And he attained the idea of 'good knowledge' that combines knowledge with action. If one makes a division between one's mind and things outside, and separates the former from the latter, and if one is concerned only with things outside of one's mind, then that concern will not be combined very easily with one's mind, that is, with one's will to act. He was a keen learner, and on his own wedding day he was so involved in discussion with a Taoist that he forgot to attend the ceremony. He was the ablest general in his age and won a high reputation as general, having quashed many rebellions. Even for him it was not a very easy task to defeat the selfish desire in his mind that beclouded the good knowledge as the Heavenly Principle. He said, 'it is easy to defeat the rebels in the mountains, but it is difficult to defeat the rebels in the mind'. In 1527 he was asked to subjugate a rebellion while he was suffering from a serious disease. After he defeated the rebellion and came back to his home, he died, aged 57; on his deathbed he must have felt that he had defeated his innate enemy, his human desire. His last words were: 'My mind is full of light; I have nothing to say any more.'

The most basic requirement of the moral philosophy of Wang Yang-ming is unity of knowledge and action. For all moral purposes the only thing which needed to be done was to bring forth 'the good knowledge' ('intuitive knowledge', or 'good conscience') of the mind. If one knows that he ought to do something and does not do it, this knowledge, for Wang, means that he does not in fact know. This reminds us of the contemporary theory of what R.M. Hare calls 'prescriptivism' in contrast with 'descriptivism'.[2] The decisive moral question in environmental ethics is not only what is the matter with the environment, but what are we to do. This is one of the reasons why Wang Yang-ming's moral philosophy is most promising when applied to environmental ethics. 'Knowing is the beginning of action, and doing is the completion of knowledge. When one knows how to attain the desired end, though one speaks only of knowing, the doing is already included; likewise, though he may speak only of action, the knowing is also implied.'[3]

Originating from some versions of Confucianism is the popular expression 'Heaven knows'. Heaven was said to watch our good acts as well as our evil acts, even if no one on earth knows. So, if someone has escaped punishment in doing some evil acts, heaven will punish him some day, because heaven was believed to be something that is completely impartial. ('Heaven's vengeance is slow but sure.') People, therefore, are recommended to 'self-care in solitude'. And also when one falls into a difficult situation, not due to their own failure, one could find consolation in such beliefs as: 'Heaven gives one severe trials, before heaven gives him a mission' or

'Sincerity can move heaven'. Wang said once that his philosophy of 'good knowledge' was born from 'a hundred deaths, a thousand difficulties'. Such ways of thinking have certainly helped people under the influence of Confucian culture in enforcing their impartialist morality.

Everyone knows the Confucian golden rule of 'Do not do to others what you do not want them to do to you' (*The Analects*, 12:2, 5:11). This is the spirit of 'jen' ('benevolence' or 'love', usually translated as 'humanity', but jen reaches far beyond humanity to all things). 'Jen' is sometimes juxtaposed with five other virtues (filial piety, loyalty, orderly love among spouses, among brothers, trust between friends). But it is often thought of not only as representative, but as fundamental, in the sense that it forms the basis by which other virtues are justified. If this interpretation is possible, what Confucius wanted to point out was a logical thesis about morality which is, in a sense, shared by recent Western moral philosophers like Hare, Peter Singer and others, who argue that 'universalizability' (but not 'universality' in the sense of 'generality') is the fundamental requirement of a moral judgement. If we interpret 'jen' as something like the utilitarians' 'impartial benevolence', then what is the difference between the two positions?

The difference is this: the utilitarian motto in the old version of the theory is 'the greatest happiness of the greatest number'. The utilitarians expanded our moral concern beyond our species to include the wellbeing of animals (Bentham and Singer), and further expanded our moral concern to future generations. Confronting environmental crisis on a global scale, people have realized that if the environment is endangered, there is no longer any happiness for any being. So an expanded utilitarianism will need to be supported by some kind of eco-holistic view. If we, as moral agents, go one step further, expanding moral subjects further than sentient beings, and include in our moral consideration the natural environment that is relative to human activities, then we will come very close to the position of Wang Yang-ming.

So far we have seen that the two logical requirements of moral judgements (i.e. prescriptivity and universalizability) which are made explicit by Western moral philosophers are already implicit in Wang Yang-ming's philosophy. However, this may seem rough and biased in the eyes of people who are experts in Chinese philosophy. If we further add an eco-holistic view to these two logical requirements, what will happen is as follows.

Wang Yang-ming had critically inherited another thesis of Ch'eng Ming-tao: 'jen is the love of all things in the universe as one body'. This thesis was related to, inherited from, the Buddhist thesis that 'Heaven and Earth have the same roots as myself and all things are one body with me' and the Taoist thesis of Chuang Tzu that 'Heaven and Earth live alongside me and all things are one body with me'. One must understand, Wang said, that 'jen' is the unity of all things. According to Wang, each and every one of us possesses the original mind, which is one with the universe. 'The man of jen regards Heaven and Earth and all things as one body. If a single thing is deprived of its place, it means that my jen is not yet demonstrated to the fullest extent' (89). Thus, jen is not only the basis of human virtue, but is the original principle according to which heaven and earth make everything live. Jen is 'the principle of unceasing production and reproduction. Although it is prevalent and extensive and there is no place where it does not exist, nevertheless there is an order in its operation and growth. That is why it is unceasing in production and reproduction' (93). 'Our nature is the substance of the mind and Heaven is the source of our nature. To exert one's mind to the utmost is the same as fully developing one's nature. Only those who are absolutely sincere can fully develop their nature and know the transformation and nourishing process of Heaven and Earth' (6).

It is in this way – that is, according to the jen, which is innate original knowledge and the principle of the universe at the same time – that a ruler is expected to rule society and the whole country. This is called 'jen-politics' or moral politics. If one somehow unifies oneself with the society that is one body, and if one knows that people are suffering, then this will become enough incentive for one to save people from suffering, because knowledge and action are united.

This element of the social philosophy of Wang Yang-ming, because it is holistic, can be extended to a view of nature. What is most important for environmental ethics is that jen is not only a matter of human concerns.

When one hears the cry of birds and animals, one will have compassion, because the jen is one with the birds and animals. If one says that animals have senses, then one will have compassion when one sees the grasses and trees faded and broken, because the jen is one with the grasses and animals. If you say that grasses and trees are animated beings, then one will regret when one sees tile-stones collapse; this is because the jen is one with tile-stones.

And yet the Grand Master [Confucius] was extremely busy and anxious, as though he were searching for a lost son on the highway, and never sat down long enough to warm his mat. Was he only trying to get people to know him and believe him? It was rather because his jen, which regarded Heaven and Earth and all things as one body, was so compassionate, keen, and sincere that he could not

> stop doing so even if he wanted to … Alas! Aside from those who truly form one body with Heaven and Earth and the myriad things, who can understand the Grand Master's intention?
>
> *(182)*

If the community is a closed one, people tend to see it as the whole and would be able to sacrifice themselves for it. If people could lift their eyes a bit higher and expand their concern to include nature as a whole, they could be prepared to devote their labours to enriching the natural environment. When people were settled and earthbound they knew their survival depended on a sound natural environment. Thus social ethics in the East was severely restricted and shaped by the limits the natural environment imposed upon the society. Yang-ming's eco-holistic tendency was for many years one of the strongest ideological backgrounds of pre-modern Japan, where the natural environment was marvellously enriched and sustainable. Kumazawa Banzan (1619–91), a samurai scholar who belonged to the Yang-ming school, is well known for his ecological policies and achievements.

Yang-ming's social ethics and the vitality of his thoughts inspired samurai revolutionaries when Japan opened its door to the West and caused the Meiji Restoration (1868). They fought for what they thought was the whole, that is, for the country to keep independence, not for the interests of their own class and, after the revolution was achieved, they eliminated their own class. On the other hand, the opinion leaders' ideological model in the period of Japanese modernization, after the long period of being a closed country, was mainly influenced by British utilitarianism. What both the Yang-ming school and utilitarianism could somehow share was their social ethics. Yet the very gap lies in their views on nature. While utilitarian concern focuses on the interests of sentient beings, Yang-ming's concern was with all beings interrelated under heaven. But such eco-holistic views were abolished and instead the Western dualistic tendency was, under the pressure from Western powers, imported and the dominion of nature had prevailed. But an industrial and economic giant means also an environment-degrading monster. The cost of the modernization has not yet been generally noticed. Only a few philosophers have started to take another look at the traditional Confucian views on nature, among them a prominent environmental philosopher, J. Baird Callicott, who classified Confucianism as a form of deep ecology.

Notes
1   Quoted on p. 179. Page references in the text are to *Instructions for Practical Living*.
2   See R.M. Hare, *The Language of Morals*, Oxford: Oxford University Press, 1952.
3   W. Liu, *A Short History of Confucian Philosophy*, p. 171.

See also in this book
Callicott, Confucius, Singer, Zhuangzi

Yang-ming's major writings
Chan, W. (trans.), *Instructions for Practical Living and Other Neo-Confucian Writings by Wang Yang-ming*, New York: Columbia University Press, 1963.
Ching, J. (trans.), *The Philosophical Letters of Wang Yang-ming*, Charleston, SC: University of South Carolina Press, 1972.
Chan, W. (ed. and trans.), *A Source Book in Chinese Philosophy*, Princeton, NJ: Princeton University Press, 1973.

Further reading
Callicott, J.B., *Earth's Insights: A Multicultural Survey of Ecological Ethics from the Mediterranean Basin to the Australian Outback*, Berkeley, CA: University of California Press, 1994.
Chang, C., *Wang Yang-ming: Idealist Philosopher of Sixteenth-Century China*, New York: St John's University Press, 1962.
Liu, W., *A Short History of Confucian Philosophy*, New York: Delta, 1964.
Tu, W., *Confucian Thought: Selfhood as Creative Transformation*, Albany, NY: SUNY Press, 1985.
Yamauchi, T., 'The Confucian Environmental Ethics of Ogyu Sorai', in Callicott & McRae eds., *The Environmental Philosophy in the Asian Traditions of Thought*, 2014, SUNY.
Yamauchi, T., 'The Agricultural Ethics of Ninomiya Sontoku', *Taiwan Journal of East Asian Studies*, 12 (2), 235–57, 2015.

T. YAMAUCHI

MICHEL DE MONTAIGNE 1533–1592

When I am playing with my cat, who can know whether she is not amusing herself with me, rather than I with her?[1]

Montaigne is the most congenial of intellectual companions. He was an exceptionally well-educated member of the local gentry, who spent most of his life on his estate near Bordeaux. A trained lawyer, he served two terms as mayor of Bordeaux and was a minor player on the national political stage, at a period when France was ravaged by religious civil war of unparalleled savagery. In 1571, he vowed to retreat to the 'peace and security' of his library and his own reflections. This retreat was much interrupted by his political responsibilities and by a tour of Italy in 1580–1, but 1571 marked the beginning of the years of study, rumination and writing that resulted in the *Essays*.

Two books of essays were published in 1580, and successive editions made alterations and additions, the most important of which was a third book, appended in 1588. Montaigne was working on the essays up to his death, and a posthumous edition in 1594 contains very extensive insertions into the body of the text. Well before then, Montaigne had discovered the subject of his writing, and that subject was himself. This was a revolution in the history of European thought. No authors prior to Montaigne had made themselves the matter of their book, except to present a partial view of themselves, as examples of God's grace or as witnesses to historical events. Montaigne's *Essays* are loosely structured, but extraordinarily intelligent and critical, ponderings on his own responses to the total diversity of his own experience, his reading, his social interactions, his habits, his environment, his mental cogitations, his sensations and his bodily proclivities. The book found avid readers throughout western Europe. John Florio published it in English in 1613, Francis Bacon's *Essays* could not have been conceived without it. Descartes, Pascal and all the major thinkers of the seventeenth century start from questions raised for them by Montaigne, even though, under the influence of the Scientific Revolution, they came to a very different world-view. The *Confessions* of Jean-Jacques Rousseau, Europe's next great essay in autobiography, is clearly of Montaigne's progeny, but so is the modern preoccupation with the self, and every essay that was ever written. As an environmental thinker (a concept he could not have recognized, though he did think about the natural environment), Montaigne may best be considered under three heads: his reaction to attitudes typical of his social class; his reflections on the 'wild'; and the place of animals in his sceptical account of human pretensions to a superior place in the natural order.

There was no appropriate language already in existence for Montaigne's novel investigation of his own psychology, and Montaigne invented one out of metaphors. One of his favourite metaphors is that of the hunt, a pastime closely linked to the social status of a gentleman. When his essay on Cruelty (Book II, no. 11) gets round to his own sense of what is cruel, he confronts us straightaway with the reality of 'the hare squealing when my hounds get their teeth into it'. As always, Montaigne mistrusts any simple analysis of his behaviour. It may be mere squeamishness, too weak to underpin a moral position. He is fully conscious that his distaste for the spectacle of the hunted beast at bay is likely to be mocked by his peers. Yet, his self-awareness forces him to stand aside from his social group and triggers more general speculation about man's natural propensity to cruelty, about the respect and affection religion and our very humanity enjoin that we give to our fellow creatures, beasts and plants, and about the lessons animals have for our presumption: 'I willingly lay aside that imaginary rule over other creatures that we have been assigned.'

Montaigne's capacity to imagine the other finds its most startling expression in his essay on Cannibals (Book I, no. 31), which also turns on questions of cruelty and presumption. This is one of two essays on the New World in which Montaigne deplores the depredations of the European conquerors with a fierce bitterness rarely articulated before him. His subject is the human inhabitants of South America, not the natural environment, but his basic distinction between the wild and the cultivated, the natural and the artificial, has implications for both. For Montaigne, the 'savagery' of Brazilian cannibals is akin to the vigour and virtue of uncontaminated nature. It is European culture that has corrupted nature by artifice. So, the refined cruelty Europeans inflict on their colonial subjects, and on each other in the name of religion, is a barbarity far in excess of the 'barbarity' we ascribe to cannibals. While condemning cannibalism in absolute terms, Montaigne contextualizes the practice in a description derived from explorers' accounts of a very simple, 'natural' society living in equilibrium with the environment, using its resources without cultivating or altering it. The Brazilians desire nothing beyond what their environment liberally provides, so they have no concept of conquest, of property, of trade or of social division. Their cannibalism derives from a competitive sense of honour, as does their polygamy. Montaigne focuses on these two practices so repugnant to his own culture to demonstrate that they are not alien to nature, that they are conceivable within a different environment and make sense within a different social structure. His capacity to imagine the other is also a tool for attacking Europeans' presumption that they are morally and culturally superior: 'There is nothing barbarous or savage about the Brazilian tribes, except that all call "barbarous" anything they are not used to.' Moreover, this view from the other side can be very disturbing. When his 'savages' visited France, they 'naively' marvelled at the disparity between rich and poor and wondered why the destitute did not 'seize the others by the throat or set fire to their houses'.

Montaigne's most sustained discussion of man's general relationship with other inhabitants of his environment is to be found in the very long Apology for [or, Defence of] Raymond Sebond (Book II, no. 12). It serves as an introduction to a translation Montaigne had been asked to make of Sebond's *Natural Theology*, written in the early fifteenth century. Sebond had argued that truth can be read in the Book of Nature, but only if those who observe nature and interpret it do so in the light of Christian revelation. His subject matter here forces Montaigne to investigate traditional theological attitudes to the natural world and to explore the cosmological, psychological and biological science of his day. The strategy of his essay requires him to be sceptical, for he has chosen to undermine those who object that Sebond's arguments are weak by demonstrating how fallible all human reasoning is. Montaigne accumulates evidence on two counts: first, to show that opinions held about the workings of nature are incoherent and self-contradictory; second, to show what a feeble creature man is, despite his much-lauded faculty of reason on which is founded his presumption to rule the rest of creation. He pursues this second theme through a copious inventory of examples where animals put man to shame, in their ability to communicate, their creative skills, their ingenuity, their powers of deduction, their memory, their moral virtues of fidelity and courage, and many more. Fact and fable are all grist to his mill. Montaigne revels in the literature from which he takes his examples, and that shows him to be a man of his time. In the middle of the sixteenth century lavish books were printed reproducing all that was known about animals, with detailed, realistic illustrations. They adhered to a basic grouping of species, but it is the profusion of animal life, rather than its taxonomy, that still entrances the browsing reader of these works, where the fabulous is interleaved with familiar creatures from the Old World and exotic beasts from the New. Their text is exhaustive about anatomy, habitat, feeding, breeding and so on, but it dwells just as much on references to the animal in historical texts, poems, fables, proverbs, sayings and emblems. Animals are literary and cultural objects, as much as objects of scientific observation. The same could be said about the study of the natural environment outside books. The sixteenth century was the great period of the curiosity cabinet, filled with a heterogeneous collection of animal and mineral objects designed to excite wonder at the uninhibited variety of natural forms rather than to initiate scientific research.

Montaigne's apparently undisciplined gathering of the more amazing feats of animals recounted in books, interspersed with the occasional personal observation, has analogies with the encyclopedism and collecting mania of his contemporaries (and exactly the same appeal as exotic 'wildlife series' on television). There are, however, features of Montaigne's discourse that betray a rather different preoccupation. In emphasizing the role of animals in human culture and in subjecting their remains to the wondering gaze of the possessor of a curiosity cabinet, encyclopedias and collections tended to promote an anthropomorphic view of the animal world just as effectively as the moral and Christological lenses through which their medieval predecessors had read the Book of Nature. Montaigne's purpose is not to show how man can know nature and therefore feel easy with it, but to discomfort man, to show that his claim to be superior to animals is undermined by counter-examples at every turn. He claimed in his essay on Cruelty that humility with respect to the rest of creation is both a proper human attitude and a Christian one, but it does go against the grain of a certain theological attitude that puts man and his immortal soul on a level above all other living things and also identifies the 'bestial' with the degenerate and morally corrupt. Moreover, there runs through Montaigne's catalogue of animal behaviour that same sense of the other that allowed him to grope towards an anthropological understanding of the alien cannibals. He does not rest in a state of wonder, but conceives imaginatively a world where other forms of language operate, other values hold, in which animals have an incomplete sense of what makes humans tick, but no less complete than our insight into them, a world where cats amuse themselves with humans no less than we amuse ourselves with cats.

Montaigne's kind and generous attitude to non-human creatures was the product of a pre-scientific mentality. For him, the natural environment was an array of mobile forms, a playground for his agile mind. When Descartes, in his *Discours de la méthode* of 1637, defined the natural world as the scientific object of man's investigating reason, he constructed it as a machine. Animals for Descartes were functioning mechanisms without thought, language or sensation. Montaigne's respect for animals survived the Scientific Revolution in the fables of La Fontaine. In La Fontaine, too, animals have their mode of communication and teach mankind a lesson.

Note
1   Book II, Essay 12 of *The Complete Essays*. All references to Montaigne in the text are to Book and Essay numbers of this work.

See also in this book
Bacon, Rousseau

Montaigne's major writings
The Complete Essays, trans. M.A. Screech, with good introduction, London: Allen Lane, Penguin Press, 1991.

Further reading

Burke, P., *Montaigne*, Oxford: Oxford University Press, 1981.

Daston, L. and Park, K., *Wonders and the Order of Nature, 1150–1750*, New York: Zone Books, 1998.

Sayce, R.A., *The Essays of Montaigne: A Critical Exploration*, London: Weidenfeld & Nicolson, 1972.

Screech, M.A., *Montaigne and Melancholy: The Wisdom of the Essays*, London: Duckworth, 1983.

ANN MOSS

FRANCIS BACON 1561–1626

Human knowledge and human power meet in one; for where the cause is not known the effect cannot be produced. Nature to be commanded must be obeyed; and that which in contemplation is as the cause is in operation as the rule.[1]

Bacon was a politician, jurist, royal councillor, natural scientist and essay writer, who spent his entire life within the highest political, courtly and intellectual circles around Queen Elizabeth and King James I. His maternal uncle, Lord Burghley, was the most powerful statesman of his age. After Cambridge University and a period in France, he became a lawyer and Member of Parliament. Despite his earlier friendship with the charismatic Earl of Essex, he was active in the prosecution of Essex for treason – a deed which has inspired some, probably unfairly, to accuse Bacon of the worst sort of betrayal. Bacon was also involved in the prosecution of Sir Walter Raleigh, once an associate of Essex.

The first edition of Bacon's *Essays and Counsels* (1591) dates from his early career, and was eventually expanded into a third edition containing fifty-eight essays in 1625. Upon the succession of King James I in 1603, Bacon moved upward in the court hierarchy even more rapidly, eventually becoming, in 1618, Lord Chancellor and Baron Verulam.

Bacon's writings during these politically active years reflect his many interests – in English Law, the Church of England and the 'Advancement of Learning', which offered a sweeping survey of the current state of knowledge in every field. In 1621, Bacon was created Viscount St Albans and finally published the *Novum Organum*, the first part of his vast systematic natural philosophy. But he had also made some serious enemies, who had him removed from office and convicted for taking bribes. Released from prison by King James, Bacon retired to his country home, where he could devote his undivided attention to carrying out many of his principal works. He died in April 1626 from pneumonia contracted while testing the preservative effects of snow on a chicken.

Perhaps no other person in the history of modern ideas has provoked such incompatible and one-sided assessments, towards which adherents maintain an almost sectarian zeal. One reason for this is that there is little agreement on what his actual intentions were or the scientific status of his achievements. The seventeenth century praised and imitated him, the eighteenth century glorified him as the precursor of Enlightenment, while the nineteenth century devoted effort to debunking him and making him the villain of the Jacobean period. The 'Secretary of Nature and All Learning' came to be despised as a charlatan, an enemy of real science, and quite recently was even described as a Satanist. In Mathews' summary of twentieth-century opinion, Bacon is dubbed an atheist and hailed as a religious thinker; acclaimed for his prophetic insights in natural history, his understanding of logic, his theory of forms, and his powerful imagination; while at the same time he is decried for his ignorance of natural history and logic, his absurd notion of forms, and his entire lack of imagination.[2]

One must resist the temptation to describe Bacon's life and works, let alone his attitude towards environmental issues, in cartoon terms. It is far too facile to condemn him for his comments about the merits of human domination and exploitation of nature. Bacon likened himself to a honey bee; the proper philosophical conduct is to work together in the accumulation of material in order to then transform it into something sweet and nutritious. The traditional thinkers, especially the medieval scholastics, were like spiders who spun intricate webs entirely from the inside and then imposed their structures on the world. The empirics, especially alchemists, astrologers and other pseudo-scientific dabblers, were like ants who merely collect curiosities and arcane lore, unable to articulate a coherent intellectual framework. Bacon described three defective methods in the pursuit of knowledge: the 'disputatious' erudition of scholastics; the 'delicate' learning which preserved the errors of revered authorities; and the 'fantastic' learning of the occultists and Hermeticists who catalogue dubious instances of isolated marvels.

Bacon attempted to address all these issues, and more, in the various parts of the *Instauratio Magna*, 'the Great Setting-Forth'. The first part appeared as a revision and expansion of *The Advancement of Learning*, while the second part, the *New Organon*, recasts Aristotle's *Organon* (the Logical Texts) in new terms and contains Bacon's most detailed though incomplete exposition of his criticisms of the false path of natural philosophy and his outline for a cooperative programme in the various natural sciences.

At the core of Bacon's notion of scientific knowledge are the doctrines of induction, hidden forms, and maker's knowledge. In his doctrine of hidden forms, Bacon resuscitated an old idea, that it is the form which gives a thing its true nature. Baconian forms are the simple constituents of matter and, though there are only a small number of them, they can be combined in an infinite number of arrangements, like the letters of an alphabet which can be combined to generate an infinite number of words. The aim of his whole project is, in his words, 'the inquisition of forms' which leads to works, the fruits of correct experimental procedures; he defines natural philosophy as 'the inquiry of causes and the production of effects'. The canon of basic physical properties is the discovery of those true forms which are 'nothing more than those laws and determinations of absolute actuality which govern and constitute any simple nature'. In Perez-Ramos' adroit words, the scientist as a human knower is first and foremost a maker or doer, and his warrant for claims to knowledge depends on his credentials as a maker: 'Bacon's idea of science … establishes that to know something (a natural phenomenon) amounts to being able to (re)produce that very phenomenon on any material substratum susceptible of manifesting it.'[3]

In order to fully comprehend the grandeur (or grandiosity) of Bacon's entire project one must realize that the doctrines of hidden forms and maker's knowledge are aspects of a scheme which concerns 'the advancement of ideas about moving and persuading things and human beings'. The overall scheme is embraced under Bacon's notion of rhetoric which combines psychological, economic and material dynamics. The very idea that one can persuade *things* seems, to modern readers, to be utterly strange, unless one bears in mind that for Bacon there are hidden spiritual forms which compose the nature of all things, including human beings. 'In the new learning, experiment is more than a method of discovery; it is an ordeal, a test of a subject's true nature. Ultimately, all experiments work upon the matter and spirit of the created world, including the minds and passions of human beings.'[4]

In Bacon's theistic picture, the Creator moves the created world in a cryptic manner; the surface language of the perceptible properties of animals, vegetables and minerals conceals a secret code, which the scientist must decipher in order to interpret the latent or deeper language. This notion helps to explain Bacon's repeated references to ciphers and codes, encryption and decryption. The New Learning endorses secrecy, an adept's privileged knowledge, and this is most appropriately expressed through aphorisms and riddles: 'God's encryption of the world is an enigma, and its maker is hidden to all but those who can discover the signs of God's wisdom by suffering the scourging of their vanities in the sweet ordeal of Solomonic inquiry.'[5] This aphoristic and riddling format is featured most prominently in the *Essays*. But if one grasps the twofold power of Bacon's rhetoric, then one can appreciate that the recommendations addressed to civil servants and power-brokers, for instance, in regard to domination of natural forces and the production of works beneficial to humankind, are designed both to persuade them in terms of their own self-interests and (more secretly) to obey the commands of nature hidden at the deepest levels:

> Each essay stands by itself as a separate counsel fitted to move those peculiarly susceptible to its appeal ... Together they are a paradigm of enlightenment. They are perhaps the classic example of the art behind the light ... 'which gradually, by imperceptible degrees, would illuminate the world'.[6]

This artful light is the philosophical force behind the statement that 'Nature to be commanded must be obeyed; and that which in contemplation is as the cause is in operation as the rule'. One can only command natural material and forces in order to shape them into works insofar as one has already understood the deeper inalterable structure of hidden forms; and further, that the 'object' of theoretical insight has its own causal dynamics which must be strictly followed in practical terms in order to (re)make the object for one's own purposes. To violate this fundamental principle, to attempt to alter nature's hidden laws and work against its intrinsic dynamics, is a dangerous enterprise. In the *Wisdom of the Ancients*, Bacon meditated on the fable of Daedalus to draw the lesson that careless fooling with mechanical techniques can have malicious and even deadly consequences. But one should be wary of an overly optimistic reinterpretation of his attitude towards scientific progress, for 'Bacon's works are various mirrors of one another, some darker than the rest'[7] – the darkness often concealed beneath the advocacy of a philanthropic practice.

Despite Bacon's awareness of the dangers of even carefully controlled experiments, he was willing to risk these for the material improvement of human life. If some of his pronouncements on these issues are obscure and ambiguous, his vision of a scientific utopia in *New Atlantis* is both unambiguous and frightening. European travellers in the South Seas are blown off course and arrive at the island of Bensalem. They are provided with the benefits of this strange welfare state – food, shelter and medical care – shown the island's indigenous customs and rituals, and given a guided tour of Solomon's House, the realization of Bacon's scientific research institute. The guide shows them through many rooms where various 'research and design' programmes are being carried out, such as the transformation of birds, beasts and plants into new, barren or super-fertile, kinds, the manufacture of more violent weapons and munitions, and 'houses of deceits of the senses'. The catalogue of twenty-four 'improvements' in scientific knowledge presages some of the most dreadful nightmares of human reason: genetic modification of living things, drug trials on animals, nuclear armaments, powerful machines for the pursuit of luxury or idleness, the ideological apparatus for the control of human behaviour, and more. Equal weight is given to the trivial and the profound, the beneficent and malevolent, reflecting Bacon's ideal of experimental inquiry focused on 'nature under constraint and vexed; that is to say, when by art and the hand of man she is forced out of her natural state, and squeezed and molded'. This utopian vision is the culmination of Bacon's twofold strategy: a physical science capable of dealing with powerful natural motions and a rhetorical 'science' capable of dealing with human emotions.

Bacon's influence on natural science and politics has been pervasive and paradoxical. An epic poem at the front of Thomas Sprat's *History of the Royal Society* (1663) treats Bacon as nearly godlike in the breadth of his vision, one whose instructions could be substituted for divine commandments. At the height of the French Enlightenment, D'Alembert thought that Bacon's grand plan was like the light after the dark, that he was 'the greatest, the most universal, and the most eloquent of philosophers'; his Great Setting-Forth was the model for the *Encyclopédie, ou Dictionnaire Raisonée*. But in Britain about the same time, Bacon's grand scheme was the object of vicious satire and ridicule in Jonathan Swift's *Gulliver's Travels*, where Solomon's House is turned into an asylum for crackpot inventors. Under the impact of the early nineteenth century's revaluation of the history of scientific ideas, Bacon was accorded a more dubious honour, having failed to realize the importance of mathematics in an understanding of physical laws. Lord Macaulay's once famous eulogy of Bacon's philosophy praised this beneficent promoter of human advancement but damned him for moral turpitude in his betrayal of close friends and pandering after honours and riches. In the twentieth century, the Frankfurt School theorists Horkheimer and Adorno denounced Bacon as the initiator of the worst forms of human domination and oppression under the aegis of an instrumental rationality in the service of a capitalist state. Perhaps one can attain a measured balance between these two extremes by considering the arguments of the eminent biologist Loren Eiseley, who arrives at an ambivalent assessment of Bacon, an Elizabethan scientist-magician who promised so much for the good of the human species, but was willing to destroy or distort at least as much to achieve this end.

Notes

1   *The New Organon*, Bk. I, p. 1.
2   N. Mathews, *Francis Bacon: A History of a Character Assassination*, chap. 1.
3   Perez-Ramos, in M. Peltonen (ed.), *Cambridge Companion to Bacon*, p. 115.
4   J.C. Briggs, *Francis Bacon and the Rhetoric of Nature*, p. 3.
5   Ibid., p. 9.
6   R.K. Faulkner, *Francis Bacon and the Project of Progress*, p. 29.
7   Briggs, op. cit., p. 12.

See also in this book
Aristotle

Bacon's major writings

The Works, 7 vols, and *The Letters*, 7 vols, ed. James Spedding, R.L. Ellis and D.D. Heath, London: Longman, 1857–74.

The Advancement of Learning and *New Atlantis*, Oxford: Oxford University Press, 1915.

The New Organon and Related Writings, Indianapolis, IN: Bobbs-Merrill, 1960.

Essays and Counsels; The Advancement of Learning, London: J.M. Dent/Everyman, 1968.

The Oxford Francis Bacon, ed. Graham Rees & others. 8 of 15 vols published. Oxford: Oxford University Press, 1996.

Further reading

Adorno, T. and Horkheimer, M., 1944, *Dialectic of Enlightenment*, trans. J. Cumming, London: Verso, 1979.

Bidhendi, M., Nigomatullina, R. and Shiravand, M. "The Role and Stance of Francis Bacon in Initiating Environmental Crisis Occurences", *Review of European Studies*; 6 (4), 2014.

Briggs, J.C., *Francis Bacon and the Rhetoric of Nature*, Cambridge, MA: Harvard University Press, 1989.

Eiseley, Loren, *Francis Bacon and the Modern Dilemma*, Lincoln, NE: University of Nebraska Press, 1962.

Faulkner, R.K., *Francis Bacon and the Project of Progress*, New York and London: Rowman & Littlefield, 1993.

Mathews, N., *Francis Bacon: A History of a Character Assassination*, New Haven, CT: Yale University Press, 1996.

Merchant, C. "Francis Bacon and the 'vexations of art': Experimentation as Intervention", *British Journal for the History of Science*, 46 (4), 2012.

Peltonen, M. (ed.), *Cambridge Companion to Bacon*, Cambridge: Cambridge University Press, 1996.

Perez-Ramos, A., *Francis Bacon's Idea of Science*, Oxford: Oxford University Press, 1988.

PAUL S. MACDONALD

BENEDICT SPINOZA 1632–1677

> The highest good is ... the knowledge of the union that the mind has with the whole of Nature.[1]

Spinoza was born in Amsterdam in 1632, the son of Jewish emigrants from Portugal; his father was a merchant and a respected member of the Jewish community's board of elders. Spinoza was brought up in the orthodox fashion, studying Hebrew, Holy Scripture and the Talmud. It is possible to trace the various cultural influences in his life through his choice of first name, from Bento (Portuguese) to Baruch (Hebrew) to Benedict (Latin), each of which means 'blessed'. Sometime about 1656, Spinoza was excommunicated from the Jewish church, on the grounds of heretical beliefs. The bookseller and freethinker Franz van den Enden played a pivotal role in his life and thought; he brought Descartes' works to Spinoza's attention, taught him Greek and Latin, and his mystical views about God or Nature as an infinite substance probably had a decisive influence on the young philosopher's unusual perspective, substance monism, as early as the treatise *On the Emendation of the Intellect*. Spinoza moved to Rijnsburg near Leiden in 1660, where his close friends persuaded him to set down his careful, though uncompleted, exegesis of Descartes' metaphysics, *Descartes' Principles of Philosophy*, published in 1663. Spinoza lived a solitary, almost reclusive life, grinding lenses and working slowly on the text of the *Theological-Political Treatise*, published anonymously in 1670. The Council of the Reformed Dutch Church condemned the book as 'a treatise of idolatry and superstition', while one professor at Utrecht wrote that it was 'the most pestilential book'.[2] In 1672, the glorious Dutch Republic came to a disastrous close with an invasion of the French and German armies. The Republic's leader, Jan de Witt, was murdered by an angry mob, and the Dutch Estate Holders brought back into power the young Prince William III. Spinoza was much distressed at the death of de Witt and the unfinished *Political Treatise* demonstrates his unyielding advocacy for the rational foundations of a legitimate state; today it shows his readers the immediate and direct manner in which a philosopher can be engaged in important social and political issues. But it is his final work, the *Ethics*, left incomplete at his premature death in 1677, which has had the most decisive influence in the history of modern philosophy.

In many respects, Spinoza's systematic philosophy in the *Ethics* is the most beautiful, perfectly ordered picture of the universe and humans' place in it. Every aspect of every dimension of human experience is consistently explained in terms of the greater whole. For Spinoza, to explain something is to know its cause, that which not only brings that being into existence but also makes that being just what it is and not something else. A cause also necessarily produces the effect that it does. Understanding, therefore, consists in showing how some feature of the universe necessarily has the role it has as some kind of essential property of substance which is the cause of all things *and* the cause of itself. In contrast with Descartes' dualism, Spinoza propounds substance monism: there is only one substance which has two principal attributes, thought and extension. In this fashion he rejects the dualisms of God and created world, and of mind and body. There are an infinite number of particulars in the world, each of which can be considered a dependent part of that one substance. There is one substance, God or Nature, with two infinite attributes.

These attributes should be thought of as different ways of 'seeing' one and the same reality. We think of extended substance as divided into separate bodies which occupy a limited area in space and time, but extension *itself* cannot be thought of as other than limitless in time and space. The way in which we think of thought will depend upon the level of knowledge which our particular finite mind has reached. The infinite and eternal mode of extension is motion-and-rest; the finite mode, which constitutes individual bodies, or the medium-scale things in our environment, are configurations of simplest particles. The configurations which compose individual physical objects are elements in a *hierarchy* of such organized systems in which there is an ascent from the simplest particles to the whole world; there is one complete cosmic substance in which all other entities are components. All individual things then are configurations of particles in a charged energy state which possess a drive (*conatus*) to maintain themselves in being. The hierarchy of beings then is a plenary order *of power*: the higher an individual is on the scale, the less it is acted on by external forces and the more its changes come from within itself. Moreover, there is an equation between being more or less active as a causal agent upon others and being more or less real. In ascending order of power, these are: the inorganic, the organic, the animal, and the human. The human body is more real than merely animal bodies because it maintains itself in being more effectively than others, does so more under its own control, and interacts with its environment with greater foresight.

The ordered arrangement of beings corresponds with the hierarchy of levels of knowledge. The highest level is intuitive knowledge which approaches the 'infinite idea of God'. 'The more each of us is able to achieve in this kind of knowledge, the more he is conscious of himself and god, that is, the more perfect and blessed he is.' Since God is the same as Nature as a whole, and since Nature is defined as perfect, every being is oriented towards its own perfection or completeness of essence. From this vantage point arises the individual's striving to unite with the source of that which causes the experience of joy or bliss.

'The mind has had eternally the same perfections which now come to it and that is accompanied by the idea of god as an eternal cause. If joy then consists in the passage to greater perfection, blessedness must surely consist in the fact that the mind is endowed with perfection itself.'[3]

One can readily appreciate how the German Romantic poet Novalis later referred to Spinoza as 'the god-intoxicated man' and Goethe dubbed him 'the most Christian one'.

The key to Spinoza's moral theory and thus to his attitude towards environmental concerns can be found in his theory of ideas. Corresponding to each level of knowledge or class of idea, there is an ideatum or 'object' of that same idea; degrees of rationality and degrees of reality must be linked at every stage. Thus, insofar as we purify our understanding in order to consider ideas of the highest order of rationality, we come close to the condition of godhood; in this way we cease to be merely parts of nature. Our status as 'natural' beings under the aspect of extension wholly depends on the class of idea (confused, adequate or intuitive) which constitutes our minds, and vice versa. Spinoza has an unusual and seemingly paradoxical claim about the union of mind and body in the human being: the complex idea which the human body has of itself is its mind. This union under two aspects which constitutes a person is only a special case of a general, uniform principle.

There is thus an equal novelty in his notion of psycho-physical causation: changes in one do not *produce* or generate changes in the other, rather every bodily change *is* a mental change and vice versa, since there is only one Nature conceived under two different attributes. Spinoza was well aware of consequent paradox in identifying mental with physical changes. The particular finite mode of extension which is my body exchanges energy with its own proximate environment, and every such 'interaction' is reflected in an idea.[4] Since Spinoza construed the moral dimension as coterminous with the perfectibility of things as parts, exchanges which diminish living beings' energy states are poisons, and thus evil, and exchanges which augment their energy states are healthy, and thus good.

Human beings maintain their identity by preserving a constant adjustment of their parts. This self-maintenance is not the result of some decision by the person, but occurs as a natural process. Other things are susceptible to fewer changes because their structure is less complex and they have less 'reality' than human beings. They can manage only a lesser field in their environment and hence the cohesion of their parts is liable to disruption by a more narrow range of external causes. Human beings have a high degree of complexity which, under the attribute of thought, is captured by saying that they have *mind* and that they are *self-conscious*. Thus, a human mind consists of ideas which reflect the effects of external causes insofar as they modify the balance of motion-and-rest which constitutes the human body. Such an alteration arises out of the body's interaction with other things and may be either an increase or a decrease in energy, its 'life-force'. There is thus a wide range of internal energy states within which a human cohesion of parts may remain united, without the individual being destroyed. These changes in state can be described both in physical terms as an increase or decrease in the organism's life-force; and they can be described in mental terms as pleasure and pain.[5] Thus, every increase in the 'life-force' is experienced as pleasure and every decrease is experienced as pain; by 'pleasure' Spinoza means 'the passion by which the mind passes to a higher state of perfection, and by pain the passion by which it passes to a lower state of perfection'.[6] Any increase in the power or perfection of the human body must be an increase in the power or perfection of the mind and vice versa. The moral principle here is that all things which contribute to one's perfection are good and all things which detract from it are evil.

The degree of power or perfection of any finite thing depends on the degree to which it is causally active in relation to things other than itself. The one infinitely powerful and perfect being is God or Nature, who is in every respect active and not passive. A human being has greater power and perfection insofar as the succession of ideas which constitutes its mind are linked together as cause and effect; a human is active insofar as the succession of ideas in its mind is a logical one. A human being has less power or perfection as a thinking being insofar as its present ideas are not explicable as the logical consequences of previous ideas in its mind. In God, there would be an infinite sequence of ideas each one of which would be logically entailed by its predecessors. But human minds, for the most part, consist of more or less *random sequences of ideas*, in the sense that the causes are *external to the sequence*. The sequence of ideas is not self-contained and hence cannot be completely intelligible – there are always gaps. The power and perfection of an individual mind is increased in proportion as it becomes *less passive* and *more active* in the production of its ideas. The equivalent for the individual human body of this increased cognitive activity is the internal stability of the organism, which enables it to carry on living without any violent perturbations produced by external causes. Thus, the mind is relatively free and active in its thinking when the body is in a relatively *constant state* vis-à-vis its own proximate environment.[7]

Human beings, unlike animals, can be *aware* of the tendency towards self-maintenance which constitutes their real 'nature'. The reflection in a *conscious idea* of this *conatus*, the drive to maintain oneself in being, is called *desire*. Spinoza defines desire as *appetite* together with conscious awareness of its occurrence and the 'object' towards which it is directed. Now pleasure and pain are not to be found in the 'objects' which desire and aversion afford, nor can they be discovered by any form of abstract reasoning. They represent a change in the psycho-physical state of the whole person; they are the mental reflection of a rise or fall in the power or activity of the organism. Which specific things will promote or depress the life-force of any organism depends on the constantly changing 'nature' of the individual organism. It may be difficult to understand how *conatus* pertains to inanimate things – how could a *stone*, for example, be said to have a *drive* to maintain itself in being? The problem for the common-sense view is that we think of a stone as an individual *thing* or substance – but of course, for Spinoza, this is incorrect. A stone, a plant, or an animal are each no more than temporary configurations of finite modifications of the infinite attributes of *one thing*, God or Nature; they are all *parts* of the cosmos which work together towards the maintenance of the whole. So, plants consume soil and water, animals consume plants and animals, and so forth, each thereby participating through exchange of energy in the greater whole, from whence all ultimately derive their life-force. This grand conception of the world-whole is perhaps more familiar to contemporary readers from Lovelock's Gaia Hypothesis.

The Norwegian 'eco-philosopher' Arne Naess proposed that Spinoza was the most important philosophical source for inspiration regarding concern for environmental issues. He claims that nature conceived by ecologists is not the passive, inert, value-neutral nature of mechanistic science, but more like Spinoza's Nature – all-inclusive, creative, infinitely diverse, and alive in the broad sense of panpsychism. Further, Spinoza's reflections on morality are 'important for striking a balance between a submissive, amoral attitude towards all kinds of life struggle, and a shallow moralistic and antagonistic attitude'. Future societies will achieve an equilibrium with their environment by following a 'third way' between the two extremes. In Spinoza's world-picture, every thing is connected with every other thing. Nothing is really causally inactive, there is nothing wholly without an essence which it expresses through a cause. And finally, every thing strives to preserve and develop its specific essence or nature, and since every thing is a part of God's perfection, this striving is an active shaping of its environment.[8] 'The highest good is … knowledge of the union which the mind has with the whole of nature.'

Notes
1   *On the Emendation of the Intellect*, p. 5.
2   Wim Klever, in D. Garrett (ed.), *The Cambridge Companion to Spinoza*, p. 40.
3   *Ethics*, V P32, 33.
4   Stuart Hampshire, *Spinoza*, pp. 72–3.
5   Ibid., pp. 98–9.
6   *Ethics*, III P11Schol.
7   Hampshire, op. cit., p. 100.
8   Arne Naess, *Freedom, Emotion and Self-Subsistence*, pp. 19–20.

See also in this book
Goethe, Lovelock, Naess

Spinoza's major writings
The Ethics and Selected Letters, ed. and trans. Samuel Shirley, Indianapolis, IN: Hackett, 1982.
The Collected Works, ed. and trans. Edwin Curley, Princeton, NJ: Princeton University Press, vol. I, 1985; vol. II, 2016.

Further reading
Delahunty, R.J., *Spinoza*, London: Routledge & Kegan Paul, 1985.
Donagan, A., *Spinoza*, Chicago, IL: University of Chicago Press, 1988.
Garrett, D. (ed.), *The Cambridge Companion to Spinoza*, Cambridge: Cambridge University Press, 1996.
Hampshire, Stuart, *Spinoza*, rev. edn., Harmondsworth: Penguin Books, 1988.
de Jonge, Eccy, *Spinoza and Deep Ecology*, Aldershot, UK: Ashgate, 2004.
Nadler, Steven, *Spinoza: A Life*, Cambridge: Cambridge University Press, 1999.

Naess, Arne, *Freedom, Emotion and Self-Subsistence*, Oslo: Universitets Vorlaget, 1975.

Sharp, Hasana, *Spinoza and the Politics of Renaturalization*, Chicago, IL: University of Chicago Press, 2011.

PAUL S. MACDONALD

BASHŌ 1644–1694

The chestnut by the eaves
In magnificent bloom
Passes unnoticed
By men of this world.[1]

Regarded by many as amongst Japan's finest literature, Bashō's work is an important development and summation of medieval Japanese cultural attitudes to the natural world, and emphasizes a heightened sense of unity with nature, much stressed in later artistic expressions of Zen Buddhism. Held in high regard in Japan, his authority as a literary figure is matched by his growing influence on religious and artistic responses to the environment, especially in contemporary Western circles.

There is a dearth of material concerning Bashō's early life. It is generally believed that Matsuo Kinsaku, who would later take the name Bashō, was born in 1644 to a samurai family in service to the lord of Ueno, south-east of Kyoto, then the capital of Japan. As a boy he was a page to Tōdō Yoshitada, the eldest son of the ruling feudal lord of the area – thus his duties as companion and page brought him into contact with the literature of the ruling classes. Both boys shared an interest in poetry, and as their friendship grew they influenced and encouraged each other, particularly in the writing of haiku. It was an old literary tradition amongst the more affluent classes of medieval Japan to engage in the team construction of *renku* (or *renga*), linked poems of thirty-five, fifty, or one hundred lines. By the end of the fifteenth century it had also become popular to generate the first stanzas alone as haiku (originally *hokku*), and haiku competitions were a fashionable leisured activity. The earliest known recorded verse by Bashō dates from his time with Yoshitada in the early 1660s.

In 1666 Yoshitada died suddenly and Bashō's grief prompted him to leave the service of the ruling family, and thereby renounce his samurai status. It is generally believed that during the period of wandering that followed he studied in Kyoto, and it is certain that he continued to write and gain a name for himself as a poet. Between 1667 and 1671 his verses were included in four anthologies, and by 1672 he was able to publish his own record of a haiku competition, *The Seashell Game* (*Kai Ōi*), which also marks the beginning of Bashō's recorded critical prose.

In 1672, at the age of twenty-eight, he left Kyoto to journey to Edo. He joined in the writing of *renku* with several local poets there, and it is assumed that around 1675 he became a professional writer, his work appearing with greater frequency in haiku anthologies of the time. He was presented with a hut or cottage surrounded by banana trees by the local people of Edo, Banana Tree Hermitage (*Bashō An*), from which Bashō took his name. However, he was not entirely satisfied with a static lifestyle and set out on what became his first major journey in 1684; he often referred to himself as homeless and certainly had few possessions. Whilst travelling he continued to compose *renku* and haiku and wrote his first travel journal, *The Records of a Weather-Exposed Skeleton* (*Nozarashi*). During the remaining years of his life he undertook several such journeys, writing journals alongside his poetry, including *A Visit to the Kashima Shrine* (*Kashima Kikō*), *The Records of a Travel-Worn Satchel* (*Oi no Kobumi*) and *A Visit to Sarashina Village* (*Sarashina Kikō*). Two poetry anthologies of 1686, *Frog Contest* (*Kawazu Awase*) and *A Spring Day* (*Haru no Hi*), include his famous frog haiku, which is often used as the paradigmatic example of Bashō's poetic style:

The old pond;
A frog jumps in –
The sound of the water.[2]

The Narrow Road to the Deep North (*Oku no Hosomichi*) is an account of a two-and-a-half-year trek, begun in 1689, taking in the villages and country north and west of Edo. It is regarded as Bashō's greatest literary achievement, combining concise, crisp prose and poetry in a breathtakingly unified piece. He wrote one other major journal, *The Saga Diary* (*Saga Nikki*), before focusing solely on poetry and the encouragement of younger writers composing in his style. Anthologies of this final period include *The Monkey's Cloak* (*Sarumino*) and *A Sack of Charcoal* (*Sumidawara*). Bashō died on a pilgrimage to Osaka in 1694.

Around one thousand haiku are attributed to Bashō. He established the haiku as a serious and deep poetic form that captures a purity and unity in the immediacy of experiences of the natural world.

The haiku translator and historian, R.H. Blyth, once commented that 'Nature *is* Japanese Literature'.[3] Although this is an exaggeration, there is much evidence to support the spirit of this claim. Japan has marked seasons, and seasonal poetry dates back to the earliest recorded anthologies, nature and natural cycles remaining key subjects for writers throughout the development of the *renku* and *hokku* and up to the present day. Many words and phrases acquired connotations of seasons and seasonal activities, bringing to mind more than the picture inspired by the literal meanings of the word. For example, 'blossom' (*hana*) in a poem means ornamental cherry tree blossom and the associated image of its fluttering in a warm spring breeze. This 'logopoeia' allowed poets to condense a great deal of imagery into a simple phrase. Sōgi (1421–1502) mastered this technique in his *renku* two hundred years before Bashō, and it subsequently became a mainstay of Japanese verse. Through use of these linked and complex images the seasons and nature remained central to later poetry as the haiku came into its own; so much so that all haiku – in their traditional form at least – must refer to a season to be complete. It is the concentration of associated 'images', across all the senses, that paradoxically makes haiku so pure. The 'plop' of the frog jumping into the old mill pond, together with the stillness, the ripples of the water and the flash of colour evoked by 'the sound of the water', brings us to an imagining of the moment that lengthier descriptions fail to evoke. So in haiku we find a supreme aesthetic expression of the experience of a thinking being in a relationship with the natural world. And Bashō was undoubtedly one of the great haiku masters. Yet his genius lies in more than the cleverness of his style.

Whilst there seem to have been no major developments in philosophy during the Tokugawa period (1600–1867) of Japanese history, the closing of the borders to foreign influences around the time of Bashō accentuated the purely Japanese aspects of art and literature produced by the 'home-grown' talent of the time. Writers and artists looked back to their own cultural forefathers, such as Sōgi, in whose writing Bashō obviously found great inspiration. But Bashō also explicitly drew on the Japanese '*conflation* of the religious and literary dimensions of human experience',[4] a deliberate refusal to separate out different aspects of an experience where a separation could be made. Consequently it is disingenuous to treat Bashō's poetry as 'pure' literature alone, separated from its religious and philosophical roots in Shintō, Buddhism and earlier Chinese thought. In this spirit of co-existence Buddhism and Shintō stood side by side in Bashō's Japan, despite differences in practice and origin, as they do today, often in the same shrine. Bashō's poetry is in tune with this tradition of 'conflation' and often incorporates into more sophisticated responses to nature an explicit animism derived from Shintō:

> Making the uguisu [warbler bird] its spirit
> The lovely willow-tree
> Sleeps there.[5]

However, to see Bashō as presenting a simple, romantic view of the living world would be to ignore a much deeper Buddhist component. Bashō almost certainly learnt to meditate under a Zen master and direct references to Buddhism are scattered throughout his verse:

> The anniversary of the Death of the Buddha;
> From wrinkled praying hands,
> The sound of the rosaries.[6]

Even when it does not point directly to Buddhism, Bashō's verse often exemplifies key components of Buddhism – awareness of impermanence, the non-existence of the self, emptiness, the suffering of all living beings, and the compassion we should feel for these beings (and that we would feel if we paid attention to our true nature):

> Singing, singing,
> All the long day,
> But not long enough for the skylark.[7]

And in reference to his dead mother's hair:

> Should I take it in my hand,
> It would melt in my hot tears,
> Like autumn frost.[8]

Above all, it is the sense of sympathetic compassion with all life that pervades Bashō's writing:

> The ancient poet
> Who pitied monkeys for their cries,
> What would he say, if he saw
> This child crying in the autumn wind?[9]

He attempts the conceptually impossible, to lead us into an unmediated glimpse of the real world of natural things in which blossom is glorious and beautiful and then fades and dies, to return the next year, each flower new and unique. It is only through reaching a state of enlightenment that such an insight could be perfected, for only then would we be aware of the connectedness and conditionality of all living beings and true Buddha-mind. And yet, Bashō strives always to point the way, showing us moments of experience of frogs and flowers and muddy roads. Bashō reminds us that the silence of an autumn moon reflected in a lake will have more to tell us about ourselves and reality than the chatter of our thoughts and theories. 'The Jewel Mirror Awareness', a Zen poem attributed to Dongshan Liangjie (807–869), a Zen master whose works Bashō would have known, states with regard to the immediate 'thusness' of experience: 'The meaning is not in the words, yet one pivotal instant can reveal it.' This is what Bashō attempts to capture. 'Although it is not created, it is not beyond words. It is like facing a jeweled mirror; form and image behold each other.'[10] A revealing of form in formlessness, and formlessness in form.

By combining his mastery of the haiku form, his Buddhist insights and his personal commitment to a life largely liberated from material concerns, Bashō produced an account of nature, in its broadest terms, that has been deeply influential in Japan's culture and literature. After Bashō, haiku writing flourished with renewed vigour, his travels having spread his teaching and style throughout the country. There were many students and imitators, but none of Bashō's contemporaries and immediate successors reached his standard of equanimous simplicity. Of later poets only Buson (1716–83) and Issa (1763–1827) can compete for his impact on modern poetry in Japan. Certainly Issa, in particular, made more of the ineliminability from pure aesthetic experience of compassion for living beings, but it was Bashō who crystallized the use of modern haiku for such a Zen purpose, superseding all earlier models for haiku writing whilst refreshing a well-established tradition.

Bashō's verse now appears in almost all inspirational Zen collections.[11] He is also quoted and used by Buddhist writers attempting to forge a connection between Zen and deep ecology, but in the end, it is the uncluttered purity of his prose and poetry that keeps his concern for the natural world alive and continually attracts new readers.

Notes
1 *The Narrow Road to the Deep North and Other Travel Sketches*, p. 108.
2 R. Aitken, *A Zen Wave*. This haiku has attracted more discussion and analysis than any other; in Japanese it is a perfect balance between sound, form and content and, as with all haiku, presents problems of translation. See Hiroaki Sato, *One Hundred Frogs*.
3 R.H. Blyth, *The Genius of Haiku: Readings from R. H. Blyth on Poetry, Life and Zen*, London: The British Haiku Society, 1994. p. 72.
4 W.R. LaFleur, *The Karma of Words*, p. 149.
5 R.H. Blyth, *A History of Haiku*, p. 111.
6 Ibid., p. 119.
7 Ibid., p. 127.
8 Ibid.
9 *The Narrow Road to the Deep North and Other Travel Sketches*, p. 52.

10 K. Tanahashi *Zen Chants, Thrity-Five Essential Texts with Commentary*, Boston and London, 2015. p. 90.
11 It would be impossible to survey them here, but see, for a good balanced example, K. Tanahashi and Tensho D. Schneider (eds), *Essential Zen*, New York: HarperCollins, 1994.

See also in this book
Buddha, Wordsworth

Bashō's major writings
There are several collections of Bashō's work, some more scholarly than others. For a survey of around 250 haiku with commentaries, see:

Uedo, M., *Bashō and His Interpreters*, Stanford, CA: Stanford University Press, 1992.

A good collection of many of the travel poems and prose is:

Bashō, *The Narrow Road to the Deep North and Other Travel Sketches*, trans. Nobuyuki Yuasa, Harmondsworth: Penguin, 1966.

Further reading
Aitken, R., *A Zen Wave: Bashō's Haiku and Zen*, New York: Weatherill, 1978.
Blyth, R.H., *A History of Haiku*, 2 vols, Tokyo: Hokuseido Press, 1963.
LaFleur, W.R., *The Karma of Words: Buddhism and the Literary Arts in Medieval Japan*, Berkeley, CA: University of California Press, 1983.
Sato, Hiroaki, *One Hundred Frogs: From Renga to Haiku to English*, New York: Weatherill, 1983.

DAVID J. MOSSLEY

JEAN-JACQUES ROUSSEAU 1712–1778

> Man's proper study is that of his relation to his environment ... this is the business of his whole life.[1]

Born in Geneva, Rousseau was raised by his aunt and eccentric watchmaker father, who instilled in him an abiding love of literature, especially classical. After an unstable childhood and several years as a vagabond, Rousseau moved in 1743 to Paris, where he met Diderot and other *philosophes* involved in the great *Encyclopédie, ou Dictionnaire Raisonée*, for which he contributed an article on music. In 1749 Rousseau experienced an overwhelming inspiration from which he later claimed all his philosophical speculations were derived. He won a prestigious prize with his *Discourse on the Arts and Sciences* in 1750, and wrote two operas. In 1754, on a return visit to Geneva, he reconverted to Calvinism and regained his citizen status, of which he was always proud. During the following eight years, living mainly in the country, he published most of his principal works, including *Émile* and *The Social Contract*. These works were condemned in Paris and Geneva, and Rousseau moved to England, on the instigation of David Hume, with whom he soon quarrelled. Returning to France in 1767, he became mentally disturbed and was always in fear of being arrested. He finally settled in Paris in 1770, where he finished work on *The Confessions*, only to have his former friend and confidante Madame d'Epinay issue a police ban against him. His final, unfinished work, before his death in 1778, was the more serene and meditative *Reveries of a Solitary Walker*.

In the *Discourse on the Arts and Sciences*, Rousseau answered the question 'Has the rebirth of the arts and sciences contributed to the purification of morals?' with an emphatic negative. In direct opposition to the view espoused by the *philosophes*, he asserted that the progress of the arts and sciences in every society has been accompanied by the corruption and diminution of morality. In this essay he broached the concept of a natural human being, characterized by simplicity, lack of vanity and basic virtue, a natural state eroded by the acquisition of politeness, superfluous ornaments and dependence on artifice, including the machinery of warfare. He drew numerous examples from ancient history to show that the arts and sciences have not inspired humans with courage or patriotism, but instead deflected their energies into unnecessary inventions, the flattery of paintings and sculptures, and the display of erudition. Even our most valued sciences have developed out of idleness and trivial pursuits: astronomy from superstition, geometry from avarice for property, and physics from excessive curiosity. Rousseau's vigorous condemnation of modern morality is drawn from a conjectural history of humanity. He argues that the human species has declined from the innocence of its original condition and the most praised civilizations are decadent under the weight of their own cultural progress.

Despite its confident tone, this first *Discourse* suffers from incoherence, lack of originality, and indecisiveness about a remedy for the parlous situation. In this essay, he is not clear whether the general decline of culture is the cause or the effect of the erosion of morality.

In the *Discourse on the Origin of Inequality*, Rousseau carries forward his central theme of the denaturation of human beings, their progressive removal from the sources of their natural being. The second *Discourse* is an ingenious, tightly argued essay which ran counter to the then-accepted view that humans in their original state were motivated solely by self-interest and aggression towards their fellows, and remained fractious until they were coerced into accepting governance under the rule of law. Rousseau distinguishes between natural inequality, which results from discrepant physical and mental abilities, and moral or political inequality, which depends on social conventions and is authorized by mutual consent. The subject of this essay then is

> the moment at which ... nature became subject to law, and to explain by what sequence of miracles the strong came to submit to serve the weak, and the people to purchase imaginary repose at the expense of real felicity.[2]

Previous political theorists, such as Hobbes, made the mistake of imputing to their hypothetical natural humans ideas which were only acquired by socialized humans. Rousseau constructs a conjectural history in order to make sense of the origins of moral and political notions such as natural right and justice. He resists the temptation to retroject notions which the civilizing process has conferred upon humans and considers instead an entirely natural human, a creature whose basic needs of hunger, thirst and sex are satisfied in the most immediate manner.

Rousseau follows Descartes in considering the animal in its bodily dimension to be an intricate machine, driven by its senses to seek what would nourish it and to guard against or avoid what would damage it. But where non-human animals carry out their actions for need-satisfaction by the internal operations of instinct, humans have a freedom to choose; they are at liberty to acquiesce or forbear to carry out what their natural desires impel them towards. 'In the power of willing or rather choosing, and in the feeling of this power, nothing is to be found but acts which are purely spiritual and wholly inexplicable by the laws of mechanism.'[3] This account of the freedom prefigures the dualism of spirit and body in the Savoyard Priest's discourse in *Émile*. Rousseau thus expressly sides with the philosophical view that only the bodily aspect of humans can be explicated in mechanistic terms. But the fact that non-human animals are sentient creatures means that *they ought to partake of natural rights*; humans are subject to an obligation even towards the brute.

> This is less because they are rational than because they are sentient beings; and this quality, being common both to men and beasts, ought to entitle the latter at least to the privilege of not being wantonly ill-treated by the former.[4]

Rousseau clearly expresses here one of the first conceptions of the intrinsic moral standing of non-human animals.

The first step beyond this entirely natural human condition was made by the first person who declared a piece of ground to be his own; civil society is founded on the notion of private property. But the satisfaction of natural humans' basic needs might not be immediate due to variations in circumstances, climate, soil and so forth which provoked the additional needs to build shelter, storage and implements. Reflection on the best way to achieve these ends would have inspired a sense of prudence which required that only in some cases would pursuit of private interest be to one's best advantage, whereas in other cases cooperation with one's fellows' pursuit of their interests would best serve one's deferred needs. Freed from the demand to be incessantly in pursuit of one's own needs, socialized humans had the opportunity to sing and dance, 'the true offspring of love and leisure', as Rousseau charmingly phrases it. It was from the desire for public esteem that the first moves towards inequality were made – on the one hand, vanity and contempt, and on the other, shame and envy. Moral sentiments are judgements conferred upon persons and actions which are deemed to endorse or contravene a suitable estimate of a person's or an action's worth.

Rousseau extols a conjectured golden age, 'the real youth of the world', whose best exemplar is the noble savage who maintains 'a just mean between the indolence of the primitive state and the petulant activity of our *amour-propre*'.[5] The next stage was the specialized labour of metal-working and agriculture, but variable distribution of natural resources ensured that those who had more property and power accumulated greater riches. It was in the interests of those with more property and power to retain the services of the poor, and for the poor to offer their labour, even their liberty, in exchange for protection. Since the rich enjoy greater physical goods and the talented enjoy greater public esteem, it becomes a new interest for those less well blessed to appear to be what they really are not. Flattery, trickery and deceit become valued skills. But since even the rich and powerful might have to contend with dangers and even rebellion from everyone else, they devised an ingenious plan: 'to make allies of his adversaries, to inspire them with different maxims, and to give them other institutions as favorable to himself as the law of nature was unfavorable'. Thus the first version of the social contract is tendered, in which the supreme power which governs everyone is invested in the rule of law. 'All ran headlong to their chains in hopes of securing their liberty'; the contract 'bound new fetters on the poor and gave new powers to the rich; which irretrievably destroyed natural liberty ... and for the advantage of a few ambitious individuals, subjected all mankind to perpetual labor, slavery and wretchedness.'[6]

This ringing denunciation of the misfortunes which result from the progressive denaturation of human beings is taken up again in *The Social Contract*: 'man is [or was] born free, and everywhere he is in chains'. *The Social Contract* portrays an association by contract which draws citizens together instead of driving them apart and protects egalitarian ideals of public engagement which enhance liberty. Rousseau argues that our proper passage from the original, natural condition to civil society must not suppress true liberty, but instead realize our freedom by transforming appetite and desire into obedience to laws which we prescribe for ourselves. His radical vision centred around the notion that this association by contract ensured that the various parties were able to fulfil ambitions which they could not have managed without the contract. By renouncing freedom from 'each other's control, ... citizens acquire moral personalities and cooperative interests unimaginable to solitary savages'.[7]

Rousseau's most complete, mature exposition of two themes little discussed in *The Social Contract* – humans' natural condition and the process of denaturation – is in *Émile*. This is divided into five books which roughly correspond with the five ages of man – infancy, childhood, puberty, adolescence and adulthood. The central theme of this convoluted work is that the proper education of children must take account of the maturation of their cognitive and affective abilities, leading their natural desires towards goals which will be of value to them as adults, and not impose adult expectations on each stage of growth. Rousseau's own experiences as a private tutor taught him that the only way to compel a child to obey one's commands was to prescribe nothing, forbid nothing, exhort nothing, and avoid boring him with useless book-work.[8]

Rousseau profoundly disagreed with John Locke's *Treatise on Education* and its numerous adherents who, he claimed, distorted the child's natural inclinations and inculcated ambitions for useless pursuits, vain conceits and superfluous social niceties. Rousseau's astonishing advice was to employ two other inborn motives for learning which do not corrupt the pupil's natural goodness. In childhood, this basic drive is for food, and after puberty it is for sex. In Alan Bloom's excellent analysis of these themes, the child seeks out desirable foods, whereas the adolescent and young adult seeks out other ideals because he does not yet know what he really longs for.

> The task is to enrich his desires before they are satisfied ... The goal is to sublimate his desires prior to his capacity to distinguish sex from love, so that when he learns about the distinction it no longer interests him.[9]

The tutor's task in the life-long education of Émile is to prepare him for his encounter with Sophie, the embodiment not merely of his sexual desires but also his longing for an ideal in this world.

Every child before the onset of education lives in the golden age of his world, a natural creature whose source of action is a surfeit of self-love. But the immediate environment does not always satisfy the child's desires, nor can the child count on the ability to manipulate persons and things to achieve its ends; however, nature has also endowed humans with imagination and this cognitive power compensates for what nature in general does not supply for the child's own existence. It is through imagination that the maturing child comes to understand that others have desires and feelings and that through compassion the child can extend its world. The adult needs other persons' compassion, their fellow-feeling for his own desires and their realization; this mutual compact with other adults is founded on an even balance between the self-serving primitive mode of human being and dependence on the esteem of others in the socialized mode.

> Man's proper study is that of his relation to his environment. So long as he only knows that environment through his physical nature, he should study himself in relation to things; this is the business of his childhood; when he begins to be aware of his moral nature, he should study himself in relation to his fellow-men; this is the business of his whole life.[10]

Rousseau is often assimilated into the broad current of the Enlightenment project, but although he concurred with the *philosophes* in their attempt to eliminate religious prejudices, he was their sharpest critic in rejecting the elitist notion that human reason should hold sway over our passions. He rejected the Baconian and Cartesian advancement of humans' dominance over the natural order and their exploitation of the precious gifts of God's creation. Rousseau argued passionately for the natural goodness of the ordinary person and championed the idea of collective self-expression and popular self-rule. His epistolary novel *Julie, or the New Heloise*, with its evocation of ideal love and an earthly paradise, was highly influential and much imitated. *Émile* became the most important treatise on education since Plato's *Republic* and the *Reveries of a Solitary Walker* became the vade mecum of the Romantic Naturalist movement. Through his entire life and writings runs one of his deepest concerns – the implacable commitment to prevent an individual's dominance or submission, which would chain him to worldly things and negate his natural liberty.

Notes
1 Émile, pp. 209–10.
2 The Discourses, pp. 44–5.
3 Ibid., p. 54.
4 Ibid., p. 42.
5 Ibid., p. 82.
6 Ibid., pp. 88, 89.
7 Robert Wokler, Rousseau, p. 61.
8 Ibid., p. 94.
9 Alan Bloom, Love and Friendship, p. 61.
10 Émile, pp. 209–10.

See also in this book
Bacon, Goethe, Kant

Rousseau's major writings

Émile, or On Education, 1911, trans. Barbara Foxley, London: Everyman, 1992.
The Social Contract and the Discourses, 1913, trans. G.D.H. Cole, London: J.M. Dent, 1973.
The Confessions, trans. J.M. Cohen, Harmondsworth: Penguin, 1954.
Julie, or the New Heloise, abr. and trans. J.H. McDowell, Pittsburgh, PA: State University Press, 1968.
The Collected Writings, ed. R.D. Masters and C. Kelly, 5 vols, Hanover, NH: University Press of New England, 1990–2000.

Further reading

Bednar, C.S. Rousseau's Counter-Enlightenment, Albany, NY: State University of New York Press, 2003. [On Ecological Economics]
Bloom, Alan, Love and Friendship, New York: Simon & Schuster, 1994.
Cranston, Maurice, Jean-Jacques, vol. I, The Noble Savage, vol. II, The Solitary Self, vol. III, Chicago, IL: University of Chicago Press, 1983, 1991, 1997.
Green, F.C., Jean-Jacques Rousseau: A Critical Study of His Life and Writings, Cambridge: Cambridge University Press, 1955.
Grimsley, Ronald, Jean-Jacques Rousseau: A Study in Self-Awareness, Cardiff: University of Wales Press, 1969.
Lane, J.H. & Clark, R.R. 'The Solitary Walker in the Political World: The Paradoxes of Rousseau and Deep Ecology', Political Theory 34 (1), 2006.
Wokler, Robert, Rousseau, Past Masters Series, Oxford: Oxford University Press, 1995.

PAUL S. MACDONALD

IMMANUEL KANT 1724–1804

> Bold, overhanging, as it were threatening cliffs, thunderclouds
> towering up into the heavens, bringing with them flashes of lightning
> and crashes of thunder, volcanoes with their all-destroying violence,
> hurricanes with the devastation they leave behind, the boundless
> ocean set into a rage, a lofty waterfall on a mighty river, etc., make
> our capacity to resist an insignificant trifle in comparison with their
> power.[1]

Immanuel Kant is one of the most influential philosophers of all time,
celebrated as a key figure of the Enlightenment and well known for his
writings across various philosophical fields. Kant's brilliant and complex
philosophical system – his 'critical philosophy' – was central to shaping
ideas in German Idealism and Romanticism. More recently, there has
been a resurgence of scholarship on Kant in philosophy, and beyond in
other disciplines. At the same time, alongside other philosophical figures
of modernity, postmodernism and poststructuralism have strongly
critiqued the centrality of reason in his philosophy.

Kant was born in 1724 in Königsberg in Prussia (now Kaliningrad,
Russia). After studying at the University of Königsberg, he became a
private tutor for several years. He became a lecturer at the university, was
promoted to Magister or doctor of philosophy (1755) and then Professor
of Logic and Metaphysics (1770), lecturing until his death in 1804. Kant
was known to be a serious, dutiful and dedicated teacher and academic,
who never married. His early work related to natural philosophy (or to
the sciences), including work on cosmology which remains significant
today. His pre-critical and critical philosophical writings addressed
epistemology, metaphysics, logic, ethics and aesthetics among other
areas. Much of his most important thinking is set out in the 'three
Critiques', the *Critique of Pure Reason* (1781), the *Critique of Practical
Reason* (1788) and the *Critique of the Power of Judgment* (1790). The
originality of Kant's approach lies in its response to the two dominant and
competing philosophical movements of his time, empiricism and
rationalism. Of particular interest to environmental thought are his
discussions in ethics and aesthetics.

Kant's most influential work in ethics appears in *Groundwork of the
Metaphysics of Morals* (1785). Kant's moral philosophy is deontological, or
an ethic based in duty. In moral decision-making and action this means
that emphasis is placed on intentions and principles or duties. His
approach contrasts strongly with consequentialism, an ethical theory
which holds that moral actions are those which produce the best

consequences. For Kant, the moral agent possesses reason and autonomy, and in deciding how to act she or he engages in rational reflection rather than being motivated by feelings, inclinations, or preferences. Reason and autonomy ground moral actions that are freely chosen for their own sake, as ends in themselves. An act is right in virtue of following a duty that can be universalized beyond oneself, rather than serving as a means to achieving a set of outcomes.

One of the key principles of Kant's moral philosophy is 'respect for persons'. As one formulation of his 'categorical imperative', it states: 'So act that you use humanity, whether in your own person or in the person of any other, always at the same time as an end, never merely as a means.'[2] Kant's understanding of 'person' has proved to be controversial when it comes to environmental thought. A person is a rational, moral agent, and Kant is careful to distinguish 'person' from other kinds of natural entities. Nonhuman natural organisms are governed only by causes and do not possess reason or autonomy. The upshot is that the principle of 'respect for persons' does not extend to nonhumans (or children), which presents the main challenge for understanding how Kant's moral philosophy might extend to the natural environment.

According to Kant, we do not have direct duties to nonpersons but we may have indirect duties to them. Although we may not have direct duties to nonhumans, our own moral perfection demands that in many cases we ought not to treat nonhumans badly. More specifically, Kant argues that in some cases we have direct duties to ourselves to treat nonhumans with respect (i.e., indirect duties to nonhumans). Discussions of these ideas have taken place within Kant scholarship, animal ethics, and environmental philosophy. Various approaches interpret or reconstruct Kant's ideas in ways that enable a deontological theory to articulate respectful treatment of nonhuman animals and other organisms. For example, Toby Svoboda has defended a Kantian environmental virtue ethic, arguing that the philosopher's ideas provide strong moral reasons to care about nonhuman organisms as having value in themselves, and as having value in their flourishing independently of human interests.[3] This interpretation rests on indirect duties and advocates that developing morally virtuous dispositions requires that humans do not unnecessarily harm other organisms.

The interesting links between Kant's moral and aesthetic philosophy help to position his ideas within environmental thought. First, arguments have been put forward that Kant prioritizes natural beauty over artistic beauty. In the 'Critique of the Aesthetic Power of Judgment', the first part of the *Critique of the Power of Judgment*, Kant's well-known and influential discussion of the 'judgment of taste', or judgments of the beautiful, begins with appreciation of nature rather than the arts. An interest in both the aesthetics of nature and of art was not unusual in the eighteenth century; however, further evidence for the significance of nature can be seen in the second part, the 'Critique of Teleological Judgment', where Kant discusses teleology in organisms, or 'purposiveness' (purposes, ends) in nature.

Second, Kant's aesthetic theory offers a conception of the non-instrumental aesthetic value of nature. Judgments of taste are grounded in a feeling of pleasure or 'liking' in the subject in response to the perception of form or appearance of some object. This feeling is 'disinterested', which is not understood as indifference, rather it identifies appreciation of something apart from practical or instrumental interests. In aesthetic appreciation, 'disinterestedness' functions to background personal preferences and utilitarian concerns and foreground the valuing of nature's qualities for their own sake. The judgment of taste is a 'free liking' arising from the mere contemplation of an object for its aesthetic qualities, rather than a liking arising from the ways in which that object might satisfy our needs, whether as a means to our own (self-interested) ends, or other ends.

Third, Kant himself draws out affinities between aesthetic and moral experience, while also maintaining the autonomy of the aesthetic. In proposing that beauty is a symbol for morality, he argues that features of aesthetic experience such as disinterestedness and the free play of imagination prepare appreciators for the freedom that characterizes moral feeling.[4] Although not presented in the context of contemporary environmentalism, these ideas provide an early foundation for discussions of 'aesthetic preservationism' within environmental philosophy today. Aesthetic preservationism holds that the sensitive perception characteristic of aesthetic attention and the discovery of beauty, majesty, and so on, can encourage the development of a moral attitude toward the natural world.[5]

Fourth and finally, Kant's theory of the sublime presents an aesthetic category linked to moral ideas, and one that is directed mainly, if not exclusively, at nature. As the quotation at the beginning of this essay illustrates, the sublime involves a response of 'negative pleasure' to qualities of great natural scale or power, where the respondent is made to feel humbled and insignificant.[6] Pleasure in the sublime response arises through an expansion of imagination and a kind of awareness of the distinctive moral capacities of persons. Feelings of displeasure are associated with frustration in the face of an inability to take in the unbounded, and there is a feeling of both attraction to and repulsion from fearsome, mighty forces in nature. Kant writes that the sublime causes a feeling of physical helplessness and insignificance in the face of such forces, yet it also brings about a sense of being able to measure one's human capacities in relation to them.

Some interpretations of Kant's theory of the sublime suggest that aesthetic and moral stances toward the natural world are also very close. If judgments of the natural sublime are understood as disinterested aesthetic judgments, then this suggests an aesthetic experience that prepares valuers for an attitude of respect for nature.[7] Because one cannot have direct duties to nonhuman nature and since the sublime is an aesthetic rather than moral experience at heart, it is not respect for nature as such. Rather, the attitude one takes toward the sublime in nature is one of 'admiration'. That admiration is directed outward at nature as part of non-instrumental, aesthetic appreciation but also inward, toward the distinctively human resources that are discovered in the face of things that appear beyond human capacities. Understanding the Kantian sublime in these ways potentially addresses objections that his theory is anthropocentric, and emphasizes the sense of humility that runs through many of his ideas. Rather than reducing sublime appreciation to some new awareness of human moral vocation (only), high mountains, thunderclouds and lightning, vast deserts, and starry skies, and so on, are also appreciated for themselves. This interpretation of Kant's theory characterizes a form of aesthetic appreciation that supports less hubristic relations between humans and the rest of the natural world.

As a philosophy of its time, Kant's approach is anthropocentric, focused on human capabilities in both moral and aesthetic experience. Nonetheless, scholars in environmental ethics, animal ethics, and aesthetics have found valuable resources in his thought either through direct interpretation of his ideas, as indicated above, or through indirect inspiration in relation to ideas of duty, respect, beauty and sublimity.

Notes
1   *Critique of the Power of Judgment*, §28, 5:261, p. 144.
2   *Groundwork of the Metaphysics of Morals*, 429, p. 36.
3   Svoboda, T., *Duties Regarding Nature: A Kantian Environmental Ethic*, New York and London: Routledge, 2015.
4   *Critique of the Power of Judgment*, §59.
5   See Carlson, A. and Lintott, S. (eds), *Nature, Aesthetics, and Environmentalism: From Beauty to Duty*. New York: Columbia University Press, 2007.
6   *Critique of the Power of Judgment*, 5:245.
7   Brady, E., *The Sublime in Modern Philosophy. Aesthetics, Ethics and Nature*, Cambridge: Cambridge University Press, 2013; Clewis, R., *The Kantian Sublime and the Revelation of Freedom*, Cambridge: Cambridge University Press, 2009.

See also in this book
Rousseau

Kant's major writings
Groundwork of the Metaphysics of Morals, trans. and ed. M. Gregor and J. Timmermann, Cambridge: Cambridge University Press, 2012.
Critique of Pure Reason, trans. and ed. P. Guyer and A. Wood, Cambridge: Cambridge University Press, 1999.
Critique of Practical Reason, 2nd edn, trans. M. Gregor, Cambridge: Cambridge University Press, 2015.
Critique of the Power of Judgment, ed. P. Guyer, trans. P. Guyer and E. Matthews, Cambridge: Cambridge University Press, 2000.

Further reading
Brady, E., *The Sublime in Modern Philosophy. Aesthetics, Ethics and Nature*, Cambridge: Cambridge University Press, 2013.
Budd, M., *The Aesthetic Appreciation of Nature*, Oxford: Clarendon Press, 2002.
Carlson, A. and Lintott, S., eds., *Nature, Aesthetics, and Environmentalism: From Beauty to Duty*. New York: Columbia University Press, 2007.
Clewis, R., *The Kantian Sublime and the Revelation of Freedom*, Cambridge: Cambridge University Press, 2009.
Guyer, P., *Kant and the Experience of Freedom*, Cambridge: Cambridge University Press, 1996.
Kuehn, M. *Kant: A Biography*, Cambridge: Cambridge University Press, 2001.
Svoboda, T., *Duties Regarding Nature: A Kantian Environmental Ethic*, New York and London: Routledge, 2015.
Zuckert, R., *Kant on Beauty and Biology: An Interpretation of the Critique of Judgment*, Cambridge: Cambridge University Press, 2007.

EMILY BRADY

JOHANN WOLFGANG VON GOETHE 1749–1832

The alarming increase in machines torments and frightens me, they are rolling down upon us like a thunderstorm, slowly, slowly, but they are on their way, they will come upon us.[1]

The Germany into which Goethe was born on 28 August 1749 was a pre-industrial collection of statelets. By his death on 22 March 1832 this pre-eminent genius, a poet, dramatist, novelist, artist, critic, lawyer, civil servant, statesman and scientist, had lived through a period which took

Germany to the very threshold of its delayed industrial revolution. After a childhood in Frankfurt, Goethe studied law at Leipzig and Strasbourg. During convalescence from serious illness he dabbled in alchemy, the influences of whose underlying philosophy are still evident in Goethe's later approaches to both science and literature. In August 1771 he began to practice as a lawyer, but the tumultuous success in 1774 of his drama *Götz von Berlichingen* and, especially, of his epistolary novel *The Sorrows of Young Werther*, catapulted him to European-wide fame as a writer. In 1776 Goethe was called to the court of the Duchy of Sachsen-Weimar, marking the start of a life-long career in Weimar as a civil servant and minister under the patronage of the Duke Carl August. In 1782 he was elevated to the aristocracy and in the same decade began to develop his interest in the natural sciences, in the course of time covering fields including geology (he was for a time Minister of Mines), botany, optics, zoology, anatomy, morphology and meteorology. His exploration of botany and geology in particular developed during a sojourn in Italy between 1786 and 1788.[2] During the 1790s Goethe not only worked on a number of long-lasting literary projects which were to become world classics (especially *Faust*, the second part not finished until 1831), but he also began a lengthy endeavour to discredit Isaac Newton's theory of optics in favour of his own chromatics. Goethe finally published his *Theory of Colours* in 1810, by which point his literary reputation was reaching new heights with the publication of the first part of *Faust* in 1806 and of the novel *Elective Affinities* in 1809. By the 1820s Goethe's fame and acknowledged importance were such that his friend Eckermann made detailed notes of his dinner conversation over several years; this along with other sources, and the huge number of words that Goethe wrote, have provided a profoundly rich source for an assessment of his views.

It is primarily Goethe's view of nature that makes him attractive to those interested in environmental thought. Having abandoned Christianity early in life in favour of a Hellenic neo-paganism (though not in any organized or evangelical manner), Goethe allowed his holistic view of nature to inform every aspect of his work. Though he himself was wary of the term pantheism, which is conventionally attributed to him,[3] there is no doubting his holistic understanding, a spiritual dimension to his approach to 'God-Nature', and above all and everywhere apparent, his passionate veneration of the natural world. Goethe rejected a view of nature which concentrated solely on the totality, however. A perception of nature as an external, complete, static given is as limiting, indeed false, as an excessively analytical, taxonomic approach which concentrates on the detailed elements in isolation. In Goethe's view the question of the whole and of the parts is inseparable; one cannot be viewed without the other, and both must be seen as part of a process, in constant change, growth, death, rebirth. He was convinced that for this reason there was an intimate relationship between 'the demands of science' and 'the impulses of art and imitation'.[4] Accordingly, in an uncanny foreshadowing of Heisenberg's uncertainty principle, and in contradiction to the secure objectivity of the eighteenth- and nineteenth-century scientific method, Goethe insisted that there can be no separation between subject and object, between observer and the observed. The interweaving of humankind and nature precludes any such division; the very act of observation affects the observed, while the observed is capable of profoundly altering the observer. The fundamental processes of nature, the polarities of bonding and separation, of breathing in and out, as he understood it, are reflected in the human spirit; Goethe's holism would admit no other. The corollary is that there must be an ethical dimension to the relationship between nature and humankind; nature demands respect, even veneration. Nature, as observed by the scientist, is imbued with values. Herein, surely, lies much of the attraction of Goethe to the modern Green movement. Writing in an age before human beings had the capacity to shape nature in a thoroughgoing post-industrial fashion (although humankind had been leaving its mark on the planet for thousands of years), Goethe nevertheless recognized that inner nature and external nature are indistinguishable, and thus came near to the concept of inner or constructed nature which was only fully developed by Horkheimer and Adorno in the 1940s and 1950s. A further dimension of the attraction Goethe's view of nature holds for modern Greens is his insistence that nature can only be properly comprehended by means of *Ahnung*, or intuition. This does not mean a rejection of science; but it does mean a rejection of the conventional scientific method; and indeed an understanding of Goethe as a scientist is fundamental to an understanding of his thought in ecological terms.

The lasting achievement of Goethe's scientific work is also his earliest in the field of natural sciences: the discovery of the intermaxillary bone in human beings. Until Goethe's discovery, the absence of a bone in the human jaw which in animals houses the canine teeth was taken as evidence of the essential distinction between the two. The suture which remains as the indication that human beings also retain such an anatomical structure bears Goethe's name still. But it is in the theological and social, not to say scientific, importance of the recognition of a relationship between human beings and animals that the importance of this discovery lies. It points to an essential cornerstone in ecological thinking; that human beings, while in Goethe's view the crowning achievement of nature and clearly distinct from animals, are a part of nature like any other.

Although with the exception of this anatomical discovery none of Goethe's scientific revelations are of acknowledged lasting significance, his writings on science nevertheless remain the subject of lively debate. Distinguished physicists including Walter Heitler, Werner Heisenberg and Max Planck have written on Goethe. The reason for this enduring interest lies in his idiosyncratic scientific methodology.

This is nowhere more clearly or fully expressed than in the substantial *Theory of Colours*, the work which he regarded as his most important.[5] On the basis of a chance observation through a prism, Goethe became convinced that Newton's spectral theory of light was wrong, in contrast to his own understanding of light as a unity of white which achieved colour by varying admixtures of shade. To his lasting chagrin, Goethe was unable to convince his contemporaries of the correctness of this thesis, partly since he was of course utterly in the wrong. It has been argued that Newton and Goethe were in fact talking about two different things; Newton about the composition of light and Goethe about the human perception of it.[6] And it is on subjective perception that Goethe's scientific method relied. The attack on Newton was anything other than objective; indeed, a 'Polemical Section' of the work is devoted in part to denigrating Newton's character in the most scurrilous fashion. In fact, the basis for Goethe's deep disquiet was Newton's analytical methodology, which allegedly embodied the nature-dominating techniques of the scientific method. Spectral analysis using optical instruments was a dispassionate dissection, objectification and subjugation of nature. For Goethe, an account of an experiment was not a formula setting out aim, method, equipment and results, but a story in itself, which included his own feelings, the origins of the experiment, the effect on his senses; in short, a contextual narrative, the whole deriving from subjective evidence. Experiment must also be experience, easily repeatable for the reader with the most rudimentary equipment. Only such '*zarte Empirie*' (delicate empiricism) could do justice to the wholeness of 'God-Nature'. Accurate detail and linear causality were of less importance to Goethe than broad-ranging context, the network of interconnections. To be absolutely clear: Goethean science is a rejection not of science, but of a science which is contemptuous of nature. The extent of Goethe's influence can be gauged by the fact that there are today scientists working in ecology and other fields who pursue their research in an explicitly Goethean fashion.[7]

Goethe is conventionally celebrated for his literary achievements, where proto-ecological elements have also been discovered.[8] Merely on the level of content, Werther's despair at the cutting down of ancient nut-trees is an emotion with which many modern Green activists could sympathize, while the fear expressed in *Wilhelm Meister* at the ubiquity of machines (quoted at the outset) also has contemporary resonances. Goethe's refusal to distinguish between art and science often led him to give literary expression to scientific results. His poem *Metamorphosis of Plants* encapsulates the results of his essay of the same name, but the poem *Metamorphosis of Animals* is even more directly relevant for our topic. The apparent foreshadowing of Darwin is so startling as to make it worth quoting (my translation):

> Thus the form determines the animal's way of living
> And the way of living powerfully affects all forms
> In turn. The ordered formation is thus clearly shown
> Which, through the operation of outside elements, tends to change.[9]

As the biochemist Friedrich Cramer argues, this does sound very much like Darwinism.[10] At the very least there is a clear recognition here of the way in which creatures adapt to their environment, and, perhaps, of the interplay between organism and environment without which any ecological view is unthinkable. But on a more fundamental level too, Goethe's assumptions concerning nature inform his literary work. In particular, his masterpiece *Faust* has been interpreted as an attempt using alchemical metaphors to show the way in which the economy depends on the exploitation of nature.[11] Similarly, Jost Hermand argues that the long-standing misreading of the text as a paradigm of technical progress and individual ambition requires correction; in fact, it is a celebration of the natural virtues of harmony, holism and mutuality. Faust's destructive drives arise, Hermand argues, because he has lost 'all sense of human solidarity or empathy with nature'.[12] Both *The Sorrows of Young Werther* and *Elective Affinities*, as well as a number of poems, have also been refracted through an ecological prism. 'The Magician's Apprentice', for example, a poem known to every German-speaking school-child, is routinely used to demonstrate the dangers of meddling with powerful forces one does not properly understand.

Goethe's influence on the history of ecological thought is manifest: Darwin, without whose work there could be no science of ecology, cites him in the *Origin of Species*. Ernst Haeckel's late-nineteenth-century fusion of science and mysticism in the form of monism, which invested a holistic nature with spiritual qualities, is explicitly derived from Goethe. Rudolf Steiner, the founder of anthroposophy and an originator of organic farming, was deeply indebted to Goethe, as are contemporary Green campaigners of the stature of Fritjof Capra.[13] Was Goethe himself an early Green campaigner? Clearly not; despite the opening quotation, the steam engine is only mentioned explicitly a handful of times in the vast number of words he wrote, though it was invented in 1776. And there is a distinct thread of anthropocentrism, to be expected in his era, running through all his work. It would be dangerous and misleading, then, to instrumentalize Goethe in the light of contemporary concerns (though each age has appropriated him for its own purposes). But it is beyond dispute that, among his many accomplishments, Goethe remains a lasting source of inspiration to the ecological imagination.

Notes

1 *Wilhelm Meister's Travelling Years*, 1829; quotation translated from the original German by Colin Riordan.

2 He was especially impressed by Rousseau's work on botany.

3 He feared its use might lead to a simplistic categorization of his views. See letter to C.F. Zelter, 31 October 1831, WA, IV, 49.

4 See WA, II, 6: 9 (my translation).

5 Indeed, Goethe felt that it was his scientific work which would be his lasting monument.

6 See H.A. Glaser (ed.), *Goethe und die Natur*, Frankfurt am Main: Peter Lang, p. 29, 1986.

7 For examples, see especially Part II of D. Seamon and A. Zajonc (eds), *Goethe's Way of Science*, entitled 'Doing Goethean Science'.

8 Non-specialists frequently and mistakenly associate Goethe with Romanticism. In fact he was an overarching figure whose relations to the Romantics were ambivalent; there were fundamental differences in philosophy.

9 WA, I, 3: 90 (my translation).

10 Friedrich Cramer, '"Denn nur also beschränkt war je das vollkommene möglich" ... Gedanken eines Biochemikers zu Goethes Gedicht "Metamorphose der Tiere"', in Glaser, op. cit., pp. 119–32.

11 See Hans-Christoph Binswanger, 'Die moderne Wirtschaft als alchemistischer Prozeß – eine ökonomische Deutung von Goethes "Faust"', in Glaser, op. cit., pp. 155–76.

12 Jost Hermand, *Grüne Utopien in Deutschland. Zur Geschichte des ökologischen Bewußtseins*, Frankfurt am Main: Fischer, p. 58, 1991 (my translation). See also Jost Hermand, 'Freiheit in der Bindung. Goethes grüne Weltfrömmigkeit', in Jost Hermand, *Im Wettlauf mit der Zeit. Anstöße zu einer Okologiebewußten Ästhetik*, Berlin: Sigma Bohn, 1991. Gerhard Kaiser makes a very similar argument in his *Mutter Natur und die Dampfmaschine. Ein literarischer Mythos im Rückbezug auf Antike und Christentum*, Freiburg im Breisgau: Rombach Verlag, 1991.

13 See especially Fritjof Capra, *Wendezeit: Bausteine für ein neues Weltbild*, Munich: Droemer Knaur, 1999.

See also in this book
Darwin, Humboldt, Rousseau, Spinoza

Goethe's major writings

Sorrows of Young Werther, 1774, Harmondsworth: Penguin Books, 1989.

Faust: Parts One and Two, 1806 and 1831, London: Nick Hern Books, 1995.

Elective Affinities, 1809, Oxford: Oxford Paperbacks, 1999.

Theory of Colours, 1810, Cambridge, MA: MIT Press, 1970.

Conversations of German Refugees I Wilhelm Meister's Journeyman Years or The Renunciants, 1829, in *Goethe: The Collected Works in 12 Volumes, vol.* 10, Princeton, NJ: Princeton University Press, 1996.

Scientific Studies, in *Goethe: The Collected Works in 12 Volumes, vol. 12*, Princeton, NJ: Princeton University Press, 1995.

Further reading

Binswanger, H.C. and Smith, K.R., 'Paracelsus and Goethe: Founding Fathers of environmental Health,' *Bulletin of the World Health Organization*, 78 (9), 1162–4, 2000.

Bortoft, H., *The Wholeness of Nature: Goethe's Science of Conscious Participation in Nature*, Hudson, NY: Lindisfarne Press, 1996.

Hoffmann, N., 'The Unity of Science and Art: Goethean Phenomenology as a New Ecological Discipline', in D. Seamon and A. Zajonc (eds), pp. 129–77.

Holdrege, C. *Thinking Like a Plant: A Living Science for Life*, Great Barrington, MA: Lindisfarne Books, 2013.

Seamon, D. and Zajonc, A. (eds), *Goethe's Way of Science. A Phenomenology of Nature*, Albany, NY: SUNY Press, 1998.

Whyte, L.L., 'Goethe's Single Vision of Nature and Man', *German Life and Letters*, 2, pp. 287–97, 1949.

Williams, J.R., *The Life of Goethe. A Critical Biography*, Oxford: Blackwell, 1998.

COLIN RIORDAN

ALEXANDER VON HUMBOLDT 1769–1859

If one destroys the forests, like the European settlers all over America are doing with careless haste, then the sources [of rivers] run dry or diminish considerably. The riverbeds lie dry for part of the year and turn into currents whenever it rains in the mountains. As along with the growth of wood, also grass and moss disappear from the mountain crests, the run-off of rainwater is no longer held back; instead of gradually seeping into the ground and feeding the streams, the heavy rains during the rainy season create furrows on the mountain slopes, wash away the loosened soil and cause flash flooding, which in turn destroys the fields. From this it follows that the destruction of the forests, the lack of continuously flowing sources [of water] and the existence of torrents are three phenomena that are causally connected.[1]

Alexander von Humboldt has been widely recognized for his universalist approach to natural knowledge, his internationalist attitude in promoting science and scientists, and his cosmopolitan humanitarianism towards cultures and peoples. As an explorer Humboldt inspired a generation of "Humboldtian" followers, among whom Maximillian Alexander Philipp, Prinz zu Wied, the brothers Adolf, Hermann and Robert Schlagintweit, Robert Hermann Schomburgk, and Charles Darwin. Humboldt's most outstanding scientific research arguably was in climatology, meteorology, and plant geography, and in studies of the distribution of heat across the globe, for the visual representation of which he devised the isotherm.

Humboldt was born on 14 September 1769, in Berlin, the second son (the older brother Wilhelm was born 22 June 1767) of the retired army major and chamberlain at the Prussian court Alexander Georg von Humboldt and Marie Elisabeth née Colomb. His youth was spent at the family Schloss Tegel, in the vicinity of Berlin. The brothers Alexander and Wilhelm von Humboldt were still teenagers – Alexander was 16 and Wilhelm 18 – when they began taking part in the salon life of Berlin's educated Jewry and became imprinted with the values of the Berlin Enlightenment. Humboldt's formal, higher education, between 1787 and 1792, consisted of relatively brief periods at four academic institutions, in Frankfurt an der Oder, Göttingen, Hamburg, and Freiberg respectively.

Humboldt's fame and his writings were to a significant extent the products of scientific expeditions. In Paris, in 1798, he teamed up with the botanist Aimé Bonpland, and together they travelled to Madrid where they received royal permission to explore Spain's American possessions. On 5 June 1799, they set sail from La Coruña in the Korvette "Pizarro", and after stop-overs on the Canary Islands, Humboldt and Bonpland arrived in Cumaná, subsequently travelling nearly five years through what later became the countries of Venezuela, Columbia, Cuba, Ecuador, Peru, and Mexico. Especially notable were the trip along the Orinoco River and the trek south along the Andes, to Quito. Humboldt climbed several volcanic peaks, and most famously, on 23 June 1802, he attempted the ascent of Chimborazo, at the time believed to be the highest mountain in the world. Some 400 m below the very top of 6310 m, he and his companions had to give up, yet it was a new mountaineering altitude record all the same. Leaving the equatorial zone, Humboldt and Bonpland set sail for Philadelphia, USA, and visited Washington where Humboldt met Thomas Jefferson and members of the cabinet. On 1 August 1804, their trans-Atlantic return journey ended in Bordeaux.

Plans for a scientific expedition to the East, conceived not long after Humboldt's return from the Americas, suffered long delays, until at long last, in 1829, he undertook a Russian–Siberian journey, supported by Czar Nikolaus I. From St. Petersburg, he set out in the company of the micropaleontologist Christian Gottfried Ehrenberg and the mineralogist Gustav Rose, almost frantically covering 15000 km in less than six months. The trip carried them across the Ural Mountains and the West Siberian Plains to as far as the Central Asian Altai Mountains, and on the return journey, down to the Caspian Sea. The voyage provided Humboldt with comparative information to the Alps and the Andes.

Humboldt's scientific travels were not in first instance journeys of discovery of unknown territories. They were primarily journeys of observation, for the purpose of which he took along a series of state-of-the-art measuring instruments, including chronometers, telescopes, sextants, instruments for measuring magnetism, atmospheric composition, and rainfall, and a variety of less complex gadgets. With the aid of these instruments, Humboldt collected a large number of data which he used to construct a generalized picture of the distribution of environmental parameters across the globe, altitudinally as well as latitudinally. He added to his *Essai sur la géographie des plantes* (1805–7) [Essay on the geography of plants] the "Tableau physique des Andes et des pays voisins" [Natural scene of the Andes and adjacent regions], an iconic cross-sectional profile of South America, from the Pacific to the Atlantic at the latitude of Chimborazo, showing the zoned occurrence of plant life at different altitudes. Later, Humboldt formulated his famous "law" of vegetational distribution which states that the changes in plant distribution by altitude matched those by latitude. Moreover, his application of the isoline technique for plotting these data proved a revolutionary contribution to the cartographic representation of spatial relationships. Thus Humboldt introduced a style of science that consisted of precision measurement of physical parameters and their visualized distribution on a global scale.[2]

At the time, Paris was the glamorous centre of the scientific world and Humboldt became one of its stars. Starting December 1807, he made Paris his domicile and remained there till 1827. Here he put together his massive American travel oeuvre which consisted of some 30 folio and quarto volumes and carried the collective title "Voyage de Humboldt et Bonpland", which contained such major works as the *Essai politique sur le royaume de la Nouvelle-Espagne* (2 vols, 1811) [Political essay on the kingdom of New Spain]; the unfinished narrative of his journey, the *Relation historique du voyage aux regions équinoxiales du Nouveau Continent* (3 vols, 1814–1825) [Narrative of travels to the equinoctial regions of the

New Continent]; and the less voluminous yet influential *Essai sur la géographie des plantes*. One of the few German publication from this period was the popular *Ansichten der Natur* (1808) [Views of nature].

Upon his return to Berlin he presented, from 3 November 1827 till 27 April 1828, his famous Cosmos lectures – a total of 77 lectures for a mixed public. Many Germans, who at times had frowned upon Humboldt's francophile life, now accepted and celebrated him as one of Germany's great sons. No sooner had he completed the "Cosmos lectures" than he planned to get an "Entwurf einer physischen Weltbeschreibung" [Sketch of a physical description of the world] in print. Yet it took nearly two decades for this plan to be realised, and the appearance of the first volume of *Kosmos* was delayed until 1845 (five volumes appeared, the last posthumously, 1845, 1847, 1850–1, 1858, 1862). The title, in addition to indicating the vast scope of his book, gave expression to Humboldt's aesthetic-holistic epistemology, as the word "cosmos" in Homeric times had meant "ornament" and "elegance", and later had come to denote the order or harmonious arrangement of the world. The way Humboldt defined climate, in the first volume of his last great book, accurately reflected this approach:

> The term climate, taken in its most general sense, indicates all the changes in the atmosphere which sensibly affect our organs, as temperature, humidity, variations in the barometrical pressure, the calm state of the air or the action of opposite winds, the amount of electric tension, the purity of the atmosphere or its admixture with more or less noxious gaseous exhalations, and, finally, the degree of ordinary transparency and clearness of the sky, which is not only important with respect to the increased radiation from the Earth, the organic development of plants, and the ripening of fruits, but also with reference to it influence on the feelings and mental condition of man.[3]

Humboldt's holistic perception of the world was given visual expression by means of isolines, a representational device that gained widespread acceptance in the context of the Geomagnetic Project of which the international phase ran from 1829 till the early 1840s. The so-called Magnetic Association, of which Humboldt was one of the leading initiators, produced a wealth of data that was plotted on global distribution maps, making use of the isoline technique. World distribution maps showing isogonics (declination), isoclines (inclination), and isodynamics (total magnetic intensity) were produced. In addition, charts became available showing temperature, rainfall, tidal movements, and, sketched on isotherm maps, the distribution of plants, animals, humans, human diseases, even levels of civilization and mental development. A veritable revolution in visual representation and communication took place. The Humboldtian charts crossed existing language barriers, allowing the new information to be grasped by all. Whereas in the past, measurements of environmental parameters in distant parts of the globe had in many instances remained isolated observations, these could now be plotted on the isoline maps and made an integral part of the international effort, and be connected to Göttingen, London, Paris, and other European centres of scientific learning. The most famous collection of Humboldtian distribution maps was the *Physikalischer Atlas* (1845–8; 2nd edn, 1852; 3rd edn., 1892) [Physical atlas] produced by the cartographer Heinrich Berghaus to accompany Humboldt's *Kosmos*.

Many of the volumes of Humboldt's "Reisewerk" were lavishly and exquisitely illustrated and the costs of production, which came on top of the costs of the journey itself, reputedly exhausted Humboldt's private fortune (c. 100.000,00 Taler). An example of such illustrated publications are the *Atlas pittoresque. Vues des Cordillères, et monumens des peuples indigènes de l'Amérique* (1810–13) (*Researches, concerning the Institutions and Monuments of the Ancient Inhabitants of America, with Descriptions and Views of some of the most striking Scenes in the Cordilleras* (1810)). From among various geomorphological features, the volcanic peaks particularly drew Humboldt's attention, and he depicted several of these, showing the stark zonal occurrence of snow caps and eternal snow. Humboldt's holistic-visual approach went hand in hand with a landscape aesthetic that influenced a generation of geographers and painters. In 1833, he gave a presentation to the Breslau "Versammlung deutscher Naturforscher und Ärzte" [Assembly of German scientists and doctors] on the topic of art-and-science. Humboldt later incorporated his views on the importance of landscape painting for the study of nature in the second volume of *Cosmos*. It influenced a generation of Romantic landscape painters, among whom Frederic Edwin Church and Johann Moritz Rugendas, and inspired such early heroes of environmentalism as Henry David Thoreau and John Muir.

More recently, from the 1960s and 1970s onwards, environmental activists and environmental historians, among whom was Donald Worster, have written Humboldt's name on their banners. A fine-grained reading of Humboldt's writings turned up further evidence of his prescient concern for the environment, in particular with respect to the devastation caused by deforestation. In addition to the quotation at the beginning of this entry, other passages from the Relation historique du voyage / Personal Narrative of Travels have been and are being highlighted, such as the one where Humboldt attributes the drying up of

Lake Valencia, in Venezuela, to the agricultural destruction of forests, continuing as follows: "By felling the trees that cover the tops and sides of mountains men in every climate cause at once two calamities for future generations: the want of fuel, and a scarcity of water."[4]

Notes
1　Humboldt, *Personal Narrative of Travels*, 7 vols, London: Longman et al., 1814–25, vol. 2, p. 72 (with minor corrections, based on the original German).
2　Since the late 1970s, in the English-speaking world, this practice of exploration has been termed "Humboldtian science", providing historians with a framework for the discussion of Humboldt's significance.
3　Humboldt, *Cosmos*, Baltimore and London: Johns Hopkins University Press, 1997, vol. 1, pp. 317–18.
4　Humboldt, *Personal Narrative of Travels*, 7 vols, London: Longman et al., 1814–29, vol. 4, pp. 142–3.

See also in this book
Darwin, Goethe, Muir, Wilson

Humboldt's major writings
[Here the original French and German titles are cited. Multiple translations in English and other languages exist.]

Essai sur la géographie des plantes, accompagné d'un tableau physique des regions équinoxiales, Paris: Schoell; Tübingen: Cotta, 1807.
"Des lignes isothermes et de la distribution de la chaleur sur le globe", *Mémoires de physique et de chemie de la Société d'Arcueil*, 3 (1817), 462–602. The isotherms illustration appeared with an abbreviated version of the text, "Sur des lignes isothermes", *Annales de chimie et de physique*, 5 (1817), 102–12.
Voyage de Humboldt et Bonpland. Première partie. Relation historique, 3 vols, Paris: Schoell, 1814–25.
Asie centrale. Recherches sur les chaînes de montagnes et la climatologie comparée, 3 vols, Paris: Gide, 1843.
Kosmos. Entwurf einer physischen Weltbeschreibung, 5 vols, Stuttgart and Tübingen: Cotta, 1845–62.

Further reading
Rupke, Nicolaas A., *Alexander von Humboldt: A Metabiography*. Chicago, IL and London: University of Chicago Press, 2008.

Sachs, Aaron, *The Humboldt Current. Nineteenth-Century Exploration and the Roots of American Environmentalism*, New York: Viking, 2006.

Walls, Laura Dassow, *The Passage to Cosmos. Alexander von Humboldt and the Shaping of America*. Chicago, IL and London: University of Chicago Press, 2009.

Wulf, Andrea, *The Invention of Nature. Alexander von Humboldt's New World*. London: John Murray, 2016.

NICOLAAS A. RUPKE

WILLIAM WORDSWORTH 1770–1850

Nature never did betray
The heart that loved her.[1]

The name William Wordsworth is almost synonymous with 'nature poet' (and with the landscape of the English Lake District); paradoxically, Wordsworth is also the 'poet of the self' (of the inner landscape). Indeed, when Wordsworth writes, 'Nature never did betray / The heart that loved her', we see him draw together his sense of external nature both as a ministering agent, one ministering 'to' the self, and as a patient recipient of the responses of the 'heart', receiving 'from' the *inner* landscape of the 'self' the promise of both their futures.[2] Here is not the science but the experience of ecology.

Wordsworth's external and internal 'natures', while literally as old as the hills (and the lakes of his native district), were startlingly new and paradoxical ones too. His reinvention of ancient nature worship or pantheism, for example, was both a challenge to and easily reconcilable with Christian humanism, Enlightenment individualism, the heady power and energy of the industrial age, and rural Toryism.

Wordsworth was born in Cockermouth in West Cumberland, just outside the English Lake District. He grew up in the Lake District in Hawkshead near Esthwaite Lake; attended St John's College, Cambridge (1787–91); spent time in France during the early part of the French Revolution; came back to England and endured an emotional crisis of some five years' duration, precipitated by severed personal relationships, confused national loyalties, and a growing disillusionment with the progress of the French Revolution; lived in Racedown, Dorsetshire with his sister Dorothy, whose own mind and writing reveal a startlingly original though usually neglected contribution to environmental thought; and then moved with Dorothy to Alfoxden, Somersetshire (1797), to be near their new friend S.T. Coleridge. There, according to one traditional account, Wordsworth recovered, in his growing sense of a personal relationship to the natural rhythms and agency of the pastoral Somersetshire landscape, his sense of purpose.

Donald Worster writes, 'The Romantic approach to nature was fundamentally ecological; that is, it was concerned with relation, interdependence and holism.'[3] For Wordsworth, these three concepts are as much psychological as ecological, a key correspondence in Wordsworth's most significant contribution to environmental thought: his anticipation of the phenomenological perspective underlying English-born, twentieth-century anthropologist Gregory Bateson's 'steps to an ecology of mind' (and feeling). Indeed, in Somersetshire, his sense of the organic wholeness of nature appears to have grown out of his sense of a need for *personal* wholeness (whole = hale = health). As some critics suggest, Wordsworth's recovery may rather have been an escape from complex political and awkward personal responsibilities than an affirmation of an intrinsic wholeness in nature itself.[4] Nonetheless, rather than undermining the centrality of Wordsworth to modern environmental philosophy, such controversy has served to keep him at its centre.

In 1798, Wordsworth and Coleridge published *Lyrical Ballads*. To speak boldly, this book instituted a Copernican-like shift in poetry and in how we think about the relationship of our inner nature to (our?) outer nature. Copernicus replaced the geocentric (and human-centred) model of the solar system with a heliocentric model. While no such absolute shift is made in *Lyrical Ballads*, Wordsworth and Coleridge seek in their early poetry to replace the anthropocentric model of experience with what today we would call a biocentric one: indeed, in this new view, 'experience' is a general biological category not just a human one. In 'Lines Written in Early Spring', from *Lyrical Ballads*, Wordsworth writes,

> The budding twigs spread out their fan
> To catch the breezy air;
> And I must think, do all I can,
> That there is pleasure there.

Here, in a key biocentric image, 'the twigs' experience pleasure! This is, of course, a far cry from the mechanistic view of René Descartes (1596–1650), who believed that animal cries are merely the organic equivalent of the squeaking gears of machines. However, even for Wordsworth, separating himself from Descartes' belief in the essential separation of matter and spirit (of 'pleasure' from 'twigs') is no easy task. When Wordsworth writes of the 'twigs' that he '*must* think' (emphasis added [line 3]) '[t]hat there is pleasure there' (line 4), such a conclusion, he tells us in the same poem, is only after he does all he can ('do all I can' [line 3]) to prevent such an irrational thought. In dramatizing his own struggle to accept the biocentric view of experience, in using the words '*must* think', Wordsworth implies that his thoughts are somehow beyond his control. In philosophical terms, he dramatizes his discovery that his thoughts are not, as in the Cartesian tradition, self-evident or immediately knowable. For Wordsworth, the mind is not fully present to itself but is always only to be understood as an encounter with the living agency of nature, an agency that Wordsworth later in *Lyrical Ballads* calls 'One impulse from a vernal wood'. Wordsworth's locating the agency of his own thoughts in part outside himself (that is, within his environment) represents a displacement of consciousness from the presumed internal locus of the rational Cartesian mind. Here is a key sense in which *Lyrical Ballads* represents a Copernican-like displacement.

Today we understand that women, minorities and children suffer disproportionately from environmental pollution and other environmental degradations. It is thus no coincidence that the poems in *Lyrical Ballads* are not only about the tenets of an emerging 'environmental' manner of knowing or being but about female vagrants, displaced pastoralists, mad women, cold and hungry people, and even an 'Idiot Boy', in other words, the dispossessed and the voiceless. Another great central insight dramatized by *Lyrical Ballads*, then, is that environmental and social issues are inseparably linked. The underlying epistemological origin of this insight may be traced to Wordsworth's brilliant 'equation' (a verbal representation of the limit of a function in calculus) from 'Book Eighth' of his *Prelude*: 'LOVE OF NATURE LEADING TO LOVE OF MAN'. Here is a key environmental idea, namely, that all subjects (especially 'man', say even a 'lowly' shepherd) may only come to approach our full understanding and appreciation (1) *over time* ('leading to', in the above quotation) and (2) *in context* ('man' understood *in* 'nature'). 'LOVE OF NATURE LEADING TO LOVE OF MAN', then, represents a kind of anti-essentialist and emergent thinking by which the dignity of anything (of any subject) is not so much 'in and of itself' as it is in its being lovingly 'shepherded' (pun intended) by each of us through time and in space.

Wordsworth settled at Grasmere, in the Lake District, with Dorothy in 1799, a move that marked a permanent return to the region, and married Mary Hutchinson (1802). Wordsworth became Poet Laureate in 1843. His places of residence and the literary landscape that he created in the Lake District became tourist attractions – the man and the place now understood as inseparable.

To continue with Wordsworth's radical displacements and replacements, in the poem 'Nutting' and elsewhere, Wordsworth takes what was for the ancient pantheist a spirit's individual embodiment in a particular object and transfers that individuality from the object itself to each human subject's (potential) individual response to that object. Wordsworth's great achievement, then, is to transform an outmoded pantheistic (spectator–spectacle) ontology of being into a modern (participant–observer) one: the spirit indwells in the mutually constituting *relationship* between nature and human beings – not in the trees themselves. This is a view that while finding some sympathy in Enlightenment sensibility anticipates twentieth-century phenomenology, the philosophy of experience.

In 'Nutting', following a boy's 'savage treatment' of a 'shady nook of hazels', those trees 'patiently g[i]ve up / Their quiet being'. But as the boy says, 'Ere from the mutilated bower I turned', 'I felt a sense of pain when I beheld / The silent trees'. For Wordsworth, as for the present-day phenomenologist Drew Leder, 'the universal or the "spiritual" need not be conceived of as something opposed to the flesh and blood. The body itself proclaims spirit in our lives, that is, transcendence, mystery, and interconnection.'[5] The body of the boy in 'Nutting' makes this proclamation. His 'pain' draws his attention to what was his own previously 'absent' body: when we do *not* hurt, our bodies often are in the background of our awareness. (Indeed, as the twentieth-century American formulator of 'ecological ethics', Aldo Leopold, later demonstrates in *A Sand County Almanac*, the hurts or 'wounds' of the body of the natural world are for many people below the threshold of their awareness.) The boy's new awareness of his own body (emerging out of the background of his self) parallels his awareness of the bodies of the trees (emerging out of the background of nature); these trees, once only a romantic 'nook' or 'bower', have become individuals. One body (the inner body of the boy) 'calls out' the other (the outer bodies of the trees), and vice versa. Importantly, the boy's inner body, his viscera, the gut from whence his pain comes, is as much a mystery to him and as outside his own control as the life forces of the 'body' of the external natural world, the 'shady nook of hazels'.[6] Wordsworth's great achievement here is that two awarenesses (the two bodies, the two mysteries) become one for the boy. Thus the boy himself concludes the poem as if nature had a 'body' sensitive to touch:

> Then, dearest maiden, move along these shades
> In gentleness of heart; with gentle hands
> Touch

Wordsworth helps us to *embody* the earth in our experience of it.

In the context of intellectual history, Wordsworth dramatizes in 'Nutting' and elsewhere what French phenomenologist Maurice Merleau-Ponty (1908–61) codifies more than a century later. As David Abram tells us, 'Merleau-Ponty sensed (1) that there was a unity to the visible–invisible world that had not yet been described in philosophy, that there was a unique ontological structure, a topology of Being that was waiting to be realized, and (2) that whatever this unrealized Being is, we are in its depths, and of it, like a fish in the sea, and that therefore it must be disclosed from *inside*'[7] – which in fact it was for the boy of 'Nutting'. From this perspective, to be put off by Wordsworth's 'egotistical sublime', as his poetic and personal (supposedly self-centred) orientation to outer nature was called in his time, is to fail to understand a great insight of Wordsworth's: environment cannot be conceived of as distinct from a unique individual, and the uniqueness (and the unity and diversity) of that environment is only revealed through a parallel revelation of the uniqueness (and the unity and diversity) of the individual.

In a related idea, in Wordsworth's 'steps to an ecology of mind', the growth of the individual (of a person) is also an environmental, an ecological, and a community phenomenon. In 'ecological succession', for example, according to twentieth-century American ecologist Eugene Odum, natural communities transform themselves from young, immature communities into 'climax', or mature, natural communities, the latter understood as places (forests or fields) of 'stability', 'protection', and 'quality'. In Wordsworth's discussion, in 'Tintern Abbey' and in *The Prelude*, of the 'growth' or *personal* succession of his own mind and person, we find the same ecological telos toward 'stability', 'protection', and 'quality' that we find in *ecological* succession: the poet's well-known personal succession from the 'glad animal movements' of his youth to the 'Abundant recompense' of a more thoughtful maturity. In this mature ecological *and* personal stage of succession, Wordsworth hears 'the still, sad music of humanity': a sign that the love of the whole, of the whole human community, has succeeded the personal or individual delights of youth. In modern ecological and ethical terms, we would say that Wordsworth has now internalized the external, the old-growth values of 'quality' (the dignity of each human being) when not separated from the natural communities that nurtured him or her – and then, of course, of Wordsworth's desire to 'protect' and 'stabilize' those communities: the beginning of his preservation ethic.

In his pastoral poem 'Michael' and in his *A Guide Through the District of the Lakes* (1835, 5th edn), Wordsworth describes the natural and cultural histories of the vale and the people of Grasmere. Wordsworth desired to preserve, as 'a sort of national property', the mature, interdependent, natural and human communities of the 'Lake District', what he calls 'a perfect equality, a community of shepherds and agriculturalists'. This community, the product of a long succession of generations (as Wordsworth also details in the *Guide*), reflects in its *social* organization the same 'old growth' or 'climax' 'values' of 'stability', 'protection', and 'quality', aspects of its *ecological* organization, that characterize Wordsworth's personal maturation. Indeed, Wordsworth creates the psychology and sociology of ecology before that science was codified.

In a kind of paradigm for tensions in social ecology today, however, those seemingly innocent ecological, community, and conservation-oriented 'values' (or parallels) that Wordsworth represents in his *Guide* and poetry may also be seen as repressively conservative and elitist – and even dangerous, if these 'values' reify 'nature', that is, if they underwrite the shrinkage of the many possible (sustainable) natures (including 'discordant harmonies'[8]) into one (imperial) Nature – if they make some communities seem more 'natural' (rather than nonjudgmentally 'sustainable') or more inevitable than others. Tim Fulford, for example, recently asks: '[C]an we derive a political lesson about the importance of ecological consciousness from a Wordsworth whose rural Toryism is included in the account?'[9] Fulford refers in part to Wordsworth's desire – expressed in his *Guide* and in two letters to the Editor of the *Morning Post* (1844–5) – to protect (and preserve) the Lake District from vacationing industrial workers and the certain commercialization that would follow.

Ultimately, then, this desire of Wordsworth's may be seen to have multiple inflections: it may be seen as a significant anticipation of the British National Trust and Park System, as Jonathan Bate argues.[10] Indeed, as Robert Hass, former American Poet Laureate writes about the history of environmental preservation, 'Thoreau read Wordsworth, Muir read Thoreau, Teddy Roosevelt read Muir, and you got national parks.'[11] However, while such parks and preserves can be seen as signs of a culture's wisdom and foresight, they may also be seen as signs of a culture's failure everywhere *except* in the park or preserve – or why else would a culture have to protect itself from itself by setting aside land from itself – as Louis Owens muses in 'The American Indian Wilderness'. Of course, Wordsworth's proposed 'national property' is more akin to what today is called a 'sustainable development reserve', one populated with people working together in a sustainable manner, not a 'wilderness' (elitist) set aside; indeed, his 'national property' is more like Yosemite National Park in the USA *before* the indigenous people who had lived there for countless generations were forcibly removed and a 'pristine' wilderness all of a sudden 'discovered'. Thus, Wordsworth's *Guide* may be seen to evince a selfish elitism, a charge sometimes justifiably levelled at upper-(middle-) class environmentalists today, or it may be seen as anticipating today's 'sustainable development reserve' or 'ecovillage movement'. In any case, Wordsworth's notion of a 'national property' and 'a perfect equality, a community' makes for a useful bridge to a discussion of the range of 'social ecologies' available today, from the conception of a sustainable 'community' as a *political* (not a 'natural') artefact (see the American leftist and anarchist Murray Bookchin) to the conception of a sustainable community as an *ecological* 'fact' (see the German 'eco-fundamentalist' and environmental 'conservative' Rudolf Bahro).

Notes

1 'Tintern Abbey', lines 122–3.
2 See Michael Polanyi's 'From-to' structure as described in Drew Leder, *The Absent Body*, Chicago, IL: University of Chicago Press, 1990, pp. 15–17.
3 *Nature's Economy: A History of Ecological Ideas*, 2nd edn, Cambridge: Cambridge University Press, 1994, p. 58.
4 See, for example, Jerome McGann, 'The Anachronism of George Crabbe', in *The Beauty of Inflections*, Oxford: Oxford University Press, 1985, pp. 310–11.
5 Leder, op. cit., p. 68.
6 Here I apply to Wordsworth another of Drew Leder's phenomenological insights.
7 'Merleau-Ponty and the Voice of the Earth', in *Minding Nature*, Guilford: The Guilford Press, 1996, pp. 98–9.
8 See Daniel B. Botkin, *Discordant Harmonies*, Oxford University Press, 1992.
9 Tim Fulford, 'Wordsworth's "Yew-Trees": Politics, Ecology, and Imagination', *Romanticism*, 1 (2), 272–88, p. 273, 1995.
10 *Romantic Ecology*, London: Routledge, 1991, pp. 10, 47ff.
11 Interview with Robert Hass in *Mother Jones*, March/April 1997 issue (Sarah Pollock, interviewer.)

See also in this book
Bashō, Clare, Darwin, Leopold

Wordsworth's major writings
Descriptive Sketches, 1793; ed. Eric Birdsall and Paul M. Zall, Ithaca, NY: Cornell University Press, 1983.
Lyrical Ballads, 1798, with S.T. Coleridge; ed. W.J.B. Owen, Oxford: Oxford University Press, 1969.
Lyrical Ballads, with Preface, 1800.
The Excursion, 1814; reprint edn, London: Cassell, Woodstock Books, 1991.

Collected Works, 1815; see *The Complete Poetical Works of William Wordsworth*, London: Macmillan & Co., 1988, on-line edn, July, 1999, Bartleby.com.

A Guide Through the District of the Lakes, 5th edn, 1835; see W.J.B. Owen and Jane Worthington Smyser, *The Prose Works of William Wordsworth*, 3 vols, Oxford: Oxford University Press, 1974.

The Prelude; or Growth of a Poet's Mind, 1805, 1850, 1933; see, for example, Jonathan Wordsworth (ed.), *The Prelude: A Parallel Text*, London: Viking Press, 1996.

The standard scholarly edition is Ernest de Sélincourt (ed.), *The Poetical Works of William Wordsworth*, 5 vols, Oxford: Clarendon Press, 1958–65.

Further reading

Bate, Jonathan, *Romantic Ecology: Wordsworth and the Environmental Tradition*. London and New York: Routledge, 2013.

Cervelli, Kenneth R., *Dorothy Wordsworth's Ecology*. London and New York: Routledge, 2007.

Garrard, Greg, 'Romantic Pastoral: Wordsworth Versus Clare', *Ecocriticism*. London and New York: Routledge, 2012.

McKusick, James, 'Wordsworth's Home at Grasmere', *Green Writing: Romanticism and Ecology*. New York: St. Martin's Press, 2000 (2010, paperback).

Morton, Timothy, *The Ecological Thought*. Cambridge, MA and London: Harvard University Press, 2010.

Rigby, Kate, 'Romanticism and Ecocriticism', *The Oxford Book of Ecocriticism*, ed. Greg Garrard. Oxford and New York: Oxford University Press, 2014.

W. JOHN COLETTA

JOHN CLARE 1793–1864

[F]ields were the essence of the song[1]

John Clare, the self-styled 'Northamptonshire Peasant Poet', was a poet of the 'fields' in more ways than one: he himself laboured in the fields; he wrote of the life of field hands; he was a superb field naturalist; he lived through and lamented the loss of the old sustainable open-field system of agriculture; he celebrated the ecology of fields, considered not only as sites of agricultural production but as habitats (homes) of mutually dependent plants and animals; and he was, like Wordsworth, but in a much more explicitly ecological way, a great poet of what may be called phenomenological ecology: the study of 'fields' of experience. That is, rather than the study of resources (plants, animals, and minerals considered with respect to their use value) distributed throughout 'space', phenomenological ecology is the study of 'lived' relationships (i.e. experience) considered with respect to a specific 'place'.

The classical definition of 'ecology' is the study of the relationships between living things and their environments. In his poem 'Shadows of Taste', written before the science of ecology was codified and even before the word 'ecology' was coined, Clare provides us with a rhymed couplet that anticipates this definition while giving it a wider experiential dimension. Clare writes:

> Associations sweet each object breeds
> And fine ideas upon fancy feeds

This is to say that the ecological 'web of *life*' (the 'associations' or 'relationships' 'bred' between things or objects) cannot be separated from the phenomenological 'web of *being*' (the perceptual and conceptual 'feeding' of 'fine ideas' upon 'fancy', a 'fancy' that itself 'feeds' upon the associations bred by natural objects in *a food chain or web of signification*). For Clare, 'all objects of all thought' (as Wordsworth's line in its context within 'Lines Composed a Few Miles Above Tintern Abbey' itself attests) are (re)charged with a significance beyond that of mere use; all objects have agency or *being*; the *objec*tive is itself subjective. As Clare also writes in the same poem,

> Flowers in the wisdom of creative choice
> Seem blest with feeling and a silent voice

Such natural objects are subjects because they have 'feeling' and 'voice'. As *subjective* ecological objects, 'birds and flowers and insects' '[a]ll choose for joy in a peculiar way': in their ability to 'choose', they also have agency. Furthermore, biological subject-objects, unlike the passive regularities of objects in Newtonian physics, are 'peculiar'; that is, they are individuals.

In contradistinction to Albert Einstein's search for a 'universal field theory' of the space–time continuum, John Clare's 'ecological field theory' of the *place–time* continuum and its great web of being was local or situated (rather than universal) and embodied (rather than abstract) – peculiar rather than regular. Though merely thought parochial in Clare's time, such a 'situated' and 'embodied' perspective now plays a key role in the work of important contemporary historians and philosophers of science such as Donna Haraway, who seeks to replace 'relativism' (again an echo of Einstein) with 'location', substituting 'local knowledges' for 'world system' and 'webbed accounts' for 'master theory'.[2]

Clare's poems (and the 'webbed accounts' and 'local knowledges' they embody) represent an explicit response to the following questions drawn from phenomenology (the 'ecology' of experience). We can perceive individual blades of grass, but can we perceive (with just our senses) a field, if by 'field' we mean not a congeries of things but a series of relationships, a living community involved in a mutually sustainable process of self-regulation? The answer is 'no': 'relationship' is not a sensory phenomenon that may be directly perceived. However, through the mediation of culture (stories, songs, art, sport, drama, taking walks, 'beating the bounds', passing by-laws, et cetera) ecological communities (such as a field, a pasture, a commons, an ecological community) may be experienced (if not directly sensed). Not all cultures, though, provide a mode of *sustainable* (or ecological) experience. Therefore, what would the songs and stories of such a sustainable culture be like? What visual images are more sustainable than others? How would songs, stories, and imagery function to provide the feedback necessary to any self-regulatory, sustainable community, constituted by both the human and the non-human? Clare provides some direction here.

As George Deacon writes, Clare was the 'earliest collector of the songs people actually sang in Southern England'.[3] Furthermore, Clare's own literary ballads show evidence of his desire not only to commemorate that (sustainable) oral tradition but to adapt it to what for Clare could only be an uncertain future community beyond his imagining. Another key question, therefore, that emerges from reading Clare today is this: is it possible for readers of Clare in our time to recover or, especially, to reinvent in a new register (in a literary and cultural critical process I call elsewhere 'renewable historicism') the lost ecological ethic and aesthetic once embodied in the folk song and ritual of Clare's rural Northamptonshire (agri)cultural tradition (and thus point towards the creation of new stories and songs of a sustainable future), a tradition that contemporary scientific 'narratives' such as Garrett Hardin's highly influential 'Tragedy of the Commons' erase or efface?[4]

Hardin argues that a 'commons', his metaphor for any ecosystem – a lake, estuary, grassland, or even ocean or atmosphere – subject to communal or unregulated use, is at risk of a tragic ecological collapse because of a virtual law of human behaviour. Consider a grassy commons used by several families of herders. Each herder will generally find it to his or her economic advantage, when the possibility arises, to add one cow to his or her herd – and thus to the commons. In the short term, the degradation of the commons will not be great, and the loss of profit that results from this general but moderate degradation – a degradation that itself resulted from the combined independent decisions of the herders – will be shared (and *experienced* therefore in 'diluted' form) by all. However, each individual herder who decided to add one cow will reap all of the *economic* gain from that cow. Of course, according to Hardin's model, in the middle or long term the ecological and economic viability of the commons will collapse. Here then is Hardin's insidious tragedy of the commons. Hardin's atomistic view, however, assumes the operation of self-interest only; it assumes that there are no community feedback mechanisms for assessing the condition of the commons and acting upon those assessments. For Hardin, the cows may feed but the herdsmen give no feedback. Clare's poems, however, are the voice (the ecological feedback mechanism) of the herdsmen – and of the other labourers whose voices parliamentary enclosure disrupted and Hardin never heard.

In the poem 'The Wild Bull', Clare begins:

> Upon the common in a motely plight
> Horses & cows claim equal common right
> Who in their freedom learn mischiveous ways
> & driveth boys who thither nesting stray ...
> & school boys leave their path in vain to find
> A nest – when quickly on the threatening wind
> The noisey bull lets terror out of doors
> To chase intruders from the cows lap [cowslip] moores.

Here, then, is a 'story' that makes the commons a place worth preserving. Clare describes the interdependent community of the commons as a self-regulating one – one that keeps 'intruders' from despoiling the nests of those birds whose habitat it is. Clare's strategy here, then, transforms the biological principle of self-regulation into the political one of self-sufficiency, which political principle is itself echoed by the 'claim' the 'horses and cows' make for 'equal common right' (line 2) and 'freedom' (line 3). The commons, then, is a place of freedom.

But here lies a terrible political irony, as we see in Clare's 'The Fallen Elm'. In this poem, Clare shows his sophistication as a writer of environmental polemic, when he writes about one of his favourite trees, felled as part of the new economics of enclosure. Speaking to the memory of the tree, Clare writes,

> Self interest saw thee stand in freedom's ways
> So thy old shadow must a tyrant be.

Here Clare shows his insight into the fact that all landscapes (even the trees in them) under enclosure's imperial gaze must themselves be made to seem tyrannical so as to justify their despoliation, ironically, in the name of *free* enterprise. But as Clare shows in 'The Wild Bull', the land is always already a free enterprise. As Robert Pogue Harrison writes: 'In an age that rallied around the cry of "freedom", that conceived of freedom as a liberation, … in short, as a freedom *from* – in such an age, then, Clare located freedom elsewhere: in what already existed in its own right.'[5]

Like the socio-economic status of the local places that Clare defends in his verse, Clare's place in the field of English literature has, until recently, been marginal. Today, however, Clare is considered the 'finest poet of Britain's minor naturalists and the finest naturalist of all Britain's major poets';[6] the 'first true ecological writer in the English-speaking world'.[7]

Ecologist Paul Sears writes that ecology is a 'subversive subject'.[8] Natural history, for Clare, could be subversive not only because it could serve to describe healthy natural communities that would themselves serve as benchmarks against which to measure environmental devastation; natural history could also help reveal the inseparability of environmental and human concerns. As James McKusick writes: 'Clare is virtually unprecedented in the extent of his insight into the complex relation between ecological devastation and social injustice.'[9] Indeed, consider the following two lines from the poem 'Remembrances', lines that illustrate how Clare's *'ecological'* argument (*'ecological'* because it sees interdependence between premises and terms that an earlier logic overlooked) subverts conventional distinctions by suggesting relationships among categories that in the nineteenth century would have been thought to belong to separate spheres, viz., natural history (ecology), religion, agricultural policy, and continental history and imperialism. Clare laments the devastation of a place he had known and loved, 'old round oaks narrow lane':

> its hollow trees like pulpits I shall never see again
> Inclosure like a buonaparte let not a thing remain.

Clare's 'hollow trees', also called den trees today, serve as homes for several species of living things. Foresters today use the number of hollow trees per acre to indicate the status of a woodland's health. Such trees are therefore also called ecological indicators. Anticipating such an indexical function for hollow trees, John Clare, in simultaneously ecological and religious terms, compares 'hollow trees' to 'pulpits', implying that such trees are sites that proclaim (or give indication of, as would a preacher from a pulpit) the status of both our spiritual and ecological health. But, Clare tells us, such trees are threatened by the politics of parliamentary enclosure, a socio-economic process of privatizing (enclosing or fencing off) the old open-fields and of industrializing the means of agricultural production. Significantly, Clare likens his local experience of parliamentary enclosure to the imperial politics of Napoleon Bonaparte. Indeed, Clare is one of the first to recognize the interdependent relationship between colonial or imperial *politics* (symbolized by Napoleon) and colonial or imperial *biologies* (symbolized by parliamentary enclosure's effect on the 'hollow trees' of Clare's 'round oaks narrow lane'). Clare also recognizes here the interdependence between *local* and *global* (or at least continental) processes. Napoleon's destruction of life on a continental scale in Europe is related to the destruction on a local scale (in and around Clare's home village of Helpstone) of both ecological *habitats* and local social *habits* (the customs in common – 'common' understood as a relationship and a place, the commons or open-fields). Plants and animals are reduced by biological imperialism to mere commodities and elevated (as in Kew Gardens or the *Jardin du Roi*) to signs of national identity.

Tim Fulford points out that Clare's poem 'The Fallen Elm' is unique in how it 'develops a discourse of political protest from a personal response to a local landscape'.[10] Even though Clare is in many ways not part of the English Romantic literary tradition to which he belongs by date of birth, in his use here and elsewhere of *personal experience* (the foundation of being and knowing for English Romantics) as a basis for *political protest* we find the origins of a Romantic style of ecological politics. For example, Clare takes the Romantic notion of the supremacy of the 'individual' (a notion criticized by some for having emerged with and being necessary to those less desirable aspects of capitalism) and uses it to make readers aware that biotic communities are individuals too. In a poem such as 'The Lament of Swordy Well', Clare has 'Swordy Well' (a once complex biotic community that has had its ecological capital nearly spent) speak for itself as an individual. In giving a voice and a face to 'Swordy Well', Clare succeeds at least aesthetically in claiming for biotic communities the moral standing that in Clare's (and even in our own) time has only been thought due to individual human beings. As James McKusick writes, 'Clare is certainly among the first to suggest that the earth itself should have the legal right to redress of environmental grievance.'[11] Not until some 150 years later, in 1972, does a law professor, Christopher D. Stone, begin to chart the legal path towards rights for natural objects.[12]

Clare also establishes 'poverty' as an environmental category or condition – not just an economic one: Robert Pogue Harrison writes that the last stanza of 'The Lament of Swordy Well' provides 'an ominous ending, for it gives the condition of poverty a broad, almost universal extension to nature as a whole'; poverty for Clare meant 'the state of defenselessness against the forces of assault and expropriation. It did not mean destitution, at least not intrinsically.'[13] Clare, therefore, makes the vulnerability of nature natural, a real possibility (in a time when extinction, for example, was still a categorical impossibility within the stability of the Natural Theological world-view). He also anticipates the philosophical basis for what today is called the voluntary simplicity movement, poverty not as destitution but as a sustainable personal alliance with the land.

Clare's natural history poetry also dramatizes the operation of natural systems in what we might today call post-modern terms: these systems are ironic agents. For Clare, natural systems are sites of resistance to the closure of science or to any other form of institutionalized thought. In 'Shadows of Taste', Clare writes of the resistance to the taxonomic scientist on the part of insects, who 'e[v]en grow nameless mid their many names'. In 'May', from *The Shepherd's Calendar*, Clare writes about a ventriloqual bird of the grasslands, a rail, that resists a swain's (a country lad's) and a schoolboy's attempts to locate it even in the most regular terrain:

> in the grass the rails odd call
> That featherd spirit stops the swain
> To listen to his note again
> & school boy still in vain retraces
> The secrets of his hiding places

The ventriloqual voice of the rail is a deferral or displacement of its identity, one that puts a stop to the boy's search for the rail's nest – the origin or centre of its environment. Clare, then, anticipates at the ecosystem level what ecologists have only relatively recently discovered: the ironic agency of the non-human biological world.

Notes

1 Clare, 'The Progress of Rhyme', in Eric Robinson, David Powell and P.M.S. Dawson (eds), *Poems of the Middle Period 1822–1837, vol.* III, line 144.
2 Donna Haraway, 'Situated Knowledges', *Simians, Cyborgs, and Women*, New York: Routledge, p. 194, 1991.
3 *John Clare and the Folk Tradition*, London: Sinclair Browne, p. 18, 1983.
4 *Science*, 162, pp. 1243–8, 1968.
5 Harrison, op. cit., p. 219.
6 James Fisher, quoted in Margaret Grainger (ed.), 'Introduction', *The Natural History Prose Writings of John Clare*.
7 James McKusick, '"A language that is ever green": The Ecological Vision of John Clare', *University of Toronto Quarterly*, 61 (2), 226–49, p. 233, 1991.
8 Paul Sears, quoted in Donald Worster, *Nature's Economy: A History of Ecological Ideas*, 2nd edn, Cambridge: Cambridge University Press, p. 23, 1994.
9 McKusick, op. cit., p. 239.
10 Tim Fulford, 'Cowper, Wordsworth, Clare: The Politics of Trees', *John Clare Society Journal*, 14 (July), p. 47, 1995. A special 'Clare and Ecology' issue.
11 McKusick, op. cit., p. 241.
12 *Should Trees Have Standing: Toward Legal Rights for Natural Objects*, special rev. edn, New York: Avon Books, 1975.

13 Robert Pogue Harrison, *Forests: The Shadow of Civilization*, Chicago, IL: University of Chicago Press, pp. 216, 213, 1992. Contains half a chapter on Clare and nature.

See also in this book
Wordsworth

Clare's major writings
Clare published, in his lifetime, *Poems Descriptive of Rural Life and Scenery* (1820), *The Village Minstrel, and Other Poems* (1821), *The Shepherd's Calendar, with Village Stories and Other Poems* (1827) and *The Rural Muse* (1835).

The Early Poems of John Clare 1804–1822, ed. Eric Robinson, David Powell and Margaret Grainger, 2 vols, Oxford: Clarendon Press, 1989.
John Clare: Poems of the Middle Period 1822–1837, ed. Eric Robinson, David Powell and P.M.S. Dawson, 4 vols, Oxford: Clarendon Press, 1996, 1998.
The Later Poems of John Clare 1837–1864, ed. Eric Robinson, David Powell and Margaret Grainger, 2 vols, Oxford: Clarendon Press, 1984.

Also of central interest to Clare's environmental thinking is:

The Natural History Prose Writings of John Clare, ed. Margaret Grainger, Oxford: Clarendon Press, 1983.
Paperback and on-line selections of Clare's poetry are readily available.

Further reading
Barrell, John, *The Idea of Landscape and the Sense of Place: An Approach to the Poetry of John Clare*, Cambridge: Cambridge University Press, 1972.
Coletta, W. John, 'Ecological Aesthetics and the Natural History Poetry of John Clare'. *John Clare Society Journal*, 14 (July), pp. 29–46, 1995. (From a special issue subtitled 'Clare and Ecology'.)
Coletta, W. John, '"Writing Larks": John Clare's Semiosis of Nature'. *The Wordsworth Circle*, 28 (3), 192–200, 1997. (From a special issue subtitled 'Romanticism and Ecology'.)
Coletta, W. John, 'Literary Biosemiotics and the Postmodern Ecology of John Clare'. *Semiotica*, 127 (1/4), 239–72, 1999.
Goodridge, John, *John Clare and Community* (Cambridge Studies in Romanticism). Cambridge University Press, 2013.
Harrison, Robert Pogue, *Forests: The Shadow of Civilization*. Chicago, IL: University of Chicago Press, 1992.
Helsinger, Elizabeth, 'Clare and the Place of the Peasant Poet'. *Critical Inquiry*, 13, 509–31, 1987.

Heyes, Robert, 'John Clare's Natural History', in Kövesi, Simon and Scott McEathron (eds), *New Essays on John Clare: Poetry, Culture and Community*. Cambridge University Press, pp. 169–188, 2015.

Karremann, Isabelle, 'Human/Animal Relations in Romantic Poetry: The Creaturely Poetics of Christopher Smart and John Clare'. *European Journal of English Studies*, 19 (1), 94–110, 2015.

McKusick, James C., 'A language that is ever green': The Ecological Vision of John Clare'. *University of Toronto Quarterly*, 61 (2), 226–49, 1991–2.

McKusick, James C., 'John Clare's Version of Pastoral'. *The Wordsworth Circle*, 30 (2), 80–84, 1999.

McKusick, James C., 'The Ecological Vision of John Clare'. *Green Writing: Romanticism and Ecology*. New York: St. Martin's Press, 2000 (2010, paperback).

Morton, Timothy, 'John Clare's Dark Ecology'. *Studies in Romanticism*, 47 (2), 179–193, 2008.

Noble, Shalon, '"Homeless at Home": John Clare's Uncommon Ecology'. *Romanticism*, 21 (2), 171–181, 2015.

Rignall, John, H. Gustav Klaus, and Valentine Cunningham, *Ecology and the Literature of the British Left: The Red and the Green*. Routledge, 2012. (Contains two articles on Clare.)

Roy, Sraboni, '"Of Green Earth's Busy Claims": Clare, Ecology and the Sense of Nature'. *Moneta's Veil: Essays on Nineteenth Century Literature*, ed. Malabika Sarkar. Dorling Kindersley (India) Pvt. Ltd, pp. 153–172, 2010. *Studies in Romanticism*, 35 (3), 1996. (Special issue on romanticism and ecology.)

Ward, Sam, '"To List the Song & Not to Start the Thrush": John Clare's Acoustic Ecologies' *John Clare Society Journal*, 29, 15–32, 2010.

Weigner, Sarah, '"Shadows of Taste": John Clare's Tasteful Natural History'. *The John Clare Society Journal*, 27, 59–71, 2008.

Williams, Raymond, *The Country and the City*, Oxford: Oxford University Press, 1975.

W. JOHN COLETTA

RALPH WALDO EMERSON 1803–1882

Ralph Waldo Emerson once penned in his *Journals*, 'Right is a conformity to the laws of nature so far as they are known to the human mind',[1] against which we can set as a retort John Stuart Mill, 'Conformity to nature has no connection whatever with right and wrong'.[2] Mill is emphatic about humans and their achievements: 'All praise of Civilization, or Art, or Contrivance, is so much dispraise of Nature.'[3] Emerson demurs, with characteristic poetic vigour: 'In their vaunted works of Art, The master-stroke is still her part.'[4] The two met once, the transcendentalist sage of New England and the British logician framing the techniques of empirical science, contemporaries setting the contrasts of their times.

Seen now, a century and a half later, Emerson was launching an ecological view, 'harmony' with nature (we might say, rather than 'conformity'), lost as this has largely been during the flowering of humanism, science and technology, the liberal 'modernism' of whom Mill is an early type specimen. What now seems clear is that humans are nowhere near a sustainable relationship with their planet Earth, and that a radical separation, humans over nature, 'dispraising' it, has been as much part of the problem as part of the solution.

Emerson was reared in nineteenth-century New England, a promising Harvard graduate, one-time Unitarian minister. He became an iconoclast critic of his establishment. He delivered a controversial Harvard Divinity School address and was not invited back for thirty years. He gained fame from his literary essays, espousing a spiritual relationship to nature, intuitively known, ultimately an idealism of self-reliance residing in a deeply sacred world. His life was spent in Concord, outside Boston, a quiet domestic life in then rural Massachusetts, but adjacent to the Boston centres of intellectual life. Over time the novelty of his views accommodated somewhat to society; society accommodated somewhat to him. Along with Henry David Thoreau, Emerson was entered among the worthy geniuses of the traditions of which he had been so critical.

Emerson is a 'Romantic', provided one correctly understands this now somewhat outmoded term. The reference is not to a suitor overly swayed by love, but to a philosophical movement, Romanticism, that reacted to an Enlightenment overemphasis on rationalism, objectivity, Cartesian dualism, and hard science, mind versus matter, the new science that was bringing increased competence at exploiting the world, and at the same time decreased confidence about the place of humans in the scheme of things. Emerson was wondering already about the negative results. Provocatively we might say that Emerson is already a 'postmodernist', or at least that he is uncomfortable with the increasingly assertive urban, urbane 'modernism' secularizing Boston life and at once civilizing it and alienating it from the New England landscape.

Keep 'romance' in life, Emerson says; or, we might say, 'love life' in its rich fullness. Enjoy life as an 'epic, adventurous narrative' (one meaning of the French *roman* and of the English *romance*). The good life is not so much reasoned analysis, dominion over nature, rebuilt environment conquering nature; rather (as the feminists would now say) life requires appropriate respect, sensitivity and 'caring' whether in culture or nature. Humans need a deep sense of engagement with the landscape. 'Nature is the opposite of the soul, answering to it part for part. One is seal, and one is print. Its beauty is the beauty of his own mind. Its laws are the laws of his own mind.'[5]

Two of Emerson's works, similarly titled, introduce his thought. The first is his earlier small book entitled *Nature*, the original transcendentalist manifesto of 1836. The second is a later essay, 'Nature', published in 1844. 'Nature' begins with a poem:

> The rounded world is fair to see;
> Nine times folded in mystery:
> Though baffled seers cannot impart
> The secret of its laboring heart,
> Throb thine with Nature's throbbing breast,
> And all is clear from east to west.

The learned seers at Harvard University (rationalists, empiricists, scientists) are 'baffled' by the developing astronomy, geology and historical biology. They puzzled over the clockwork heavens, the rock strata, the fossil record. Asa Gray was filling his herbarium with strange plants from around the world. Science was upsetting old world-views; but an attuned heart throb understands. The sciences cannot teach us all we need to know about nature; indeed they cannot teach what we need most to know: how to value it. The wise person needs to 'transcend' this cold, mechanistic universe, known by reason and observation in its causal sequences, and to realize deeper truths.

Nature cannot be understood merely as a commodity, a *resource;* it can only be understood in *romance.* So Emerson revels in nature's 'sanctity', in the 'spell' of nature; its 'enchantments'. 'We ... make friends with matter', reconciling mind and matter. Nature 'shames us out of our nonsense'. 'Cities give not the human senses room enough.'[6] Richer aesthetic experiences are possible in forest and field – more to see, smell, touch, taste, more sense of space, time, place, proportion.

Less than a quarter of a mile away, at Walden Pond, Henry David Thoreau agreed: 'In Wildness is the preservation of the World.'[7] Socrates claimed: 'I'm a lover of learning, and trees and open country won't teach me anything, whereas men in town do.'[8] But Emerson and Thoreau objected.

In *Nature* Emerson argues that nature yields: Commodity; Beauty; Language; and Discipline. The planet's endless circulations give us sustenance, life, life-support, and prosperity. All the human useful arts but further embellish these natural cycles. As we now say, an ecology underlies every economy – a fact Bostonians were increasingly inclined to neglect. Nobler wants are served by the beauties of woods and sky. 'There is ... the necessity of being beautiful under which every landscape lies.'[9] Such beauty is reciprocal and ancillary to human character and intellectual life. In current vocabulary, Emerson has a 'virtue ethics'.

Nature's function is linguistic or sacramental. 'Every natural fact is a symbol of some spiritual fact.'[10] Rivers speak of the flux of things; rocks speak of permanence. Nature equally offers stability and dynamism – the everlasting hills, the timeless natural givens, wind, rain, sea, sky, land. Language, indeed all wisdom, roots in these earthy, proverbial symbols, as when we say that what you sow you reap, or that into each life some rain must fall. Nature disciplines, schools the will. As nature confronts us, and we figure life out, character unfolds.

'There are all degrees of natural influence', from the commodity of 'the bucket of cold water from the spring', across a spectrum to the sacramental and 'sublime moral of autumn and of noon'. 'We nestle in nature, and draw our living as parasites from her roots and grains.' 'It seems as if the day was not wholly profane, in which we have given heed to some natural object.'[11] We never have a bad day if we have enjoyed a snowfall, a field of waving grain, or wildflowers. 'He who knows the most, he who knows what sweets and virtues are in the ground, the waters, the plants, the heavens, and how to come at these enchantments, is the rich and royal man.'[12]

Nature has correlate aspects. *Natura naturata* (borrowing from medieval Scholasticism) is particular separated objects, passive, inert. These result from *natura naturans* – active energetic, the restless processes generating such objects, expressing itself in diverse and varied forms.[13] In myth, this is Mother Nature; etymologically, the root meaning of 'nature' is to give birth or spring forth. In science, this is creative natural history.

Though pre-Darwinian, Emerson is already accepting an evolutionary advance over long timespans:

> Geology has ... taught us to ... exchange our Mosaic and Ptolemaic schemes for her larger style. We knew nothing rightly, for want of perspective ... What patient periods must round themselves before the rock is formed, then before the rock is broken ... into soil, and opened the door for the remote Flora, Fauna ... How far off yet is the trilobite! ... It is a long way from granite to the oyster; farther yet to Plato.[14]

There are two faces – 'secrets', as Emerson calls them – of nature:

1   Motion, process, the flux of things, an *élan vital*, catches the element of change and development. Nature is always moving on: a 'system in transition', breaking through to new achievements in know-how and power. 'Plants ... grope ever upward toward consciousness; the

trees are imperfect men.'[15] Nature, to use current vocabulary, is 'self-organizing'.

2   Rest, changelessness or identity catches a complementary dimension. The same laws and materials are present in all its forms – from stars to men. Matter is conserved, as is energy; there is homeostasis and re-cycling. 'From the beginning to the end of the universe, she (Nature) has but one stuff.' 'The direction is forever onward, but the artist still goes back for materials, and begins again with the first elements on the most advanced stage.'[16] 'Nature is a mutable cloud, which is always and never the same.'[17] Nature's diversity and unity, its stability and spontaneity, are dialectical and complementary values.

Emerson sees wisdom in what we now call co-evolution. An animal is armed, given a niche, yet checked by its predators. An animal lives in an environment, yet has to maintain itself against that environment. So birds have feathers. Nature's order is enthusiastic and extravagant; nature seems to overdo it, but thereby succeeds.

> Exaggeration is in the course of things. Nature sends no creature, no man into the world, without adding a small excess of his proper quality ... Nature ... makes them a little wrong-headed in that direction in which they are rightest.[18]

We first think that an oak tree makes too many acorns or that the squirrel in the oak is too nervous. But the seeming waste of seeds and the squirrel's instinctive fear, usually groundless, ensures the propagation of their species. In the checks and balances of an ecosystem, this results in beauty and integrity in the biotic community, as Leopold later termed it. The 'calculated profusion'[19] adds excitement, efficiency, creativity and diversity.

There are similarities here to recent thought about the spontaneous generation of integrated order in decentralized systems, as happens in society with language and markets, or in nature with ecosystems. Such decentralized order is not low quality; to the contrary, it is richer and more diverse than centralized order.

Human life and society are, or ought to be, lived in continuity with nature. 'A man does not tie his shoe, without recognizing laws which bind the farthest regions of nature.' 'We talk of deviations from natural life, as if artificial life were not also natural.' Yet in humans there is novelty added to identity. We are not simply to 'camp out and eat roots; but let us be men and not woodchucks'.[20] Still, we should not look for the meaning of life in technological advances – hoping to use electromagnetism to grow salads quickly (or, we might say, microwave ovens to cook chicken instantly). Such accomplishments will never replace living out our threescore and ten years with roots in the soil, enjoying the seasons, spring salads included. 'Nothing is gained: nature cannot be cheated: man's life is but seventy salads long.'[21]

Homo sapiens is a microcosm, an epitome or compendium of nature, in whom nature comes to completion. At times, Emerson can seem anthropocentrist: 'All the facts in natural history taken by themselves, have no value, but are barren like a single sex. But marry it to human history, and it is full of life.'[22] Natural phenomena have their glory unrealized, until humans wake up to this, and this is a principal destiny of humankind.

Emerson closes 'Nature' trying to make sense of a certain 'deceit' in the 'face of external nature', in contrast to his opening revelry in its beauty. We travel hopefully and never arrive. 'There is throughout nature something mocking, something that leads us on and on. All promise outruns performance.' There is 'friction' and 'inconvenience'. 'Must we not suppose somewhere in the universe a slight treachery and derision?'[23]

At first yes, but ultimately no. A better perspective sees a creative discontent in which nature satisfies, but never quite fully. She is ever 'inaccessible', always remaining at an unconquerable 'distance'. We never arrive at possessing nature – 'always a referred existence, an absence, never a presence and satisfaction'. Nature is 'a vast promise, and will not be rashly explained'. She is 'fathomless'. We only touch her 'outskirts'. We never reach the end of the rainbow. This may overwhelm us with 'uneasiness' and 'helplessness', but rightly understood this should give a sense of transcendence, a higher power, a spiritual universe.[24] If, in these secular years, this seems overly romantic, consider Loren Eiseley's exclamation, as a paleontologist: 'Nature itself is one vast miracle transcending the reality of night and nothingness.'[25]

Emerson concludes his brooding over nature in the philosophical idealism that underlies all his thought: 'Nature is the incarnation of a thought ... The world is mind precipitated.'[26] But it takes long insight to see this.

Notes

1 *Collected Works, vol.* 3, p. 208.
2 John Stuart Mill, 'Nature', 1874, in *Collected Works, vol.* 10, Toronto: University of Toronto Press, p. 400, 1963–77.
3 Ibid., p. 381.
4 'Nature II', p. 226.
5 'The American Scholar', p. 55.
6 'Nature' (1844), p. 382.

7   Henry David Thoreau, 'Walking', 1862, in *The Portable Thoreau*, ed. Carl Bode, New York: Penguin Books, p. 609, 1980.
8   *Phaedrus*, 230d.
9   'Nature' (1844), p. 386.
10   *Nature* (1836), p. 18.
11   'Nature' (1844), p. 382.
12   Ibid., p. 384.
13   Ibid., p. 388.
14   Ibid.
15   Ibid., pp. 389–90.
16   Ibid., p. 389.
17   'History', p. 8.
18   'Nature' (1844), p. 392.
19   Ibid., p. 393.
20   Ibid., pp. 390–1.
21   Ibid., p. 400.
22   *Nature* (1836), p. 19.
23   'Nature' (1844), pp. 396–8.
24   Ibid., pp. 398–9.
25   Loren Eiseley, *The Firmament of Time*, New York: Atheneum, p. 171, 1960.
26   'Nature' (1844), p. 400.

See also in this book
Berry, Carson, Darwin, Griffin, Lovelock, Thoreau

Emerson's major writings
The Collected Works of Ralph Waldo Emerson, Cambridge, MA: Harvard University Press, 1971.
Nature, 1836, in *Collected Works, vol.* 1, pp. 7–45.
'The American Scholar', 1837, in *Collected Works, vol.* 1, pp. 49–70.
'Nature', 1844, in *Emerson's Essays*, New York: Thomas Crowell, pp. 380–401, 1961.
Journals, Boston, MA: Houghton-Mifflin Company, 1910.
'Nature II' (a poem), in *The Complete Works of Ralph Waldo Emerson*, vol. 9, Boston, MA: Houghton-Mifflin Company, p. 226, 1918.
'History', in *Collected Works*, vol. 2, pp. 3–23.

Further reading
McMurry, A., *Environmental Renaissance: Emerson, Thoreau and the Systems of Nature*, Athens, GA: The University of Georgia Press, 2003.

HOLMES ROLSTON III

CHARLES DARWIN 1809–1882

The plough is one of the most ancient and most valuable of man's inventions; but long before he existed the land was in fact regularly ploughed, and continues to be thus ploughed by earth-worms. It may be doubted whether there are many other animals which have played so important a part in the history of the world, as have these lowly organised creatures.[1]

Charles Robert Darwin was born on 12 February 1809 in Shrewsbury, England, fifth child of Robert Waring Darwin and Susannah Wedgwood. From the ages of eight to sixteen, he attended schools in Shrewsbury before going to Edinburgh University to study medicine. After two years he transferred to Cambridge to study divinity, where, however, he continued to indulge his boyhood passion for collecting beetles and was increasingly drawn into the company of scientists. He was a constant companion of John Henslow, Professor of Botany at Cambridge and undertook an expedition in 1831 with Adam Sedgwick the Professor of Geology. Being fit and athletic in his youth – he once felled a hare with a marble – he was gifted also with a capacity for close observation and had presented his first scientific paper at the age of seventeen on the larvae of the sea-mat *Flustra*. At the suggestion of Henslow, Darwin was taken on as naturalist aboard HMS *Beagle*, undertaking a scientific expedition to South America (1831–6). The voyage was to prove a life-shaping experience and on his return he set to work writing up his observations. In 1839 he married his cousin Emma Wedgwood. The couple first settled at Upper Gower Street in London and were soon expecting the first of their ten children. In 1842 the family moved to Down House in Kent where Darwin lived for the rest of his life. His output was prolific in the circumstances. For although he claimed to have had ample leisure 'from not having to earn my daily bread', he was plagued by illness throughout his latter years and was often incapacitated for weeks on end. He died on 19 April 1882 at Down and was buried in Westminster Abbey.

After an early focus on geology Darwin had, from 1837, begun to collect notes towards what he called his 'species theory' which was to culminate in the publication in 1859 of his most celebrated book *The Origin of Species by Natural Selection*. Eight of the intervening years – from 1846 to 1854 – had been devoted to a study of barnacles and almost all of his work following publication of *The Origin*, covering plants, animals and humans, served as corroboration for his theory of natural selection.

The central proposition of *The Origin* was that presently existing species are not independent creations but are descended from other and generally extinct species as a result of modifications brought about by chance variation and natural selection – the so-called 'theory of natural selection'. The theory of the mutability of organic species was not new. Darwin's contribution was to present a credible and purely naturalistic account of the *mechanism* by which modifications might occur, and then to amass a vast quantity of evidence which was *best explained* on the supposition that such modifications actually had occurred. He was greatly assisted in amassing this evidence by a large network of informants from all over the world and from all walks of life.

Central to the proposed mechanism was the 'struggle for existence'[2] that ensued when the population of a given species that was in principle capable of a geometrical rate of increase encountered a limited supply of resources. This is said to result in the 'survival of the fittest' (a phrase that Darwin borrowed from Herbert Spencer), where the fittest are simply those whose individual difference brings them some slight advantage as compared with fellow members of their species. These survive to reproduce, and their offspring inherit the advantageous difference. Darwin used the term 'natural selection' to describe this process – a term he chose both to contrast, and connect, with the artificial selection practised by breeders, whose effectiveness he saw as constituting strong support for his theory.[3] 'I do not pretend to adduce direct evidence of one species changing into another' he wrote in a letter to Frederick Wollaston Hutton, 'but ... I believe that this view in the main is correct, because so many phenomena can be thus grouped together and explained'.[4] Among the phenomena thus grouped and explained were the fossil record, affinities both between existing species and between existing and extinct species, and the geographical distribution of species.

In considering the circumstances which led Darwin to his theory, it is clear that his researches during the five-year voyage aboard the *Beagle* were of crucial importance, especially his observations of the affinities between existing and extinct species and of the affinities and differences between existing species such as those he found inhabiting the Galapagos islands. But even though he was aware that many of his observations could be explained by 'the common descent of species', he was not ready to embrace the idea until he could explain the mechanism. It was only in October 1838, when reading Thomas Malthus's *Essay on Population* 'for amusement', as Darwin says, that he became convinced that the struggle for existence might actually be a strong enough force to account for the modification of species. Moreover, it was only in the 1850s, after Darwin had established to his satisfaction that there were adequate means both of geographical isolation and of geographical dispersal, that he was in a position to explain the diversity, as well as the modification, of species in terms of their adaptation to increasingly contrasting environments.

Of major importance also in explaining how Darwin arrived at his theory were certain intellectual influences, chiefly that of the geologist, Sir Charles Lyell, whose *Principles of Geology* Darwin had taken with him on his long voyage. Lyell's work was crucial in several ways. Besides dramatically increasing the time available for the existence of life on earth, it demonstrated above all the power of small incremental changes to effect large results – an insight Darwin first applied to great effect in his work on *The Structure and Distribution of Coral Reefs*, published in 1842, and which arguably achieved its greatest expression in *The Origin*.

Darwin never actually wrote the major work on species that he intended to write. Instead, we have, first, the 1844 pencil sketch that he arranged to have published posthumously in the event of his early death. Then we have *The Origin* itself which was but a preliminary sketch for the 'big book', and is referred to by Darwin as an 'Abstract'. *The Variation of Animals and Plants under Domestication* (1868) was thought of as at any rate a partial fulfilment of the original plan, and shows the immense importance that Darwin attached to the evidence from artificial breeding. *The Descent of Man* (1871) in turn, grew out of a proposed chapter in *The Variation* and *The Expression of the Emotions in Animals and Man* (1872) grew out of a proposed chapter in *The Descent*.

Nor did Darwin publish his major work at a time or in circumstances of his own choosing. For that work was interrupted in June 1858 by the arrival of the manuscript from Alfred Wallace which anticipated Darwin's conclusions in almost every detail. As a result, and thanks to the intercession of Lyell and the botanist Joseph Dalton Hooker, who became a lifelong friend and supporter, the new theory was first presented as a joint paper to the Linnaean Society on 1 July 1858 entitled: 'On the Tendency of Species to form Varieties and on the Perpetuation of Varieties and Species by Natural Means of Selection'.

On its publication *The Origin* encountered fierce opposition, and even dismayed some of his friends, of whom Adam Sedgwick was the most forthright: 'I have read your book with more pain than pleasure ... other parts I read with absolute sorrow, because I think them utterly false and grievously mischievous.'[5] Many of the younger generation, on the other hand, led by Hooker and Thomas Huxley, were immediately won over. So far as religious opposition was concerned, Darwin always insisted that the theory of natural selection was compatible with belief in a God. Nevertheless, he was convinced that it thoroughly undermined the argument to design presented in Paley's *Natural Theology* which, despite the devastating attentions of David Hume, remained the argument of choice for many believers.

Many scientific objections were raised to Darwin's theory both immediately upon its publication and in subsequent years. In the period after Darwin's death, the popularity of his ideas was somewhat eclipsed, partly because of the doubt sown by the physicist, Lord Kelvin, on the likely date of the formation of the earth, and partly in the wake of Fleeming Jenkin's objection that any favourable variation would be likely to be diluted over time. Eventually however, Darwin's theory of natural selection, together with Gregor Mendel's work on genetics, which served to mitigate Jenkin's objection, were synthesized to form what came to be known as 'neo-Darwinism'. This version of the theory also finally jettisoned Darwin's earlier reliance on the inheritance of acquired characteristics and the direct effects of the environment as supplementary causal factors.

The extent and nature of Darwin's influence is still very much work in progress. While the main thrust of his theory of evolution has come to be generally accepted, there remain a number of unresolved and keenly debated issues. One is whether evolution is fuelled in the main by small changes or whether dramatic changes play a key role; also whether change is fairly constant or whether it is punctuated by periods of relatively rapid change. Another is how far the operation of natural selection is open to a variety of structural and developmental constraint. Although his work is often perceived as 'materialist', or 'reductionist', both of these designations are unhelpful, mainly because natural selection precisely points towards the plurality of explanations required to explain the variety of natural phenomena. Indeed, on one view, Darwin's genius is simply to show how all the facets of life on earth as we know it can in principle be explained as a result of the accumulation of countless haphazard and everyday events.

In placing humans at a twig's edge of a branch of the tree of life, Darwin effected a profound change in our view of the human species and of man's place in nature – a 'Copernican' displacement of humans from their place at the centre of the biosphere. He also established a view of life on earth as an open-ended and irreversible historical process, lacking any pre-determined purpose. He set an entirely new agenda for the life sciences in general, and wrote several mature ecological treatises at a time when that discipline had scarcely been conceived and was not yet named (it was named in 1863 by Ernst Haeckel, a leading champion of Darwin's views in Germany). But possibly of greatest significance is his demonstration of how it might be possible to give a purely naturalistic account of the living world. He changed forever the way in which we view the natural world, and profoundly affected the basic assumptions, modes and contexts of inquiry of both the natural and social sciences. Darwin did not so much *contribute to* environmental thought and practice as create the framework within which almost all contemporary environmental thought and practice is conducted.

But notwithstanding this huge intellectual achievement, perhaps the most abiding picture we ought to retain is of the man, on his knees, studying the earthworm to see how it drags a leaf into its burrow, or the orchid, as it fires its pollinium at the touch of an alighting insect, in a pose of humility, or indeed reverence, before the natural world that was his passion.

Notes
1   *The Formation of Vegetable Mould*, p. 288
2   *The Origin*, 6th edn, p. 46.
3   *The Origin*, 6th edn, p. 60
4   F. Darwin (ed.) *Charles Darwin*, p. 250.
5   F. Darwin (ed.) *Charles Darwin*, p. 216.

See also in this book
Callicott, Carson, Goodall, Midgley, Wilson

Darwin's major writings
Journal of Researches into the Geology and Natural History of the Various Countries Visited by H.M.S. Beagle, 1839, ed. J. Browne and M. Neve, Harmondsworth: Penguin Classics, 1989.
On the Origin of Species by Means of Natural Selection, or the Preservation of Favoured Races in the Struggle for Life, London: John Murray, 1859 (2nd edn, 1860; 3rd edn, 1861; 4th edn, 1866; 5th edn, 1869; 6th edn, 1872).
The Various Contrivances by which Orchids are Fertilised by Insects, London: John Murray, 1862.
The Variation of Animals and Plants under Domestication, London: John Murray, 1868.
The Descent of Man, and Selection in Relation to Sex, London: John Murray, 1871 (2nd edn, 1874).
The Expression of the Emotions in Man and Animals, London: John Murray, 1872.
The Formation of Vegetable Mould through the Action of Worms, with Observations on their Habits, London: John Murray, 1881.
The Correspondence of Charles Darwin, ed. F.H. Burkhardt et al., Cambridge: Cambridge University Press, 1985.

Further reading

Browne, E.J., *Voyaging. A Biography of Charles Darwin*, London: Jonathan Cape, 1995.

Darwin, F. (ed.), *Charles Darwin: His Life Told in an Autobiographical Chapter, and in a Selected Series of His Published Letters*, London: John Murray, 1902.

Dawkins, R., *The Blind Watchmaker*, London: Longman, 1986.

Desmond, A.J. and Moore, J.R., *Darwin*, London: Michael Joseph, 1992.

Eldredge, N. and Gould, S.J., 'Punctuated Equilibria: An Alternative to Phyletic Gradualism', in T. Schopf (ed.), *Models in Paleobiology*, San Francisco, CA: Freeman Cooper, pp. 82–115, 1972.

ALAN HOLLAND

HENRY DAVID THOREAU 1817–1862

The West of which I speak is but another name for the Wild; and what I have been preparing to say is, that in Wildness is the preservation of the world.[1]

No doubt there would have been an environmental movement without Thoreau, but it is hard to imagine such a movement without the rhetorical fire of his words or the inspirational force of his actions. Thoreau embodied his actions in powerful and incisive language, crystallizing concepts that shaped generations to come. His imprisonment for nonpayment of taxes that supported the Mexican War and Southern slavery kindled his protest essay 'Civil Disobedience', which inspired Mohandas Gandhi and Martin Luther King, Jr.;[2] his two-year sojourn at Walden Pond, a glacial kettle lake just a mile from town, suffused *Walden* with poetic energy. *Walden* moves from caustic criticism of American society to a lyrical intimacy with nature, as Thoreau teaches himself, and his readers, how the spirit of the one can redeem us from the evils of the other. In the decades after his death, Thoreau's new and deeper valuation of nature led to the beginnings of the environmental movement in the USA, starting with Ralph Waldo Emerson and John Muir. As Lawrence Buell writes, thousands of devotees have made pilgrimages to Walden Pond and Thoreau has become our 'environmental hero',[3] the father of American nature writing.

Thoreau was hardly born a naturalist. As a child, he joined his family's outings into the countryside around their home in Concord, Massachusetts, a county seat set in a rolling landscape of farms, lakes, rivers and second-growth woodlands. Apart from these rambles, Thoreau showed no special disposition towards nature study. His education at

Harvard turned him into an accomplished scholar of Greek and Latin, well prepared for his intended profession of schoolteaching. When his notions proved too progressive for the established schools, Thoreau opened a school of his own, soon joined by his brother John; the Thoreau brothers' school flourished until John's ill health forced them to close it in 1841. Henry's life took a further turn when John died suddenly of lockjaw (or tetanus), in January 1842. Thereafter, Henry tried various ways of making a living: as handyman, inventor and assistant in his father's pencil factory, and land-surveyor; but with the encouragement of his friend Ralph Waldo Emerson and the 'Transcendentalist' movement he inspired, Thoreau set his sights on literature as his true vocation. In 1844, Emerson bought land on Walden Pond, and early in 1845, with Emerson's blessing, Thoreau built his house there. When he moved in – on Independence Day, 4 July 1845 – Thoreau took with him the materials for his first book, *A Week on the Concord and Merrimack Rivers* (1849), a meditative re-telling of a two-week journey he and John had taken in 1839. While at the Pond, Thoreau also gathered materials for his next project, *Walden*, which grew over the years to encompass not only the events of his stay at the Pond but also the philosophy of living he learned to practice on its shores.

In July 1846, when he was living at the Pond, Thoreau was seized and jailed while running errands in town. The ensuing controversy sharpened his political thought; already a vocal abolitionist and a modest success on the lecture circuit, from the 1840s onwards Thoreau was increasingly prominent as an anti-slavery speaker and activist. Two other events at the Pond also shaped his future career. First, a few weeks after his arrest, Thoreau travelled to Maine, where on Mount Katahdin he encountered true wilderness. This experience, as he narrated it in 'Ktaadn', shattered his image of nature as a safe and nurturing mother: here, 'vast, Titanic, inhuman Nature' seemed to corner him and query, 'why came ye here before your time? This ground is not prepared for you.' It was difficult, Thoreau pondered, 'to conceive of a region uninhabited by man', for we presume our presence 'everywhere. And yet we have not seen pure Nature, unless we have seen her thus vast, and drear, and inhuman ... Here was no man's garden, but the unhandselled globe.'[4] After this revelation, Thoreau could see that even Walden's peaceful landscape held its terrors, for some element in nature was always and irreducibly Other: or, as he would soon call it, Wild.

The second event suggested a way to approach, if not fully comprehend, that otherness. As Thoreau increasingly turned to nature, he also turned to writings about nature, especially to works of natural history. After the arrival in Boston of Louis Agassiz, the famous Swiss natural scientist,

Thoreau became a naturalist himself, joining Agassiz's collecting network; by April 1847 he was shipping specimens of fish, turtles, mice and even a fox to Agassiz, who declared some of the species Thoreau collected new to science. Soon afterwards, Thoreau came to the writings of Agassiz's mentor, Alexander von Humboldt, as well as Charles Darwin and Charles Lyell, also deeply influenced by Humboldt. While Thoreau had dismissed natural history surveys as 'inventories of God's property, by some clerk',[5] here was something else again, a cosmic vision of nature as one great whole to be approached through the loving and exacting study of its myriads of details, not in the laboratory but out in the wild. Thoreau caught the Humboldt wave just as it was cresting in the USA, where Humboldtian science was stimulating the organization and funding of government-sponsored Exploring Expeditions to the American West and beyond. Humboldt promoted a global science that included organism and environment in one interconnected web, a synthesis later named 'ecology'. Thoreau's discovery of proto-ecological science was of tremendous importance to his development, for it gave him tools and models for conducting his own 'ecological' studies of the Concord environment. By the early 1850s, this new vocation absorbed most of his productive hours, including the records in his *Journal*, which eventually totalled over two million words. Under the excitement of his emerging passion, *Walden* – which had languished in manuscript form since the commercial failure of *A Week* – grew to maturity.

Published at last in 1854, *Walden* remains the classic text at the head of all American nature writing since. It is directed to all those who recognize that, like the 'mass of men', they too 'lead lives of quiet desperation'.[6] Thoreau's 'experiment' at Walden Pond sheds all but the essential trappings of 'civilized' life to reveal a more truly civil life of the mind, lived close to nature's rhythms and attentive to her creatures, which, of course, included humans. 'Not till we are lost, in other words, not till we have lost the world, do we begin to find ourselves, and realize where we are and the infinite extent of our relations', Thoreau wrote.[7] Walden is above all a place to dwell and 'find' both oneself and one's relations, and so the emphasis in *Walden* is on domestic nature. Two other works, which overlap in the time of their composition but which were not published in final form until after Thoreau's death, take up the nature of wilderness and of those whose lives border civilized and wild. *The Maine Woods* (1864) collects the narratives of Thoreau's three trips to Maine: 'Ktaadn' was followed by 'Chesuncook', in which Thoreau joins a moose hunt, and 'The Allegash and East Branch', in which he considers the mind and life of the Indian through his friendship with the Penobscot guide Joe Polis. In *Cape Cod* (1865), Thoreau visits the men and women who live in the dunes with the sea at their backs, and here, facing that sea, Thoreau considers that 'wilderness reaching round the globe, wilder than a Bengal jungle, and fuller of monsters'. Thoreau's beach delineates, like Mount Katahdin, the outermost edge of humanity and holds similar terrors: 'It is a wild rank place, and there is no flattery in it ... There is naked Nature, – inhumanly sincere, wasting no thought on man, nibbling at the cliffy shore where gulls wheel amid the spray.'[8]

Thoreau's early death, at age 44, cut short his developing projects; he was well on his way to a unique synthesis of scientific precision with a poet's love of metaphor. Most notably, 'The Succession of Forest Trees' (1860) – written in the wake of his reading of Darwin's *Origin of Species* – presents both a scientific theory accounting for patterns of forest succession and a passionate argument for intelligent forest management.[9] Not only had 200 years of human use created a patchwork landscape which Thoreau untangled with verve and brilliance, but the coming of the railroad to Walden Pond alerted him to the long-term consequences of the Industrial Revolution. The railroad had cut across a corner of Walden just before he moved there, and by the 1850s, the demand for ties and fuel had nearly levelled the surrounding forests. Once-familiar species like turkey, deer and beaver had long since been hunted out of Thoreau's neighbourhood, and his critique of capitalism included the fear that soon all open land would be fenced and posted against trespassers, outlawing the kind of long cross-country walks he took daily.[10] Yet he did not counsel despair, but began to work out solutions whereby the community would combine to create local parks and 'national preserves',[11] taking selected lands out of the system of private property and holding them in trust for all, 'a common possession forever, for instruction and recreation'. Such land, if forested, was not to be cut but to 'stand and decay for higher uses', suggesting an ethic of preservation;[12] in another late manuscript, Thoreau suggested that 'forest wardens should be appointed by the town' to oversee the management of private woodlots. He believed that Americans had much to learn from the English, who 'have taken great pains to learn how to create forests', where Americans still bushwack the infant trees or plough them into the ground.[13] Thus the seeds of the American environmental movement's two contending sides – preservation of resources and their conservation or managed use – may both be found in Thoreau's late writings.

Though he was active in educating his townspeople about better ways to live with the land, Thoreau resisted joining a 'movement', in environmental activism or anything else. He presents his reasoning in 'Civil Disobedience', where he argues that political change emerges from the convergent actions of all persons with a conscience who, based on their independent moral reasoning, resist participation in social injustice. For Thoreau, nature, too, has the power to 'resist' humankind. That is, nature is not plastic in our hands, to mould as we wish; physical nature has the power to push back, against our designs, or is simply indifferent to them, like the Titan of 'Ktaadn' or the world-circling ocean. When Thoreau looked at wild creatures, they looked back at him, and what he saw in their eyes was not his own reflection but something alien, 'wild.' Thus for Thoreau nature had its own moral standing: 'Who hears the fishes when they cry?', he asked of the shad trapped before the Billerica dam; and he went on to warn, 'It will not be forgotten by some memory that we were contemporaries.'[14] Thoreau understood that were humans removed, nature would still exist and she would not mourn. That insight, astonishing for its time, both fascinated and frightened him, for he was fundamentally a humanist in his outlook; as a result, Thoreau attended to the relationship between humans and nature in a way that few were yet capable of imagining.

Overall, Thoreau believed that power flowed from the individual to the collective. Emerson entertained this idea, but like most Romantics he was even more taken by its complement, the way in which power flowed from the whole organization through the individual. Thoreau's stubborn way of living his independent convictions unnerved his friends, but this was his way of joining his political ideals – his vision of the ultimate democracy – with his understanding of how nature worked: through a creative harmony of independent agents, each seeing to their own ends, but in their purposes borrowing each other to combine towards a higher whole. Thoreau's intellectual convictions also shaped his literary style: since the individual initiated social change, Thoreau sought to move each single reader. By turns he shocks, insults, mocks, jokes, disarms, reasons, preaches, contradicts and sings, knowing that while some readers will shake him off, others will be provoked and inspired to change themselves and, thence, society. Knowing the power of a good story, in *Walden* Thoreau tellingly offers a narrative of his own narrow escape from bondage to freedom. Of course, the point is lost if readers cannot imagine recreating the story in their own lives, and so he invites his readers to follow him, not to Walden Pond but to our own 'Walden', from which we might find our way to a life lived not in desperation but in wisdom.

For Thoreau, such a goal was inconceivable apart from nature: 'culture', the definitive characteristic of humanity, was a process of self-growth or 'cultivation' which joined human labour with the natural landscape, changing both together – like Thoreau in his famous Walden bean-field. We are not set into our environment; rather, we and our environment grow together into an interlinked whole, such that a careful look around will tell who, and what, we truly are. Thoreau's exacting observation of the landscape of Concord told him America still had a long way to go, that most human possibility still lay unrealized. If we are any closer to the civil society he imagined, it is partly because he spoke, in a way that made us listen.

Notes

1 'Walking', in *Excursions*, p. 202.
2 This essay often appears under its original title, 'Resistance to Civil Government'. After Thoreau's death this essay was reprinted under the new title, which a preponderance of evidence suggests was authorial.
3 Lawrence Buell, *The Environmental Imagination: Thoreau, Nature Writing, and the Formation of American Culture*, pp. 315–16.
4 *Maine Woods*, pp. 64, 70.
5 *A Week on the Concord and Merrimack Rivers*, p. 97.
6 *Walden*, p. 8.
7 Ibid., p. 171.
8 *Cape Cod*, pp. 148, 147.
9 The ideas in 'The Succession of Forest Trees' were developed at length in Thoreau's unfinished manuscript, 'The Dispersion of Seeds', recently published in *Faith in a Seed*, pp. 23–173. Other previously unpublished late natural history writings appear in *Wild Fruits*.
10 See Thoreau's essay, 'Walking', in *Excursions*, pp. 185–222, or in any of its many reprints.
11 *Maine Woods*, p. 156.
12 'Huckleberries', in *Wild Apples*, p. 201.
13 *Faith in a Seed*, pp. 173, 172.
14 *A Week on the Concord and Merrimack Rivers*, p. 37.

See also in this book
Darwin, Emerson, Jeffers, Lovelock, Muir

Thoreau's major writings
The standard edition of Thoreau's writings is now being issued by Princeton University Press as *The Writings of Henry D. Thoreau*, Editor-in-Chief Elizabeth Hall Witherell, 17 vols to date, 1971–. This is replacing the earlier edition, *The Writings of Henry David Thoreau*, 20 vols, Boston, MA: Houghton, Mifflin & Co., 1906.

A Week on the Concord and Merrimack Rivers, 1849, ed. Carl Hovde et al., Princeton, NJ: Princeton University Press, 1980.
Walden, 1854, ed. J. Lyndon Shanley, Princeton, NJ: Princeton University Press, 1971.
Excursions, 1863, ed. Joseph J. Moldenhauer, Princeton, NJ: Princeton University Press, 2007.

The Maine Woods, 1864, ed. Joseph J. Moldenhauer, Princeton, NJ: Princeton University Press, 1972.

Cape Cod, 1865, ed. Joseph J. Moldenhauer, Princeton, NJ: Princeton University Press, 1988.

Faith in a Seed: The Dispersion of Seeds and Other Late Natural History Writings, ed. Bradley P. Dean, Washington, DC: Island Press, 1993.

Wild Fruits, ed. Bradley P. Dean, New York: Norton, 1999.

Wild Apples and Other Natural History Essays, ed. William Rossi, Athens, GA: University of Georgia Press, 2002.

Further reading

Buell, L., *The Environmental Imagination: Thoreau, Nature Writing, and the Formation of American Culture*, Cambridge, MA: Harvard University Press, 1995.

Robinson, D.M. *Natural Life: Thoreau's Worldly Transcendentalism*, Ithaca, NY: Cornell University Press, 2004.

Richardson, R.D., Jr, *Henry David Thoreau: A Life of the Mind*, Berkeley, CA: University of California Press, 1986.

Walls, L.D., *Seeing New Worlds: Henry David Thoreau and Nineteenth-Century Natural Science*, Madison, WI: University of Wisconsin Press. 1995.

Walls, L.D. *Henry David Thoreau: A Whole Human Life*, Chicago, IL: University of Chicago Press, 2017.

LAURA DASSOW WALLS

KARL MARX 1818–1883

> The bourgeoisie, during its rule of scarce one hundred years, has created more massive and more colossal productive forces than have all preceding generations together. Subjection of nature's forces to man, machinery, application of chemistry to industry and agriculture, steam navigation, railways, electric telegraphs, clearing of whole continents for cultivation, canalisation of rivers, whole populations conjured out of the ground – what earlier century had even a presentiment that such productive forces slumbered in the lap of social labour?[1]

Karl Marx, economist and philosopher, is generally regarded as the founder of modern communism as well as a major influence on socialist theory. He was born in Trier, the son of a lawyer, and studied law and philosophy at Bonn and Berlin. After a lively and short-lived career in political journalism he sought refuge first in Paris and then in London, where he was supported financially in a life of impoverished scholarship by Friedrich Engels, with whom he collaborated extensively in his writings. He worked in the British Museum on his great study of the principles of capitalism, *Das Kapital*; it was unfinished at his death and completed by Engels from the notes that Marx left.

At the heart of Marx's thinking lies an acute sense of the damage done to human life and the human spirit by social and economic conditions, conditions which were not new but which had been exacerbated by the Industrial Revolution, as the quotation above indicates. Marx saw the rapid growth of capitalist economy as achieved by exploitation: the exploitation of one social class (the proletariat, roughly the 'working classes') by another (the bourgeoisie or owners of capital, such as the owners of mills and factories). Under these conditions all values and relations, including environmental ones, become subordinated to monetary or commercial ones: there occurs what we would now call the triumph of market values. Marx regarded this as the cause of *alienation*, of a great gulf that estranges human beings from nature, from themselves and their own vitality, and from their fellows. His ambition was to free humankind from narrow utilitarian and commercially inspired desires and help us to 're-humanise' our senses.

This acute sense of the alienating properties of capitalism, which requires us to engage in what Marx called *labour*, as opposed to productive and fulfilling *work*, remains one of his most enduring achievements. The connection he drew between these properties and the estrangement of humankind from the natural world is the principal reason for his continuing importance to thinking about the environment.

Marx's view of the moral standing of the natural world (a concept that would probably have struck him as wholly obscure), and of our relationship to it, is equivocal. In places he directly criticizes the exploitation of nature by humankind. In one essay[2] he writes that 'The view of nature attained under the domination of private property and money is a real contempt for, and practical debasement of, nature', and in the same essay he approvingly quotes Thomas Munzer as declaring it intolerable 'that all creatures have been turned into property, the fishes in the water, the birds in the air, the plants on the earth; the creatures, too, must become free'. It is usual to attribute this kind of view exclusively to the early Marx; nevertheless, in the relatively late third volume of *Das Kapital*, written between 1863 and 1883, he is still insisting that we are not the owners of the planet, whether 'we' here are construed as a society or a nation or 'even all simultaneously existing societies taken together'. We are only 'its possessors, its usufructuaries', and 'must hand it down to succeeding generations in an improved condition'.

When he writes like this Marx can appear to hold a 'stewardship' view of our responsibilities to the ecosphere. Sometimes he sounds as holistic as any modern Green could wish:

> Man lives from nature – i.e., nature is his body – and he must maintain a continuing dialogue with it if he is not to die. To say that a person's physical and mental life is linked to nature simply means that nature is linked to itself, for a person is a part of nature.[3]

The method of 'dialectical materialism' typical of later Marxist thought (it is necessary to be cautious here: Marx never used the phrase, though he often wrote of 'dialectics') also appears to promise a kind of holism. Thinking that is dialectical, in Marx's sense, is impressed by the non-static nature of things and the propensity of any state of affairs to generate contradictions and opposite states, a tension out of which new and often better conditions emerge. Hence later Marxists often welcome contradictions and conflicts as a sign that social evolution is occurring; sometimes their welcome extends to denying that it is logically impossible to maintain directly contradictory propositions (for example, that nothing can be both completely white and completely black at the same time). There is a clear connection here with the modern complaint that binary thinking (yes *or* no, black *or* white, 1 *or* 0) is at the root of the techno-rationalism that fuels our ecological ills. Engels, rather than Marx, makes explicit the relationship between dialectics and respect for nature, meaning here by 'metaphysics' roughly what we would now call binary thinking: 'Dialectics, on the other hand, comprehends things and their representations, ideas, in their essential connection, concatenation, motion, origin and ending ... Nature is the proof of dialectics ... Nature works dialectically and not metaphysically.'[4]

Marx himself sometimes appears to regret the 'disenchantment' of the world that comes from the increasing gulf between the natural world and humankind: it is this gulf that he believes communism will bridge, this conflict (among many others) that it will resolve.

> This communism, as fully developed naturalism, equals humanism, and as fully developed humanism equals naturalism; *it is the genuine resolution of the conflict between man and nature*, and between man and man, the true resolution of the conflict between existence and being, between objectification and self-affirmation, between freedom and necessity, between individual and species. It is the solution of the riddle of history.[5]

It would however be a mistake to attribute to Marx, on the basis of remarks such as these, any great degree of environmental sensitivity as we would now understand it. First, this almost romantic strain in his thinking is at odds with the far more central and dominant materialist strain. It is more typical of the mature Marx to repudiate any notion of mystical or spiritual unity between humankind and nature as an expression of false consciousness, a manifestation of the 'superstructure' put in place by priests and others in order to secure their own power base. He writes that nature 'first appears to men as a completely alien, all-powerful and unassailable force, with which men's relations are purely animal and by which they are overawed like beasts'[6]: thus we need to be liberated from such a superstitious view of nature as much as from any other kind of mystification. Disenchantment then is the name of our cure, not of our disease.

Second, Marx's labour theory of value makes it clear that nature is not to be understood as having any intrinsic worth: nature acquires worth insofar as it is transformed by human work. It is, otherwise, simply nothing: 'nature, too, taken abstractly, for itself, and fixed in its separation from man, is nothing for man'.[7] Although Marx comments here and there on the importance of respecting nature and not 'appropriating' it, the importance lies in the benefits for humankind and not in any sense for nature itself. The fundamental outlook is thoroughly anthropocentric and often Marx writes as if nature exists simply in order to be used: 'The worker can create nothing without nature, without the sensuous external world. It is the material in which his labour realizes itself, in which it is active and from which, and by means of which, it produces.'[8] The danger of mastering nature is not simply that we shall lose our awe for the natural world, but that we shall do so only to replace it with awe for the man-made one. 'What a paradox it would be', he writes,

> if the more man subjugates nature through his labour and the more divine miracles are made superfluous by the miracles of industry, the more he is forced to forgo the joy of production and the enjoyment of the product out of deference to the power of technology and those miracles of the industrial process.[9]

Third, Marx has a pronounced tendency to deprecate peasant communities and those who work on the land in traditional ways – people who are, we might think, significantly in touch with nature – as reactionary and superstitious. He placed his hopes of progress in the urban proletariat, whom he expected to form the backbone of revolution. He saw industrial capitalism, whose effects he criticized so eloquently, as nevertheless a necessary phase in sweeping away the old peasant economies and moving humankind forward into a new age when the limitations of capitalism would be transcended in turn.

What can we say about the environmental legacy of Marx and Marxism? Certainly the architects of Soviet Marxism showed no sign of nostalgia for the peasant way of life: they collectivized it at enormous human cost, introducing a form of factory-farming that mirrored in the countryside the industrialization of the cities. But it is simplistic to blame Marxism for the environmental shortcomings of the few socialist republics that have taken Marx's writings as doctrine. For example, the devastating pollution of parts of the former Soviet bloc, or the disaster of Chernobyl, can probably be traced to a significant degree to the over-rapid industrialization of backward economies and to a host of other factors, including the reluctance of many Western countries to share the benefits of advanced technology with regimes to which they are ideologically opposed.

Notes
1   Karl Marx and Friedrich Engels, *The Communist Manifesto*, 1848.
2   *On the Jewish Question*, 1844.
3   *Economic and Philosophical Manuscripts*, first manuscript.
4   *Socialism: Utopian and Scientific*, 1892.
5   *Economic and Philosophical Manuscripts*, third manuscript, emphasis added.
6   *The German Ideology*, 1846.
7   *Economic and Philosophical Manuscripts*, first manuscript.
8   Ibid.
9   Ibid.

See also in this book
Bahro, Bookchin, Passmore

Marx's major writings
Economic and Philosophical Manuscripts, 1844.
The Communist Manifesto, with Friedrich Engels, 1848.
The Eighteenth Brumaire of Louis Bonaparte, 1852.
Grundrisse, 1857.
A Contribution to the Critique of Political Economy, 1859.
Kapital, 1867–94.
These texts can be found in numerous editions. They are conveniently available on the world-wide web at the Marx/Engels Library, whose homepage can be found at www.marxists.org/archive/marx/works/index.htm. All quotations in the article above are taken from this electronic edition.

Further reading
Burkett, P., *Marx and Nature*, New York, St. Martin's Press, 1999.
Fromm, E., *Marx's Concept of Man*, New York: Frederick Ungar, 1973.
Foster, J.B., *Marx's Ecology: Materialism and Nature*, New York: Monthly Review Press, 2000.
McLellan, D., *The Thought of Karl Marx*, London: Macmillan, 1980.
Parson, H.L. (ed.), *Marx and Engels on Ecology*, Westport, CT: Greenwood Press, 1977.
Schmidt, A., *The Concept of Nature in Marx*, London: New Left Books, 1971.
Singer, P., *Marx*, Oxford: Oxford University Press, 1980.

RICHARD SMITH

JOHN RUSKIN 1819–1900

[Building a city fit for people to live in, Ruskin wrote, means] remedial action in the houses that we have; and then the building of more, strongly, beautifully, and in groups of limited extent, kept in proportion to their streams, and walled round, so that there may be no festering and wretched suburb anywhere, but clean and busy street within, and the open country without, with a belt of beautiful garden and orchard round the walls, so that from any part of the city perfectly fresh air and grass, and sight of far horizon, might be reachable in a few minutes' walk.[1]

John Ruskin was born of a possessive mother and wine-merchant father. Instead of being sent away to school he was tutored at home; at Oxford University he won the prestigious Newdigate prize for poetry in 1839. He published the first volume of his series *Modern Painters*, establishing the importance of the painter Joseph Turner, when he was 24 years of age. He rapidly established a reputation as the foremost art critic of his time, later holding a Chair of Fine Arts at Oxford. His personal life was not wholly happy: his wife Euphemia ('Effie') Gray divorced him, on the grounds that the marriage had not been consummated after five years, to marry the painter Millais who had been an intimate friend of the couple. Three years later Ruskin met and fell in love with the young Rose La Touche; when he eventually proposed to her, Rose's parents opposed the marriage. In late life he experienced periods of mental illness, exacerbated perhaps partly by this disappointment and partly by the strain of the libel action brought against him by the painter Whistler (whose *Nocturn in Black and Gold* he had accused of 'flinging a pot of paint in the public's face'). He resigned his Oxford Chair in 1879, four years after Rose La Touche's death and two years after Whistler's technical victory in his lawsuit (Ruskin was ordered to pay damages of one farthing). A connection between his resignation and Oxford University's proposal to sponsor vivisection is unproven. His house, Brantwood, overlooking Lake Coniston in England's Lake District, still stands as a memorial to many of his artistic and environmental ideals.

By the 1860s Ruskin was drawing significant connections between art and architecture on the one hand, and the natural world and social and economic conditions on the other. He criticized the economic thinking of his day for emphasizing material wealth at the expense of social welfare, and insisted on the moral basis of any true economics:

> The idea that directions can be given for the gaining of wealth, irrespectively of the consideration of its moral sources, or that any general and technical law of purchase and gain can be set down for national practice, is perhaps the most insolently futile of all that ever beguiled men through their vices.[2]

The conditions of industrial mass production, he argued, were destructive of human sensibility and of a harmonious relationship with nature. They involved making the worker into a tool, his fingers like cog-wheels and his arms like compasses. Demanding 'engine-turned precision' of human beings is a degradation of them.

More than that, to demand precision or perfection goes against what we understand of the natural world of which we are part. Nature teaches us that imprecision and imperfection are essential if anything is to be good. This is what might be called 'The Foxglove Principle':

> Nothing that lives is, or can be, rigidly perfect: part of it is decaying, part nascent. The foxglove blossom, – a third part bud, a third part past, a third part in full bloom, – is a type of the life of this world.[3]

It is this principle that Ruskin believed was enshrined in the 'rude and wild' Gothic architecture that he revered. It displayed a 'look of mountain brotherhood between the cathedral and the Alp', its ruggedness and even crudeness paying homage to its models in nature. Since no truly great man stops working until he has reached his point of failure, Ruskin notes, it follows that 'no good work whatever can be perfect, and *the demand for perfection is always a sign of a misunderstanding of the ends of art*' (ibid., emphasis in original). He is prepared to point up the paradox: 'Of human work none but what is bad can be perfect' (ibid.). It would be interesting to hear his comments on that tawdry educational slogan of our times, 'Excellence'.

Gothic architecture, furthermore, is the product of the medieval guild system, which Ruskin viewed with romantic eyes as embodying 'healthy and ennobling labour'. It rejected the idea of the division of labour with its excessive specialization and repetition, and thus it involved work which was intrinsically satisfying, as opposed to work which merely makes possible the acquisition of satisfactions through the wages it commands. It fostered creativity, or *Invention*, as Ruskin calls it, never demanding exactness for its own sake but only where there is a practical or aesthetic need for exactness; it discouraged mere imitation. The cathedral's gargoyles are 'signs of the life and liberty of every workman who struck the stone' (ibid.). Here Ruskin's ideas are remarkably similar to those of Marx and Engels, who shared his sense of the damage done by industrialization; the language of his denunciation of the evils of industrialization is considerably more vehement and impassioned even than theirs.

Ruskin's reputation now is not primarily that of an environmental thinker. Yet his formulation of the connections between social and economic, artistic and what we would now call environmental questions is important and humane. It is not misleading to call his thinking holistic in its lively sense of those interconnections. At the same time he repudiates any easy distinction between anthropocentrism and the idea of intrinsic value in nature, in a way that many later writers might learn from:

> The desire of the heart is also the light of the eyes. No scene is continually and untiringly loved, but one rich by joyful human labour; smooth in field; fair in garden; full in orchard; trim, sweet, and frequent in homestead; ringing with voices of vivid existence ... As the art of life is learned, it will be found at last that all lovely things are also necessary: – the wild flower by the wayside, as well as the tended corn; and the wild birds and creatures of the forest, as well as the tended cattle.[4]

Ruskin was a powerful influence on the development of socialism, on the arts and crafts movement of the later nineteenth century, and on a diverse range of thinkers. For example, Gandhi reported that he discovered some of his deepest and most life-transforming convictions from reading *Unto This Last* on an overnight train from Johannesburg to Durban. William Morris wrote the utopian *News from Nowhere* in 1890: it is a vision of a pastoral and ecologically harmonious England that we would now perhaps call 'ectopian'. It has been claimed that 'the most important period of green politics before 1980 lay between 1880 and 1900'.[5] During this twenty-year period were founded many environmental and conservation groups, such as the Edinburgh Environment Society and the Coal Smoke Abatement Society. Ruskin's ideas inspired many of these, as well as the founding of the National Trust and the Society for the Protection of Ancient Buildings.

Notes
1   *The Mystery of Life and its Arts*, 1868.
2   *The Veins of Wealth*, 1862.
3   *The Stones of Venice*, 1851–3.
4   *Unto This Last*, 1887.
5   P. Gould, *Early Green Politics*, Brighton: Harvester, 1988.

See also in this book
Marx

Ruskin's major writings
The Stones of Venice, Orpington, UK: G. Allen, 1886.
Unto This Last and Other Writings, ed. C. Wilmer, Harmondsworth, UK: Penguin, 1985.
Seven Lamps of Architecture, New York: Dover Publications, 1995.

Further reading
Anthony, P., *John Ruskin's Labour: A Study of Ruskin's Social Theory*, Cambridge: Cambridge University Press, 1983.
Craig, D.M., *John Ruskin and the Ethics of Consumption*, University of Virginia Press, 2006.
Landow, G.P., *Ruskin*, Oxford: Oxford University Press, 1985.
Wheeler, M. (ed.), *Ruskin and Environment: The Storm-cloud of the Nineteenth Century*, Manchester: Manchester University Press, 1995.

RICHARD SMITH

FREDERICK LAW OLMSTED 1822–1903

The dominant and justifying purpose of Central Park was conceived to be that of permanently affording, in the densely populated central portion of an immense metropolis, a means to certain kinds of REFRESHMENT OF THE MIND AND NERVES which most city dwellers greatly need and which they are known to derive in large measure from the enjoyment of suitable scenery.[1]

Among the many fields in which he excelled, Frederick Law Olmsted, as a noted journalist, travelled through the pre-Civil War Southern states from 1852 to 1856, reporting on the social abuses of apartheid to the *New York Daily Times*. In several books, such as *A Journey in the Seaboard Slave States* (1856), Olmsted exposed the social and economic deprivation of negroes in America. His writings became rallying documents for the repeal of slavery. In 1855, Olmsted became the managing editor of *Putman's Monthly Magazine*, and in 1866 he was to become one of the founders of *The Nation*, a national intellectual monthly.

Olmsted also was a 'scientific' farmer, from 1844 to 1852, utilizing new agricultural methods and advancing horticultural cultivars on his successive farms at Hartford, Connecticut, and Staten Island. In his travels, he observed the latest innovations and recorded them in his many writings for the *Horticulturist*, a monthly journal.

Olmsted too was a great reformist public administrator. He became the first Superintendent of Central Park in New York City in 1856 and during the Civil War he founded and directed the American Sanitation Commission, which became the blueprint for the American Red Cross. Olmsted was also a social critic of America's cities. Joining the Century Association in New York City in 1856 and while living on the Lower East Side of Manhattan, Olmsted banded together with a group of radical artists, writers and religious leaders – including William Cullen Bryant, Washington Irving, and Andrew Jackson Downing – to discuss strategies to alleviate poverty, poor sanitation and lack of organization in services to the poor.

As significant as these achievements were, Olmsted's most noted accomplishments were the creation of public parks and the establishment of a new profession – landscape architecture. Landscape architecture, which he founded with Calvin Vaux, an architect who had trained under the landscape artist A.J. Downing, was developed for the purpose of creating a specific type of urban open space. Olmsted and Vaux, in 1857, entered the design competition for the new park for New York City. They named the project by its advocacy and location, 'The Central Park of New York'. Of thirty-two entrants, their 'Greensward Plan' won.

The design of Central Park was uniquely American. Central Park broke with all precedents. It was revolutionary in social response, power and control, in layout and organization and in emotional content. Until then, no other city had such a park. In Europe, parks were generally either remnants of royal preserves or parks built for the privileged few, with limited access. Central Park broke all traditions in size alone. Its 770-acre expanse was enormous, greater than any park that had ever been proposed and was a huge undertaking in terms of expenditure of money and manpower. It was promoted for many reasons: scientifically, for the prevention of malaria and for clean air; economically, to provide employment at low wages at a time of recession; and to increase land values for real estate profit. It provided the city with a new reservoir and water delivery system and converted polluted and derelict pig farms with clean fill and erosion control vegetation. It provided improved positive drainage and storm water management through new streams and seemingly natural water courses. And it provided much-needed public infrastructure for the future growth of the city. In its infrastructure, it was visionary, providing for grade-separated crosstown through traffic and grade-separated internal park circulation for carriages, pedestrian and equestrian traffic. Like all parks since Central Park, the project was proffered to the public on issues of health, safety and welfare.

So revolutionary was the design that it was often criticized by public officials as being too ambitious, but as it was built, enthusiastic approval attracted unprecedented numbers in great social and economic diversity. Its landscape character replicated, in well-defined areas, the very landscapes that the Hudson River School painters, such as Frederick E. Church and Asher B. Durand, had captured on their canvases. Through this new park development, and for the very first time, a park was designed for the average citizen. It was democratic. Even the most indigent of New York City's citizens were able to experience the beauty and pleasure of the scenic natural settings available previously to only the most wealthy. Central Park was accordingly dubbed the People's Park in which, as William Cullen Bryant put it, citizens could 'forget for a time the rattle of the pavements and the glare of brick walls'.

The construction of Central Park proved enormously beneficial. New York gained in reputation in Europe where, previously, the city was thought to be foul and unrefined. The park proved very useful to politicians as well, as the recession of 1857 had left many people unemployed and park building provided employment for many at low wages. The park provided an upgraded image and viewing space for the city's cultural institutions, which competed for a place within, or on its perimeter. It accreted New York's most significant civic institutions: the Metropolitan Museum, the City of New York Museum and many others. It proved very successful for the property owners adjacent to the park, as their property values escalated overnight.

Central Park's reputation quickly spread world-wide. Every American city wanted a park of this type. The American Park Movement was born. The city of Brooklyn was the second to commission Olmsted and Vaux to design a new public park on an abandoned brick quarry. Olmsted refashioned the central part of the quarry into the Long Meadow, an undulating sloping green expanse with an axis of a curvilinear valley similar to those of the Hudson River paintings. All walkways were designed along the perimeter of the space. Plantings were added for depth and layering of distant views. It was a magnificent composition in total. Years later, after completing hundreds of parks, Olmsted stated that his singular most successful landscape space was that of the Long Meadow of Prospect Park, as it encompassed all the attributes of the picturesque style.

Olmsted's ability to make every natural feature a design asset enabled him to produce brilliant regional parks and solve many urban problems at once. As his public works commissions grew larger and more extensive, they became park systems organizing an entire city. Boston's 'Emerald Necklace' of 1875, for example, strung together widely differing parks. These included the Commons, a traditional New England pastoral space; the Garden, a Victorian public park; the Muddy River Run; Jamaica Pond; and the tree-lined Massachusetts Avenue. The Emerald Necklace established America's first green corridor, created a unified park space – a serpentine connection from natural feature to natural feature, through existing regional settlements, and tying Boston proper with its newly annexed suburbs. It has guided all urban development throughout Boston's hundred years of growth. During the 1890s, Olmsted's disciple and partner, Charles Elliot, expanded the park system as proposed by Olmsted and developed new parkland acquisition criteria based on five new scientific principles: safeguarding drinking watershed; providing tidal estuaries to protect the urban populace from diseases; preserving unique scenic resources; designing for river flood planes; and establishing barrier beaches. The full concept of a uniquely American urban park system was formulated around naturalistic, scenic and conservation design parameters.

From this beginning, the profession of landscape architecture grew in multiple directions, justifying its broad definition as 'architecture of the land', or 'design of land and the objects placed upon it', or 'design of all exterior spaces'. For some, it was the beginning of a new career in a profession that had no name, definition or direction for future growth. Olmsted himself groped for a name before combining two words, prevalent at the time, to best describe the profession – 'architecture' from the art of building, and 'landscape' from the art of painting. Olmsted thought that building parks was closely related to the art of building the landscape scenes of paintings and gave the profession its name, 'landscape architecture'.

The most written-about project in America after Central Park must surely have been the great Chicago's World Columbian Exposition. It was a watershed for architecture, planning and landscape architecture. In his 1894 Chicago Fair plan, Olmsted managed to create a master plan that responded perfectly to the widely divergent design philosophies of the Prairie and Classical schools of design. For the Classical, eclectic buildings of the east coast's group, Olmsted proposed a central formal water basin around which the building would form an imposing unified urban/civic space. For the organic American Prairie Architecture School, Olmsted proposed the Lagoon area, with softened edges and romantic islands. And for his own naturalistic and American park landscape, he proposed the lake-front barrier beach areas. Together, the Fair's landscape composition embodied not just answers to specific problems Olmsted encountered, but it created a system of connected park types which had applications to many varied urban conditions. The principal positive contribution of the Fair was that it represented a total cohesiveness of design from section to section in spite of varying architectural styles, uses, commercial enterprises and land forms. The Fair was distinctly urban and presented a new urban ideal of beauty, codified in America as the City Beautiful Movement – a movement continued by, among others, his son, F.L. Olmsted jr, who had worked with his father on the Chicago Fair plan and was to become a distinguished landscape architect in his own right.

In the City Beautiful philosophy, formal landscape architecture became a powerful counterpoint to the informal pictorial, naturalistic landscapes. Its design philosophy drew from the architecture and art of the Classical periods and used art construction terms as the basis for its rational design methodology. It captivated many landscape architects who collaborated with these architects or with landscape architects who were engaged in planning the urban expansions of America's cities. The resultant combination of Olmsted's urban parks combined with City Beautiful boulevards, local parks and natural riverway conservation corridors have given these cities a uniquely American green infrastructure.

The designs of all American parks were a distinct break from the European parks of England, France and Italy. Olmsted loved the natural scenery and began to introduce systematically ecological processes within his parks. Olmsted's parks were works of nature. In place of water basins, for example, there were natural water bodies such as lakes and cascades, all representative of natural water courses. The trees of Central Park were not trimmed bosques of singular types of trees nor ornamentals as in European parks, but were complete collections of species, of differing vegetation types and differing ecological associations, all representative of the region's ecology. An Olmsted park always provided a wilderness area – the Ramble in Central Park, for instance, an area where the forces of nature were left to define the parkland in complete, perfect representation of the natural environment, or the Bramble in Brooklyn's Prospect Park.

Olmsted, in the prime of his professional life, approached the problem of the conservation of scenic areas. He realized the uniqueness of America's majestic landscapes and witnessed the encroachment by commercial developments. In 1865, while in California working as manager of the Mariposa mines, Olmsted visited the scenic valley of Yosemite Falls. Included in the tour was the stunning Mariposa Big Tree Grove featuring some of the most mature Giant Sequoia trees in America. Enthralled by the majestic scale of the valley and its delicate waterfall suspended high above the valley floor, he envisioned its despoilment by commercial loggers, miners and other resource-extracting enterprises. With support from leading American conservationists, he successfully petitioned the United States Congress to set these lands aside, 'granting the Yosemite Valley to the State of California as a public park' and to create a commission to manage this 'land grant'. In turn, he became the preserve's first Commissioner and set in course the concepts and the basis for America's National Park System. This action is considered to be the centrepiece of the American Conservation Movement, which is generally placed in the period 1850 to 1920. This movement, led by such American conservation notables as Henry David Thoreau, John Muir and George Perkins Marsh, began the unprecedented public and private initiatives intended to insure the wise and scientific use of natural resources, and the preservation of wildlife, forestry and landscapes of great natural beauty.

In 1880, Olmsted visited Niagara Falls with the intent of rekindling boyhood memories. He was shocked and disheartened by the rampant commercialization of both the American and Canadian sides. Armed with the support of Canadian colleagues, he strove for the first international park to organize the visitor experience of this scenic wonder. In 1887, Olmsted submitted his plan to remove all commercial enterprises from the reserve and provide visitor facilities open to the general public. His plans for parklands on the American embankment and on Goat Island were carried out and the natural ecology restored. While the Canadian side was approved quickly and the removal of commercial enterprises swift, their landscape development was that of the formal European Park. Olmsted's effort at Niagara Falls was the first park of a state-wide park system established for a state, New York. It was also instrumental in establishing the need for regional and state parklands across America.

To Olmsted no land-oriented problem seemed out of bounds for his professional interest. Olmsted, the social reformer, took on the design of numerous new social institutions. In Buffalo, for example, he planned the State Asylum, a mental health facility, and in Boston the Massachusetts General Hospital. He worked on America's most distinguished university campuses. He remodelled many of these, such as Yale University, planned the expansion of others, and designed whole new institutions, including Stanford University in California and the University of Florida at Gainesville. Olmsted rethought the typical American residential subdivision, giving new order to commercial centres, street patterns, housing mixes by densities and, most importantly and unique for its time, Commons as public open space. These projects, along with parks and park systems, city plans and private estates, formed the scope of landscape architecture for the professionals of his day to those of the present.

Note

1   F.L. Olmsted and C. Vaux, *The Conception of the Winning Plan Explained by its Authors, Part Two; The Greensward Plan*, Central Park Competition, New York: New York, pp. 1–6, 1856.

See also in this book
Muir, Nash, Thoreau

Olmsted's major writings

Walks and Talks of an American Farmer in England, 2 vols, New York: G.P. Putnam, 1852.

A Journey in the Seaboard Slave States: With Remarks on Their Economy, New York: Dix & Edwards, 1856.

A Journey through Texas: or a Saddle-Trip on the South-western Frontier; with a Statistical Appendix, New York: Dix & Edwards, 1857.

A Journey in the Back Country, New York: Dix & Edwards, 1860.

The Frederick Law Olmstead Papers, 12 vols., various editors, Baltimore, MD: Johns Hopkins University Press, 1972–2016.

Further reading

Beveridge, Charles E. and Rocheleau, Paul, *Frederick Law Olmsted: Designing the American Landscape*, New York: Rizzoli, 1995.

Fabos, Julius, Milde, Gordon T. and Weinmayer, V. Michael, *Frederick Law Olmsted, Sr.: Founder of Landscape Architecture in America*, Amherst, MA: University of Massachusetts Press, 1968.

Hall, Lee, *Olmsted's America: An 'Unpractical' Man and His Vision of Civilization*, Boston, MA: Bulfinch Press, 1995.

Newton, Norman T., *Design on the Land: The Development of Landscape Architecture*, Cambridge, MA: Harvard University Press, 1971.

Roper, Laura Wood, *FLO: A Biography of Frederick Law Olmsted*, Baltimore, MD: Johns Hopkins University Press, 1973.

R. TERRY SCHNADELBACH

JOHN MUIR 1838–1914

> In God's wildness lies the hope of the world ... The great fresh
> unblighted, unredeemed wilderness. The galling harness of
> civilization drops off, and words heal ere we are aware.[1]

Of his beloved wild Sierra, John Muir wrote, 'mountains as holy as Sinai
... they are given, like the gospel, without money and without price. 'Tis
heaven alone that is given away.'[2] Like the mountain creatures he so
admired, ranging from prophets of old to grizzly bears, Muir was the
mountain embodied: 'I am hopelessly and forever a mountaineer',[3] he
wrote, and it was in mountains that he found meaning and metaphor,
glory and imaginative possibility.

'The mountains are fountains of men as well as of rivers, of glaciers, of
fertile soil. The great poets, philosophers, prophets, able men whose
thoughts and deeds have moved the world have come down from the
mountains.'[4] Like Moses and visionaries of ancient Christianity such as
Augustine and John of Damascus, Muir delivered his message from the
mountains with prophetic purity and power. Key events in Muir's life,
documented in his voluminous journals and recollections and in public
records of his fame, have explanatory power in the raising of this mighty,
righteous voice of the mountains.

Born to the family Muir, meaning 'a wild stretch of land', in Dunbar,
Scotland on 21 April 1838, son of Daniel and his second wife Anne
Gilrye, John Muir spent a lifetime living up to the name and to his
father's stern expectations. Daniel was a convert to evangelical
Presbyterianism, and a strict, dour man who beat John throughout his
childhood. Biographer Stephen Fox writes: 'John read his Bible and grew
pious beyond his years, but he could never please his father. The endless
scoldings and beatings made his adolescence a grimly unequal contest of
wills with a tyrant blinded by his own righteousness.'[5]

According to Edwin Way Teale, young Muir was

> repelled by the harsh fanaticism of his father's religion ... he affiliated
> himself with no formal creed. Yet he was intensely religious. The
> forests and the mountains formed his temple. His approach to all
> nature was worshipper. He saw everything evolving yet everything
> the direct handiwork of God. There was a spiritual and religious
> exaltation in his experiences with nature.[6]

Muir's dutiful and passionate engagement with learning led him from a
Wisconsin farm to which his family had emigrated, to the University of

Wisconsin where he took no degree but took the courses he felt he needed. Avoiding the American Civil War and often depressed and lonely, Muir wandered and worked in Ontario and Wisconsin.

Muir came to a turning point in his life when, while working at a wagon factory, he was blinded by a file flying into his right eye and by a sympathetic reaction in his left eye. Struck into abject fear at the prospect of never again seeing natural beauty, he later wrote: '[M]y days were terrible beyond what I can tell, and my nights were if possible more terrible. Frightful dreams exhausted and terrified me every night without exception.'[7]

Recovering his vision, Muir determined to have a three-year-long 'sabbatical' to store, he wrote, 'a stock of wild beauty sufficient to lighten and brighten my after life in the shadow'.[8] Muir's own Sierran baptism and mountain enlightenment climaxed this long search for self-understanding – an odyssey of spiritual and intellectual searching that took place largely out of doors across the North American continent.

In 'First Glimpse of the Sierra' he begins:

> [W]hen I set out on the long excursion that finally led to California, I wandered, afoot and alone, from Indiana to the Gulf of Mexico, with a plant-press on my back .... I crossed the Gulf to Cuba, enjoyed the rich tropical flora there for a few months ... but I was unable to find a ship bound for South America ... therefore I decided to visit California for a year or two.[9]

Arriving in San Francisco by steamer on 1 April 1868, he set out to meet his destiny in the Yosemite Valley:

> A landscape was displayed that after all my wanderings still appears as the most beautiful I have ever beheld. At my feet lay the Great Central Valley of California, level and flowery, like a lake of pure sunshine, forty or fifty miles wide, five hundred miles long ... from the eastern boundary of this vast golden flowerbed rose the mighty Sierra, miles in height, and so gloriously colored and so radiant, it seemed not clothed in light but wholly composed of it, like the wall of some celestial city.[10]

Muir saw in such wilderness the source of humanity's spiritual health and wholeness. His philosophy of nature as the glorious handiwork of a God who created a democracy of life forms has inspired the post-modern deep ecology movement. Muir was keenly aware of the anthropocentric character of human attitudes towards nature, including the values embedded in utilitarian conservation. In his mind, a different ethic was at work – one which was to inspire Aldo Leopold, Arne Naess, John Seed and contemporary deep ecologists.

> The world we are told was made especially for man – a presumption not supported by the facts ... why should man value himself as more than a small part of the one great unit of creation? And what creature of all the Lord has taken the pains to make is not essential to the completeness of that unit – the cosmos? The universe would be incomplete without the smallest transmicroscopic creature that dwells beyond our conceitful eyes and knowledge ... [P]lants are credited with but dim and uncertain sensation, and minerals with positively none at all. But why may not even a mineral arrangement of matter be endowed with sensation of a kind that we in our blind exclusive perfection can have no matter manner of communication with? ... [B]ut glad to leave these ecclesiastical fires and blinders, I joyfully return to the immortal truth and immortal beauty of Nature.[11]

For him such truth and beauty as one can know in nature answered his questions. Through immersion in wild nature one could know how best to live. As Michael P. Cohen puts it, 'ecological consciousness would generate an ecological conscience'.[12] Muir moved from his own profound spiritual experiences in wilderness to preaching action to a nation. According to Cohen:

> His vision, he now felt, must lead to concrete action, and the result was a protracted campaign that stressed the ecological education of the American public, government protection of natural resources, the establishment of National Parks, and the encouragement of tourism.[13]

He was much ahead of his time in promoting action based on ecological responsibility. Many have called Muir the voice of the wilderness and his passion to protect it from destruction gave birth to the popular conservation movement. In 1898 he founded the Sierra Club for these purposes.

Within his own historical context Muir had remarkable influence – literary, political and philosophical – on those who were to follow him in environmental ethics and environmental education. Inspired from an early age by the Bible, Shakespeare, Milton, Scott and Burns, he later discovered Thoreau and Emerson.[14] He kept journals with no intent to publish and his first book was not printed until he was aged 56. Literary fame came fast though – the result of a turn-of-the-century love of nature and the urgent need to conserve America's vast natural resources from the unbridled rapaciousness of her maturing capitalism.

His political influence grew as he devoted himself to proselytizing the grandeur of the American West and the vital importance of protecting it. He led an array of important figures from Ralph Waldo Emerson to Theodore Roosevelt on excursions in the Sierra. Some of these camping trips had an enormous effect, such as that upon Robert Underwood Johnson, editor of the influential *Century* magazine, who subsequently launched a campaign to create Yosemite National Park, while President Roosevelt ordered his Secretary of the Interior to extend the Sierra Reserve one day after emerging from his sojourn with Muir.

For generations his work inspired not only the movement to conserve nature but the impetus to appreciate it. His journals brim with the power of his experiences which could and ought to be accessible to all. He thought if only people would save the land and take the time to saunter on it, then would come wisdom. His encounter with the rare orchid *Calypso Borealis*, later famous as marking the beginning of his evolution into pantheism, is recorded in such a journal entry. The entry was written in 1864 near Lake Huron. Muir was in Canada to avoid being drafted into the American Civil War:

> I never before saw a plant so full of life; so perfectly spiritual. It seemed pure enough for the throne of its Creator. I felt as if I were in the presence of superior beings who loved me and beckoned me to come. I sat down beside them and wept for joy.[15]

His philosophical contributions to the conception of wilderness, to the democratic ethical responsibility of humans towards all life forms, and to the ecological consciousness of a vast eternal unity are immense. Earlier, among Americans, only Thoreau spoke with such moral authority; later only Carson had such an influence on environmental thinking.

Notes
1 'Alaska Fragments, June–July (1890)', in *John of the Mountains*, p. 317.
2 Quoted in 'Chronology' by William Cronon, *John Muir: Nature Writings*, New York: The Library of America, p. 839, 1997.
3 Edwin Way Teale, *The Wilderness World of John Muir*, Boston, MA: Houghton-Mifflin Company, p. 143, 1954.
4 'The Philosophy of John Muir', in Teale, op. cit., p. 321.
5 Stephen Fox, *John Muir and His Legacy: The American Conservation Movement*, Boston: Little, Brown & Co., p. 31, 1981.

6　Teale, op. cit., p. xiii.
7　Fox, op. cit., p. 48.
8　Ibid.
9　Teale, op. cit., p. 99.
10　Ibid, p. 100.
11　Ibid, p. 318.
12　Michael P. Cohen, *The Pathless Way: John Muir and the American Wilderness*, Madison, WI: University of Wisconsin Press, 1984, the quote appears on the dust jacket.
13　Ibid.
14　Cronon, op. cit., p. 836.
15　Fox, op. cit., p. 43.

See also in this book
Carson, Leopold, Lopez, Naess, Nash, Thoreau

Muir's major writings
The Mountains of California, New York: Century, 1894.
Stickeen, Boston, MA: Houghton-Mifflin Company, 1909. This book is in the public domain and can be found on the web at the following URL: http://vault.sierraclub.org/john_muir_exhibit/bio/stickeen.aspx
The Story of My Boyhood and Youth, Boston, MA: Houghton-Mifflin Company, 1913.
Letters to a Friend, Boston, MA: Houghton-Mifflin Company, 1915.
Travels in Alaska, Boston, MA: Houghton-Mifflin Company, 1915.
A Thousand Mile Walk to the Gulf, Boston, MA: Houghton-Mifflin Company, 1916.
John of the Mountains: The Unpublished Journals of John Muir, ed. Linnie Marsh Wolfe, Boston, MA: Houghton-Mifflin Company, 1938.
John Muir in His Own Words: A Book of Quotations, compiled and ed. Peter Brown, Lafayette, CA: Great West Books, 1988.

For more writings by John Muir visit the on-line John Muir exhibit: http://vault.sierraclub.org/john_muir_exhibit/writings/books.aspx

On-line text versions of many of Muir's works are available at http://vault.sierraclub.org/john_muir_exhibit/

Further reading
Kimes, William F. and Kimes, Maymie B., *John Muir: A Reading Bibliography*, Fresno, CA: Panorama West Books, 1986.
Miller, Sally M. (ed.), *John Muir: Life and Work*, Albuquerque, NM: University of New Mexico Press, 1995.

Nash, Frederick, *Wilderness and the American Mind*, New Haven, CT: Yale University Press, 1967.

Wolfe, Linnie Marsh, *Son of the Wilderness*, New York: Alfred A. Knopf, Inc., 1945.

PETER BLAZE CORCORAN

ANNA BOTSFORD COMSTOCK 1854–1930

> In order to appreciate truly his farm, the farmer must needs begin as a child with nature-study; in order to be successful and make the farm pay, he must needs continue in nature-study; and to make his declining years happy, content, full of wide sympathies and profitable thought, he must needs conclude with nature-study; for nature-study is the alphabet of agriculture and no word in that great vocation may be spelled without it.[1]

A serious agricultural depression in the northeastern United States drove people from rural landscapes to burgeoning cities in the late nineteenth century. Such a migration took place in New York 1891 to 1893. Anna Botsford Comstock wrote in the Preface to her *Handbook of Nature-Study*:

> [T]he charities of New York City found it necessary to help many people who had come from the rural districts – a condition hitherto unknown. The philanthropists managing the Association for Improving the Condition of the Poor asked 'What is the matter with the land of New York state that it cannot support its own population?'[2]

In response, a movement was created to interest 'the children of the country in farming as a remedial measure', being that 'the first step toward agriculture was nature-study'.[3]

From such a utilitarian concern for the future of rural life grew the American nature-study movement. The centre of the movement towards reiteration of the importance of agriculture and country values was Cornell University in Ithaca, New York, which from its founding in 1865 had been committed to the problems of agricultural extension.

The leader of this movement was Liberty Hyde Bailey (1858–1954), the great communicator of the idealistic, progressive, romantic beliefs of the Cornell school of thought. The practical purpose of this effort was 'making children sympathetic with nature-study so that they would truly enjoy rural life and be happy on the farm'.[4]

Working with Bailey at Cornell, and the spiritual leader of the nature-study movement, was the great proselytizer of happy, intimate contact with the earth, Anna Botsford Comstock. Born into a Quaker family in rural Cattaraugus County of upstate New York in 1854, she lived until she was three years old in a log cabin which she remembers well enough to describe in her autobiography, *The Comstocks of Cornell*. Farm life and a mother named Phoebe who loved nature made indelible impressions on the young Anna. She quotes her mother as saying one day at sunset: 'Anna, heaven may be a happier place than earth, but it cannot be more beautiful.'[5]

An educated female neighbour, Mrs. Ann French Allen, was an important influence in directing Anna Botsford towards higher education. She chose new, nearby Cornell, which had opened its doors to women. Zoology study with Professor John Henry Comstock led to long walks and courtship. Marriage interrupted her formal education but led to a decades-long partnership in scientific research, teaching and entomology illustration.

Her path of inquiry seems to have been selected by both cultural limits and personal choices. Biographer Pamela Henson writes that she 'entered science through the "back door" as many female relatives of scientists did, and she always worked on the "peripheries" of science in art, popularization, and children's education',[6] Henson cites Evelyn Fox Keller and the gendered theory of masculinist objective science in explaining Comstock's choices.

> Comstock found it more comfortable to incorporate her aesthetic appreciation of nature into scientific interpretation for children and a popular audience. This appreciation was part of her overall sense of subjective connectedness to the world around her. Anna Comstock experienced the natural world in emotional terms and felt a sense of personal relationship and responsibility to living things around her.[7]

Comstock also faced sexist societal barriers as a member of the first generation of American women with university educations. Often cited as the first woman professor at Cornell, appointed in 1898, it is less often noted that the Board of Trustees revoked the title and did not allow women professors until 1911, and only in home economics. Comstock was finally reappointed in 1915.

During her Cornell career, Comstock worked with other founders of the nature-study movement. She called Wilbur Samuel Jackman of Chicago the father of nature-study. Jackman's belief that children derived intellectual benefit, as well as personal satisfaction, from the formal study of their immediate environments seems to have been one of the most influential ideas in the history of nature-study. Comstock carried this idea in the period from 1900 to 1920 – the zenith of the nature-study movement and the era of Anna's leadership. She edited the *Nature-Study Review* and served as president of the American Nature Study Society, now more than a century old.

Nature-study was to a large extent a reform movement which rejected the methodologies of schools in the late nineteenth century. Although the movement was to become fractured at a later date by conflicting purposes, there was a common purpose at the outset. According to Richard Raymond Olmsted in his dissertation 'The Nature-Study Movement in American Education',

> like most curriculum movements, the nature study agitation developed into a complex phenomenon. The leaders of this movement found initial agreement, however, in the assumption that elementary school children should be taught about nature, defined usually as the immediate countryside, through field trips and other direct experiences.[8]

The relationship of events in the nature-study movement to social conditions is vital to an understanding of the controversy which surrounded its introduction into the schools. The historical period from the mid-nineteenth century to 1880 was one of signal change. The Civil War, westward expansion, immigration of millions of new citizens and rapid industrial growth altered the nature of American society. Education was influenced by the introduction of universal schooling, the publication of Charles Darwin's *The Origin of Species* and the growth of child psychology.

Comstock and her heroes Jackman and Bailey saw nature-study as a pedagogical ideal and social reform initiative with roots in the work of Johan Amos Comenius, Heinrich Pestalozzi, Jean-Jacques Rousseau and Friedrich Froebel.[9] She was able to channel her own feeling for nature and her progressive social ideals into an educational and environmental philosophy much needed in her cultural period in America.

Her philosophy was that at the heart of a fully human existence is the cultivated imagination and insight for truth and beauty, as found in nature. In her seminal essay 'The Teaching of Nature-Study' she wrote:

> [N]ature-study cultivates ... a perception and a regard for what is true, and the power to express it. All things seem possible in nature; yet this seeming is always guarded by the quest of what is true. Perhaps half the falsehood in the world is due to lack of power to detect the truth and to express it. Nature-study aids both in discernment and in expression of things as they are. Nature-study cultivates in the child a

love of the beautiful; it brings to him early a perception of color, form, and music. He sees whatever there is in his environment, whether it be the thunder-head piled up in the western sky, or the golden flash of the oriole in the elm, whether it be the purple of the shadows on the snow, or the azure glint on the wing of the little butterfly. Also, what there is of sound, he hears; he reads the music score of the bird orchestra, separating each part and knowing which bird sings it. And the patter of the rain, the gurgle of the brook, the sighing of the wind in the pine he notes and love becomes enriched thereby.[10]

She also believed nature was a nurse for human health, an elixir of youth for the teacher, and a cure for problems of school discipline. Her reverence for the power of nature in strengthening human nature was reiterated throughout her writing.

A respected scientist, she published in 1911 what was to become the classic, *Handbook of Nature-Study*; since reprinted in many editions. She advocated direct observation and contact and made great, even extravagant, claims for the mental and physical wellbeing of students and teachers.

'Nature-study is nature love taught in the schools',[11] she wrote. She advocated, without apology, love of the world through harmonious relationship with it. Nature-study is the vehicle for such love – for student and teacher, in school and out. In a speech in Philadelphia in 1914 she said:

If nature-study as taught does not make the child love nature and the out-of-doors, then it should cease. Let us not inflict permanent injury on the child by turning him away from nature instead of toward it. However, if the love of nature is in the teaching heart, there is no danger; such a teacher, no matter by what method takes the child gently by the hand and walks with him in paths that lead to the seeing and comprehending of what he may find beneath his feet or above his head. And these paths, whether they lead among the lowliest plants, or whether to the stars, finally converge and bring the wanderer to that serene peace and hopeful faith that is the sure inheritance of all those who realize fully that they are working units of this wonderful universe.[12]

In her retirement speech as president of the American Nature Study Society, she said:

[T]he nature-study idea almost from the first overflowed the school boundaries to enrich and make happier the lives of those who loved the life of the woods and fields, and who would fain know something of the mysteries and wonders therein hidden.[13]

The advocacy of nature-study made her well-read and well-regarded. She lectured widely in the Chautauquan movement and published science writing for the public. According to Pamela Henson, 'Comstock's popularity was built on a melding of accurate science with popular sentimentality and her aesthetic talents.'[14]

Called the Dean of American Nature-Study and finally promoted to full professor in Entomology and Nature-Study, she was admitted to Phi Kappa Phi, the honorary society. In 1923 the League of Women Voters elected her one of the twelve greatest women in America. She remained energetic over a long productive career. She was not, however, tireless. When asked why she did not actively fight for women's suffrage, she said: 'I had been using all of my strength to fight narrowness, prejudice, and injustice, in the curriculum of the common schools, and I was weary with fighting.'[15]

She was at her best, as she humbly proclaimed herself, as an interpreter of science. Keller and others have said this was so that she need not assume the objectivist perspective of Western male science. She was an artist *and* a scientist. She educated about the complex power of nature in symbolic forms.

This style also enabled her to advocate for educational reform and nature conservation. She saw the power of the human spirit and of love of nature as the best motivator, putting aside the more typical American concern with practical benefit. In 1914, she said:

> With a fatuity that our descendants of three centuries hence will characterize a criminal stupidity we have exterminated many species of birds, destroyed many interesting and harmless wild animals, hacked down our trees ruthlessly and cleared our streams of valuable fish. Men of science had remonstrated in vain. It was not until the nature-study movement permeated the people throughout the land that they came to resent this extermination; and not until then was there a sufficiently strong popular opinion created to establish and carry out protective laws ... It should be remembered that in all history crusades have been born and led of the spirit.[16]

Her own leadership in science education and in environmental thinking was an inspiration to the American conservation movement. Also, critically important was her gender. She helped make possible the later environmental leadership of many American women from Alice Rich Northrup to Edith M. Patch, from Rosalie Edge to Rachel Carson. And she made legitimate advocacy for nature on spiritual and emotional grounds by both women and men.

Notes

1 *Handbook of Nature-Study*, p. ix.
2 Ibid.
3 Ibid.
4 Leo E. Klopfer and Audrey B. Champagne, 'Six Pioneers of Elementary School Science', University of Pittsburgh, Manuscript Draft, p. 299, 1975.
5 *The Comstocks of Cornell*, p. 57.
6 Pamela M. Henson, 'Through Books to Nature: Anna Botsford Comstock and the Nature Study Movement', in T. Gates and Ann B. Shteir, *Natural Eloquence: Women Reinscribe Science*, Madison, WI: University of Wisconsin Press, p. 116, 1997.
7 Ibid., pp. 118–19.
8 Richard Raymond Olmsted, 'The Nature-Study Movement in American Education', Indiana University, Dissertation, p. 2, 1967.
9 Liberty Hyde Bailey, *The Nature-Study Idea: Being an Interpretation of the New School-Movement to Put the Child in Sympathy with Nature*, New York: Doubleday, Page & Company, p. 7, 1903.
10 *Handbook of Nature-Study*, p. 4.
11 Ibid., p. 3.
12 Speech delivered at Philadelphia, 30 December 1914, entitled 'The Growth and Influence of the Nature-Study Idea'.
13 Comstock, 'The Attitude of the Nature-Study Teacher toward Life and Death', *Nature-Study Review*, 5 (May), p. 121, 1909.
14 Henson, op. cit., p. 128.
15 Marcia Myers Bonta, *Women in the Field: America's Pioneering Women Naturalists*, College Station, TX: Texas A & M University Press, p. 164, 1991.
16 Speech delivered at Philadelphia, 30 December 1914.

See also in this book
Darwin, Emerson, Rousseau

Comstock's major writings
Manual for the Study of Insects, Ithaca, NY: Comstock Publishing Company, 1895.
Ways of the Six-Footed, Ithaca, NY: Cornell University Press, 1903.
How to Know the Butterflies: A Manual of the Butterflies of the Eastern United States, with John Henry Comstock, New York: D. Appleton Publishing Company, 1904.
Confessions of a Heathen Idol, originally published under the pseudonym Marian Lee, New York: Doubleday, Page, & Company, 1906.
Handbook of Nature-Study, Ithaca, NY: Comstock Publishing Associates, 1911.
The Comstocks of Cornell, with John Henry Comstock, Ithaca, NY: Comstock Publishing Associates, 1953.
Comstock also wrote many essays for *The Nature-Study Review*, 1904–23.

Further reading
National Society for the Scientific Study of Education, *The Third Yearbook, Part III, Nature Study*, Chicago, IL: University of Chicago Press, 1904.

PETER BLAZE CORCORAN

RABINDRANATH TAGORE 1861–1941

I still remember the very moment, one afternoon, when I ... suddenly saw in the sky ... an exuberance of deep, dark clouds lavishing rich, cool shadows on the atmosphere. The marvel of it ... gave me a joy which was freedom, the freedom we feel in the love of our friend.[1]

Rabindranath Tagore was a great poet and profound thinker. He was born in Calcutta on 6 May 1861. He belonged to a family which is the most gifted in Bengal in the realm of religion, philosophy, literature, music and painting. Although he was not educated in any college or university, he was clearly a man of learning. He had his own original ideas about education which led him to establish an educational institution at Shantiniketan in December 1901 following the model of the forest hermitages of ancient India. He named it Viswa Bharati with the intention of re-opening the channel of communication between the East and the West. He was a versatile genius. There is no aspect of literature – poetry, short story, novel, drama – which he has not enriched. He was awarded a Nobel Prize in 1913 in recognition of his outstanding literary activities. Equally important are his innumerable essays and many books which reveal his deep socio-political as well as spiritual commitments. He was also a most original composer of music. He travelled extensively in different countries of the world, and was a successful mediator between Western and Eastern cultures. He died on 7 August 1941.

Crucially, Tagore's poems, short stories and novels, as well as books and essays, exhibit his love and concern for nature, for land, sea, air, plants and animals that constitute the 'environment' around us. His concern or thinking about the environment is not, however, activated by any pragmatic or utilitarian consideration. Rather it grows on a different – non-utilitarian – ground. And here we may profitably utilize his idea of 'surplus'. The surplus in man which, according to Tagore, constitutes his spiritual make up, overflows pragmatic need, the stage of pure utility, and 'extends beyond the reservation plots of our daily life'.[2] This surplus indicates an aspect of human being, 'a fund of emotional energy' which is 'useless' or 'superfluous' in the sense that it is not regulated by self-interest, by any moral or other practical ends. Thus the point is that we no doubt have one side which is governed by pragmatic necessity, but parallel to it we have also another side – a spiritual one – which requires fulfilment of our creative urge, our capacity to appreciate and enjoy. And our life cannot be meaningful in the strict sense of the term by pragmatic fulfilment alone, without this spiritual fulfilment. This is what Tagore wants to convey by his notion of 'surplus'.

This insistence on surplus gives us the clue as to why the environment matters to Rabindranath; why he would want to see it defended against any unnecessary tampering. Nature is dear to him, since with all its enthralling beauty it can evoke our appreciation, and thus fulfil the demand of the surplus in us. To put it in a different way, he entertains nature in terms of the aesthetic appreciation or delight that it prompts.

> Would they not attract me from all sides –
> These trees, creepers, rivers, mountains and woods
> The deep blue eternal sky?[3]

This explains clearly why the natural environment with its 'special harmony of lines, colours and life and movement' should be preserved.[4] It should be preserved, for it gives us aesthetic joy, and thereby a bond of love is established between it and us.

But this defence of the environment on aesthetic ground will not enjoy the approval of all ecologists even in India. Some will condemn it as an anthropocentric denial of the intrinsic value of nature. Let us ponder how far it is fair to bring this charge against Rabindranath. Strictly speaking, the use of the words 'intrinsic' and 'anthropocentric' is infected by ambiguity: 'An object X has intrinsic value' may be understood in at least two senses:

1   X has intrinsic value' may be understood to mean that 'X has non-instrumental value', i.e. the value of X does not consist in its being a means to some end. So 'X has intrinsic value' will denote that X is an end-in-itself. This is the sense in which many environmentalists consider the value of nature. Hence it will be wrong, in their opinion, to view nature only as an instrument for serving some end of man. That would be anthropocentric imperialism. I call this anthropocentrism in the first sense.

2   'X has intrinsic value' refers to what may be designated as 'objective' value. An objective value is that which X possesses independently of any human evaluation. The denial of objective value in this sense will amount to what I call anthropocentrism in the second sense.

The view of Rabindranath indeed has an anthropocentric flavour, at least in the second sense, since he thinks that no account of value can be isolated from all relations to human being. That is why he observes, 'What we call nature is what is revealed to *man* as nature.'[5] Or, 'Reality is ... [that] by which we are affected, that which we express.'[6] Evidently Rabindranath wants to emphasize that even to say that nature has value must involve some reference to man, to his being affected by it. There is nothing wrong in highlighting this human reference. It does not mean that values are conferred on things by man. What it implies is the crucial truth that even if, like the ecologists, we grant values to nature on account of the qualities it has, this has no real sense or bearing unless we are able to understand 'why something with those qualities should matter to us, how it might fit into the orbit of our concerns'.[7]

But to hold that Rabindranath takes an anthropocentric attitude in the second sense to nature (which sounds quite reasonable) is not to hold that he is inclined towards anthropocentrism in the first sense, towards the stronger claim that nature is only a means for satisfying human purpose. Rabindranath's point that value presupposes human evaluation is only a 'formal' one about how value is to be understood; but from this does not follow the stronger claim, which is a 'substantial' one about what makes something valuable.[8]

That Tagore would not endorse any instrumentalism is strengthened by another consideration of his when, like Kant, he employs, as already suggested, the concept of 'disinterestedness'. He talks about aesthetic enjoyment – 'the enjoyment which is disinterested'.[9] The disinterestedness of aesthetic contemplation can be made explicit by the idea of an 'alternative world'.[10] The same forest which is the source of one's livelihood can open a different horizon – an alternative world – which is unconnected with any question of livelihood, with any pragmatic concern or interest. Then the smell of grass, the graceful movement of boughs of trees, the sweet melody of birdsongs begin to move us in a new way. Thus emerges the aesthetic moment when the forest is imaginatively explored and when any thought of using it for our interest or personal benefit becomes completely redundant.

This comes out more clearly from Tagore's insistence, as indicated in the opening quotation, on the relation of love we enter into with nature in our aesthetic contemplation of it. Inspired by the teachings of the *Upanishads*, he holds that when I love anyone, I cannot think of seeing my beloved in the light of any usefulness. On the contrary, I find in my beloved an extension of my own being which gives me the feeling of real freedom. It is this relation of love or of heart that we have with nature in our aesthetic experience of it. Hence this relation must be 'superfluous', i.e. beyond the bounds of any interest or satisfaction of practical purpose. 'There is an element of [the] superfluous in our heart's relation with the world.'[11]

Incidentally, but very crucially, this disinterestedness will also enable Rabindranath to meet the challenge often made by ecologists that aesthetic appreciation, since it admits of variations, cannot be effectively utilized as the ground for environmental preservation. Even if we concede that aesthetic appreciation is variable, there is yet a very good sense of it that we can hopefully attend to in the context of environment protection. The concept of disinterestedness helps us extract this good sense. As Kant puts it: '[W]here anyone is conscious that his delight in an object is with him independent of interest, it is inevitable that he should look on the object as one containing a ground of delight for all men.'[12] In other words, if aesthetic appreciation is based on disinterestedness, as Tagore thinks it is, we can very reasonably be assured that nature can give rise to the same appreciation or delight in others as it does in my case. And then it can well provide a formidable reason in favour of environment preservation.

If you visit Shantiniketan (in Birbhum district, West Bengal), you will find a concrete example of Tagore's concern for the beauty of nature. Beautiful gardens with varieties of flowers of different colours, a small river flowing nearby which glitters with sunshine, rows of long green trees – this is Tagore's Shantiniketan. It is the place which will refresh your soul. And you will get here an aura of infinity. Obviously, this aesthetic concern of Tagore can by no means be set aside as being trivial. With this in mind, I have tried to explain and defend Tagore's thinking about the environment on aesthetic and spiritual grounds. True, some environmental thinkers would not receive him well. Yet it is also true that his emphasis on the beauty of nature endeared him to many of his eminent contemporaries both in India and abroad. Note how D.R. Bhandarkar, a great Indian thinker, approves of and admires his sensitivity to nature: 'Everywhere in his poems and songs you see sunshine ... still night and various aspects of nature ... His is a mind most responsive to nature.'[13] Similarly, another eminent writer, Lim Boon Keng from the University of Amoy, China, writes: 'His soul seems at once to vibrate in full harmony with the orchestra of melodies and echoes reflected from the sound of rushing waters, from the songs of birds, from the rustling of leaves ...'[14] And it cannot be denied that caring for nature on aesthetic grounds, as Tagore did, has now become one of the major environmental concerns in the developed countries of the world.

Notes
1 *The Religion of an Artist*, pp. 16–17.
2 *The Religion of Man*, p. 33.

3 Tagore, 'Vasundhara', trans. in Rabindranath Choudhury, *Love Poems of Tagore*, Delhi: Orient Paperbacks, p. 55, 1975.
4 *The Religion of Man*, p. 85.
5 Ibid., p. 72, original emphasis.
6 Ibid., p. 83.
7 D.E. Cooper, 'Aestheticism and Environmentalism', in D.E. Cooper and J.A. Palmer (eds), *Spirit of the Environment*, London: Routledge, p. 103, 1998.
8 Ibid., p. 102.
9 *Lectures and Addresses*, p. 79.
10 D.E. Cooper, op. cit., p. 109.
11 *Lectures and Addresses*, p. 93.
12 Immanuel Kant, *The Critique of Judgement*, trans. J.C. Meredith, Oxford: Clarendon Press, p. 50, 1928.
13 D.R. Bhandarkar, 'My Impressions about the Poet', in R. Chatterjee (ed.), *The Golden Book of Tagore*, p. 36.
14 Lim Boon Keng, 'The Beauty and Value of Tagore's Thoughts', in Chatterjee (ed.), op. cit., p. 125.

See also in this book
Gandhi, Kant

Tagore's major writings
Creative Unity, 1922, New Delhi: The Macmillan Company of India Limited, 1980.
The Religion of Man, 1970, paperback edn, London: Allen & Unwin, 1970.
The Religion of an Artist, 1936, Calcutta: Viswa-Bharati, 1988.

The standard edition of Tagore's writings is *The Works of Tagore* in 15 volumes, Calcutta: West Bengal Government, 1961.

A one-volume selection of Tagore's important lectures is *Lectures and Addresses*, ed. Anthony X. Soares, New Delhi: Macmillan Pocket Tagore Edition, 1995.

Further reading
Chatterjee, Ramananda (ed.), *The Golden Book of Tagore*, Calcutta: The Golden Book of Tagore Committee, 1990.
Ghosh, S.K., *Rabindranath Tagore*: New Delhi: Sahitya Academy, 1986.
May, Larry and Sharratt, Shari Collins (eds), *Applied Ethics: A Multicultural Approach*, Englewood Cliffs, NJ: Prentice-Hall, 1994.
Sen Gupta, Kalyan, *The Philosophy of Rabindranath Tagore*, Aldershot, UK: Ashgate, 2005.

KALYAN SEN GUPTA

BLACK ELK 1862–1950

> Birds make their nests in circles, for theirs is the same religion as
> ours.[1]

Black Elk was born in 1862 on the banks of the Little Powder River, a
tributary of the Yellowstone River in what is now the state of Wyoming.
Then it was in the westernmost territory of the Lakota. Black Elk belonged
to the Oglala Band. His father and grandfather – both also named Black Elk
– were medicine men. He followed them in this calling. Black Elk was
born into a world radically different from the one in which he would die.
It was a sacred world in which 'the two-leggeds and the four-leggeds lived
together like relatives, and there was plenty for them and plenty for us'.[2] By
the time of his death, at the age of 88, on the Pine Ridge Indian Reservation
in South Dakota, the vast herds of game, especially bison, that his people
hunted for their subsistence were a fading memory; the faces of four United
States presidents had defaced Mount Rushmore in the Black Hills, which
were sacred to the Lakotas; the Yellowstone Plateau was a National Park;
and the prophecy of Drinks Water, a contemporary of Black Elk's
grandfather, had been fulfilled: '[Y]ou shall live in square gray houses, in a
barren land, and beside those square gray houses you shall starve.'[3]

Trouble began the year after Black Elk was born. As a young child, he
never saw a 'Wasichuo' (the name means not 'white', but 'too-many-to-
count'), but he grew up hearing of them. Black Elk's mother would
invoke the name as a bugbear: 'If you are not good the Wasichus will get
you.'[4] His father was wounded fighting the Wasichus when Black Elk was
only three. Black Elk later fought for his people and saw their defeat and
dispossession. He was a cousin of the great Lakota warrior, Crazy Horse.
He was an eye witness of Custer's Last Stand at the Battle of Little Big
Horn: 'These Wasichus wanted it, they came to get it, and we gave it to
them.'[5] Black Elk participated in the Ghost Dance, a millenarian pan-
Indian religious revival. Although at first sceptical, it was, indeed, he who
dreamed of and reproduced the famous Ghost Shirt that was supposed to
protect its wearer from bullets. Black Elk was present at the slaughter of
more than 300 Lakota men, women and children at Wounded Knee
Creek, the last 'battle' of the 'Indian wars' in the USA. He travelled to
England and France as a dancer in Buffalo Bill's Wild West show. In short,
Black Elk lived through the transformation of the central plains of North
America from their aboriginal condition inhabited by indigenous peoples
to a land of Wasichu farms, ranches, railroads, highways, power lines,
towns, motels, monuments, parks, diners, movie theatres and all the other

trappings of modern American civilization. And he participated in some of the most legendary events in the history of the American West.

After the murder of Crazy Horse and the pacification and reservationization of the Plains Indians, Black Elk undertook a vision quest, and began his career as a Thunder-Being medicine man, age 17. As a condition of employment in Buffalo Bill's troupe, he converted to Christianity in his mid-20s, and seems, during his three years abroad (1886–9), to have been a sincere and devout convert. Then, the fervour of the Ghost Dance, which swept the country in 1889 and 1990, encouraged him to return to his native religious beliefs. Wounded Knee ended the Ghost Dance episode in American history on 29 December 1990, and embittered and demoralized the Lakota, who were the sole victims of the massacre. Afterwards, Black Elk, like most of the Lakota, turned his back on European-American culture, and defiantly continued to practise traditional medicine, which put him in conflict with the missionaries on his reservation. As the psychic and spiritual wounds of the tragedy at Wounded Knee scarred over and the nineteenth century gave way to the twentieth, Black Elk slowly abandoned his traditional medical practice and the religious world-view in which it was embedded in favour of Catholicism and modernity. In this transformation, he may been encouraged by his first wife, Katie War Bonnet. He was baptized in 1904 and given the name Nicholas.

Above all else, Black Elk was a religious genius, and he turned this genius into a career as a catechist in the Catholic Church's St Joseph Society, spreading the gospel to other Lakotas in their own language. For the next ten years, he travelled the Great Plains as something of a Native evangelist. With fragile health (he suffered from tuberculosis) and failing eyesight, Black Elk quit travelling and settled on the Pine Ridge reservation, the head of a large family, a pillar of the Church. His humble home was a centre of Catholic social life, and he was a man to whom the missionaries pointed with pride as a model of their success in leading the Lakota from the darkness of heathenism into the light of Christianity and civilization.

In August 1930 John G. Neihardt came to Pine Ridge looking for informants on the Ghost Dance and the massacre at Wounded Knee for the final volume of his epic poem *Cycle of the West*. He was directed to Black Elk, who seemed to be expecting him, in the traditional manner of a prescient shaman recruiting a spirit-designated apprentice. The two immediately discovered they had an extraordinary rapport. At the end of the day Black Elk said: 'There is so much to teach you. What I know was given to me for men and it is true and it is beautiful. Soon I shall be under the grass and it will be lost. You were sent to save it, and you must come back, so that I can teach you.'[6] Neihardt did return the following spring, not for the purpose of fulfilling his own agenda, but Black Elk's. A special teepee was erected. In it, Black Elk spoke for many days to Neihardt in Lakota; Black Elk's son Benjamin interpreted; and Neihardt's daughters, Enid and Hilda, recorded the translation, from which they later made typescripts. Neihardt then drew upon his literary skills to craft these interviews into *Black Elk Speaks*, one of the greatest achievements of American letters, and a genre exemplar in post-colonial American Indian literature. According to Vine Deloria, Jr, a Lakota philosopher and activist, the book has realized Black Elk's intent and more: 'The most important aspect of the book ... is not its effect on the non-Indian populace who wished to learn something of the beliefs of the Plains Indians, but upon the contemporary generation of young Indians who have been aggressively searching for roots of the structure of universal reality. To them the book has become a North American bible of all tribes ... So important has this book become that one cannot today attend a meeting on Indian religion and hear a series of Indian speakers without recalling the exact parts of the book that lie behind contemporary efforts to inspire and clarify those beliefs that are "truly Indian".'[7]

It is a mistake to suspect that *Black Elk Speaks* is solely the product of Neihardt's romantic imagination. The typescripts of the 1931 interview were preserved among Neihardt's papers in the archives of the University of Missouri and were published in 1985. Comparison with these shows the book to be a faithful rendition. Neihardt's contribution was in fact purely literary, editing the narrative, simplifying and stylizing the prose. Indeed, in Neihardt's own estimation, *Black Elk Speaks* is 'the first absolutely Indian book thus far written ... all out of the Indian consciousness'.[8] How Black Elk's poignant account of the 'truth' and 'power' of his Great Vision – vouchsafed to him when he was only a 9-year-old boy, as innocent of missionary propaganda as he was of all things Wasichu – may be reconciled with his later and never-recanted devotion to Christianity remains unclear. In response to Neihardt's question about that, he said simply: 'My children had to live in this world.'[9] *Black Elk Speaks* should, therefore, be taken at face value – as an authentic window into the traditional Lakota world-view (if not that 'of all tribes').

And when we look through that window, what do we see? Many wonderful things, including a powerful environmental ethic.

The Lakota world-view, although thoroughly indigenous, is hardly aboriginal. As late as the eighteenth century, the Lakota were a woodland people living in the region of the western Great Lakes. They were pushed out onto the plains by the Algonkian-speaking Ojibwa in a kind of domino-effect of expanding European settlement of the American Eastern Seaboard. They quickly adopted the mounted bison-hunting plains culture that was already established, and which was itself a post-Columbian phenomenon. Although evolved in North America, the horse, upon which reliable bison hunting depended, had been extinct in the Western hemisphere for ten thousand years. It was reintroduced by the Spanish, and the domesticated species re-established feral populations on the vast grasslands of North America. It was welcomed by the Indians of the interior, not, as formerly, a game animal, but as a beast of burden and a companion in war and in the chase. Further, the Lakota themselves recognized that their sacred-pipe religion is of recent historical origin in the myth of White Buffalo Cow Woman, who gave it to them.

The Lakota world-view grew out of and reflected the relatively featureless, open spaces of the Great Plains. Its parameters are six in number – sky, earth, and the cardinal directions: west, north, east and south – each personified as a 'power'. *Black Elk Speaks* opens with an invocation and an explanation of the symbolism of the sacred pipe:

> These four ribbons hanging here on the stem are the four corners of the universe. The black one is for the west where the thunder beings live to send us rain; the white one for the north, whence comes the great white cleansing wind; the red one for the east, whence springs the light and where the morning star lives to give men wisdom; the yellow for the south, whence come the summer and the power to grow.[10]

Either the traditional collective Lakota world-view is very abstract and sophisticated or Black Elk's own personal version of it is, for there is a unity within this multiplicity that one scholar compares to the concept of *Brahman* in Vedic Hindu philosophy, to the mystery of the Trinity in Christian theology (one God, three persons), and to the monism of the early modern European philosopher Benedict Spinoza.[11] The unifying concept is *Wakan Tanka*, the 'Great Spirit', whom Black Elk often refers to as 'Grandfather':

> But these four spirits are only one Spirit after all, and this eagle feather here is for that One … Is not the sky a father and the earth a mother, and are not all living things with feet or wings or roots their children? And this hide upon the mouthpiece here, which should be bison hide, is for the earth, from whence we came and at whose breast we suck as babies all our lives, along with all the animals and birds and trees and grasses.[12]

So, in brief, the sky is a universal father; the earth, a universal mother; each of the four quarters (sometimes also called winds and each associated with its distinctive colour) is a spirit with a peculiar power. All are united, however, in the Grandfather (as distinct from the Father) Spirit, *Wakan Tanka*, the Great Spirit.

This world-view is the foundation of an environmental ethic, which is quite expressly stated in *Black Elk Speaks*, albeit with characteristic simplicity and brevity: after invoking each of these spirits individually and the Great Spirit, of which they are all particular manifestations, Black Elk prays: 'Give me the strength to walk the soft earth, a relative to all that is!'[13] Black Elk's rhetoric routinely implies a familial egalitarianism among all the children of Father Sky and Mother Earth – human animal, non-human animal or plant. Human beings differ from other living beings only in number of legs, or the absence of wings or roots. Again, this egalitarianism is expressly stated briefly and simply: 'All over the earth the faces of living things are all alike'.[14] In bad things as well as good, the native two-leggeds and four-leggeds share a common destiny: 'The Wasichus came, and they have made little islands for us and other little islands for the four-leggeds, and always these islands are becoming smaller, for around them surges the gnawing flood of the Wasichu.'[15]

The Lakota environmental ethic is similar to, but, in important ways, also differs from the familiar 'land ethic' formulated by Aldo Leopold in 1949. The Leopold land ethic is based on a social model of nature, which is similarly egalitarian – in which a human being is but a 'plain member and citizen' of the 'biotic community'.[16] But nature in the land ethic is represented as one big *society*, while in the Lakota environmental ethic nature is portrayed as one big *family*. According to the ecological 'community concept', each species occupies a niche, role or profession in the economy of nature. Just as in the human social microcosm there are farmers, truckers and doctors, each specializing in a particular task, so in the natural macrocosm there are producers (the green plants), consumers (animals of all sorts) and decomposers (fungi, bacteria and the like). And just as our non-privileged membership of human communities generates our human-to-human ethics, so our 'plain' membership of biotic communities generates land ethics, according to Leopold. In the Lakota environmental ethic, however, the relationship of human beings to nature seems closer, warmer – just as our relationship to a family member is more intimate and our obligations more compelling than to a fellow citizen of our municipality or country. Instead of a 'land' environmental ethic, perhaps we could call Black Elk's a 'family' environmental ethic.

Notes

1 *Black Elk Speaks, Being the Life Story of a Holy Man of the Oglala Sioux*, p. 199.
2 Ibid., p. 9.

3 Ibid., p. 10.
4 Ibid., p. 13.
5 Ibid., p. 127.
6 Ibid., p. 10.
7 Vine Deloria, Jr, 'Introduction', *Black Elk Speaks, Being the Life Story of a Holy Man of the Oglala Sioux*, Lincoln, NE: University of Nebraska Press, pp. xii–xiii, 1979.
8 John G. Neihardt to Julius T. House, 3 June 1931, *The Sixth Grandfather*, p. 49.
9 Ibid., p. 47.
10 *Black Elk Speaks*, p. 2.
11 *The Sacred Pipe*.
12 Ibid., pp. 2–3.
13 Ibid., p. 6.
14 Ibid.
15 Ibid., p. 9.
16 Aldo Leopold, *A Sand County Almanac and Sketches Here and There*, New York: Oxford University Press, p. 204, 1949.

See also in this book
Callicott, Leopold

Black Elk's major writings

Neihardt, John G., *Black Elk Speaks, Being the Life Story of a Holy Man of the Oglala Sioux*, New York: Morrow, 1932.

Neihardt, John G., *When the Tree Flowered: An Authentic Tale of the Old Sioux World*, New York: Macmillan, 1951.

Brown, Joseph Epes, *The Sacred Pipe: Black Elk's Account of the Seven Rites of the Oglala Sioux*, Norman, OK: University of Oklahoma Press, 1953.

DeMallie, Raymond J. (ed.), *The Sixth Grandfather: Black Elk's Teachings Given to John G Neihardt*, Lincoln, NE: University of Nebraska Press, 1984.

Further reading

Deloria, Ella C., *Dakota Texts*, Publications of the American Ethnological Society, no. 14, New York: G.E. Stechert, 1932.

Lame Deer, John (Fire) and Erdoes, Richard, *Lame Deer: Seeker of Visions*, New York: Simon & Schuster, 1972.

Luther Standing Bear, *Land of the Spotted Eagle*, Boston: Houghton Mifflin, 1933.

Rice, Julian, *Black Elk's Story: 'Distinguishing Its Lakota Purpose'*, Albuquerque, NM: University of New Mexico Press, 1991.

Walker, James R., *Lakota Belief and Ritual*, ed. Raymond J. DeMallie, Lincoln, NE: University of Nebraska Press, 1980.

J. BAIRD CALLICOTT

JAKOB VON UEXKÜLL 1864–1944

Only the knowledge that [...] all environments are composed into the world-score opens up a path leading out of the confines of one's own environment. Blowing up our environmental space by millions of light-years does not lift us beyond ourselves, but what certainly does is the knowledge that [...] the environments of our human and animal brethren are secured in an all-encompassing plan.[1]

Since his works of the 1910s, in which he first developed the concept of species-specific Umwelt, the German-Baltic biologist Jakob Johann von Uexküll has exerted a deep and lasting influence on many disciplinary fields, from biology to philosophy and from ethology to semiotics and biosemiotics. Born in Estonia in 1864, he spent his childhood years directly experiencing nature at the family estate of Keblaste (today Mikhli). He attended the German Gymnasium of Coburg and then spent his university years at the Faculty of Zoology in Dorpat. Until World War I, the family patrimony allowed him to study and take research trips to France and Italy (in particular to the Zoological Station of Naples). The Russian Revolution and the consequent loss of the family fortune put an end to this fundamental period.

In 1927, Uexküll succeeded in founding in Hamburg an *Institut für Umweltforschung* (Institute of Environmental Research), which would be one of the places for the elaboration of the emerging discipline of ethology. Despite frictions with the Nazi regime, due among other factors to his anti-Darwinism, his years guiding the institute were productive and rich in satisfaction. During World War II, he moved for health reasons to Italy with his wife. Uexküll died in 1944 on the isle of Capri.

The young Uexküll's predominant interests were in zoology and physiology. His main research field was the empirical study of the neurophysiology of marine animals with particular attention to their sense-perception apparatus. Two remarkable elements of this research were a prelude to the making of an original biological theory pivoting on the concept of Umwelt. The first is Uexküll's early adherence to the Kantian transcendental approach, later reinforced by the study of the physiology of perception developed by Hermann von Helmholtz (1821–1894). Through Helmholtz, the study of the transcendental structures of the mind allows, according to Uexküll, for the field of pure logic to enter that of natural sciences – that is, zoology and what we today would call ethology and cognitive sciences. The second element is Uexküll's rejection since his university years of the mechanistic and deterministic version of Darwin's thought that was commonly taught at the turn of the nineteenth century; he instead adopted a form of vitalism with strong teleological traits.

The core of Uexküll's theoretical biology is the idea of the subjective constitution of the species-specific environment (or Umwelt) of an animal. This process can be divided into two subsequent phases, the first of which takes place in the physiological sphere of the animal. In the central nervous system connecting receptors and effectors, there are quanta of excitation that the organism spontaneously transforms in environmental 'marks'. Such marks are immediately transposed outward (*hinausverlegt*) and experienced by the subject as part of the objective reality.

To be more specific, this outward transposition transforms series of undifferentiated nervous excitations into two different external spheres: the sphere of perception (*Merkwelt*) and the sphere of action (*Wirkwelt*). The sphere of perception is made up of consciously re-perceived sensory marks (*Merkmale*), that is, the visual, acoustic and olfactory elements of the environment. The sphere of action includes the intraorganic, behavioural impulses and the external counterparts they meet – the so-called operative marks (*Wirkmale*) – as when, for example, the skin of a mammal receives the mosquito's sting. Frequently, the elements of the sphere of action remain unperceived; they play a decisive role in the behaviour of the animal without entering its consciousness.

In this way, a behavioural sequence appears as a continuous, smooth passage from one environmental mark to another. Let us consider Uexküll's best known example, the feeding behaviour of the tick (*Ixodes ricinus*). The first *Merkmal*, the smell of butyric acid, makes the tick fall from the bush where it was waiting onto a passing mammal, colliding with its body (*Wirkmal*). At this point, according to Uexküll's idea that 'the operative mark extinguishes the [previous] perception mark',[2] the tick finds the next adequate mark, the warmth and the hair of the mammal's skin. That allows it to move around until it finds the first accessible bare spot, where it can begin to feed on the mammal's blood. It is very important not to insert the notions of *Merkmal* and *Wirkmal* into a realist or objectivist theoretical frame. Both perception and operative traits are, at their source, qualitatively undifferentiated waves of excitation that arise in connection with external events but are not a faithful representation of them. If it is true that Uexküll frequently defines such traits as 'signs' (*Merkzeichen* and *Wirkzeichen*), it also is true that he denies to them any mimetic or representative function.[3]

The first, physiological phase of the constitution of the species-specific Umwelt constitutes around the organism a complex but unitary environment composed of perception and operative marks. This first phase is performed by every animal species, even if sometimes in a very elementary way.[4] The second phase, which not every animal species shows, is linked to the capacity to give further meaning to some first-level environmental elements.

This higher form of Umwelt-constitution is based on a transcendental function that Uexküll defines as the 'colouring' (*Färbung*) or 'tone' (*Ton*) of an element. According to the situation of the organism, some element of the environment appears as charged with a particular qualitative 'aura' which overlaps the content. For instance, a hungry predator doesn't perceive a prey as something neutral but rather as an object 'coloured' with a particular feeding tone (*Freßton*), which can be regarded as the *meaning* of the object in the particular living situation.

One of the clearest examples of the semiotic variability of the Umwelt is provided by the behaviour of the hermit crab (*Pagurus bernhardus*) faced with a sea anemone (*Anemonia sulcata*). If the hermit crab has been kept without nutriment for a long time, the anemone assumes a 'feeding tone'; if the animal is deprived of the anemones which crabs usually have on their shell as a defence from predators, the anemone undertakes a 'defensive tone'; if the hermit crab has lost both its shell and the mimetic anemones, the anemone acquires a 'dwelling tone' (as attested to by the hermit crab's vain efforts to enter it).[5] In this semiotic variability of the environmental elements Maurice Merleau-Ponty would see 'a beginning of culture' or 'a species of preculture within Nature',[6] thus adhering to Uexküll's decidedly gradualist position as far as the difference between human and non-human animals is concerned.[7]

At a first sight, Uexküll's concept of Umwelt seems of little utility for contemporary ecological thought. His Kantian, transcendental approach is likely to make the constitution of the Umwelt a solipsistic or merely subjectivist process – even if one should not forget that the physiological starting point of the process, the undifferentiated waves of excitation, are provided by the receptors of the nervous system, which inevitably are connected with the external world. Moreover, Uexküll's vitalism makes him see nature as a harmonic and powerful whole and consequently undervalue the risks of the extinction of single animal species. Even in his consideration of the Umwelt of man – which is seen as more comprehensive and rich than those of non-human animals, but not at all opposed to them – there is no indication that the way in which humans give meaning to their living sphere could interfere in a destructive manner with the Umwelt-formation performed by non-human animals.

If we put Uexküll's thought together with the contemporary, scientific and realist vision of nature, we can however find in his work a valuable contribution to the protection of biodiversity. Both the theoretical importance of the Uexküllian concept of Umwelt and its utility for ecological and conservationist thought, in fact, are directly linked to the high degree of spontaneity and 'cognitive creativity' it gives to every animal species. As a transcendental constitution, the Umwelt is spontaneously developed around each animal. It can be limited to the physiological sphere, and consequently rather poor in meaning, or it can be rich in meaning thanks to the aforementioned elaborations of the perceptive sphere, but in any case the Umwelt is a sort of extension of the organism itself. And, in a conservationist perspective, this 'field of meaning' displayed by each organism should be protected as an essential part of its biodiversity. In other terms, if an animal lives in a situation that doesn't allow it to constitute its species-specific Umwelt (for instance, in captivity), its integrity is only apparent. And, if we consider that very frequently the more meaningful elements in a species-specific Umwelt are other organisms (or the signs of their presence), it is evident that the preservation of the 'Uexküllian' Umwelt (the perceptive-operative sphere of the organism) depends on the preservation of the environment meant as complex and balanced ecosystem.

Notes

1   *A Foray into the Worlds of Animals and Humans; with, A Theory of Meaning*, p. 200.

2   *A Foray into the Worlds of Animals and Humans; with, A Theory of Meaning*, p. 49.

3   For a more complete evaluation of this point, see C. Brentari, *The Discovery of the Umwelt. Jakob von Uexküll between Biosemiotics and Theoretical Biology*, Springer, Dordrecht, 2015, pp. 85–89 and pp. 225–228.

4   It is the case of the Umwelt of medusa Rhisostoma, which is made up of a single intraorganic stimulus that cannot even be transposed to the outside; see *A Foray into the Worlds of Animals and Humans; with, A Theory of Meaning*, p. 75.

5   *A Foray into the Worlds of Animals and Humans; with, A Theory of Meaning*, pp. 92–93.

6   M. Merleau-Ponty, *Nature. Course Notes from the Collège de France*, Evanston, IL: Northwestern University Press 2003, p. 178.

7   If it is true that Uexküll supports a gradualist conception of the relationship between human and non-human animals, yet it should be said that (in Uexküll's conception as well) in the environment of human beings there is a sort of leap in quality. In humans, the transcendental constitution of the environment depends on individual variables in a measure unequaled in the rest of the animal kingdom; for different human subjects the same environmental element can assume an utterly different meaning, and can thus confer a different coloring to the environment. Uexküll's favorite example of this phenomenon is based on the interpretative relation that different people (a frightened little girl and a professional forester) can have regarding an oak; see *A Foray into the Worlds of Animals and Humans; with, A Theory of Meaning*, p. 126–129.

See also in this book
Heidegger, Merleau-Ponty, Wilson

Uexküll's major writings
'Die Umwelt', *Die neue Rundschau*, 21, 638–649, 1910.
Umwelt und Innenwelt der Tiere. 2. vermehrte und verbesserte Auflage, Berlin: Springer, 1921.
Theoretical Biology, London: Kegan Paul/Trench, Trübner and Co., 1926.
A Foray into the Worlds of Animals and Humans; with, A Theory of Meaning, trans. J.D. O'Neill, Minneapolis, MN and London: University of Minnesota Press, 2010.

Further reading

Agamben, Giorgo, *The Open. Man and Animal*. Stanford, CA: Stanford University Press 2004.
Brentari, C., *The Discovery of the Umwelt. Jakob von Uexküll between Biosemiotics and Theoretical Biology*, Dordrecht: Springer, 2015.
Buchanan, Brett, *Onto-Ethologies: The Animal Worlds of Uexküll, Heidegger, Merleau-Ponty and Deleuze*, Albany, NY: SUNY Press, 2009.
Canguilhem, Georges, 'The Living and Its Milieu', in *Knowledge of Life*, New York: Fordham University Press 2008, pp. 98–120.
Deleuze, Gilles and Guattari, Felix, '1837: On the Refrain', in *A Thousand Plateaus. Capitalism and Schizophrenia*, London and Nex York: Continuum 2004, pp. 342–386.
Heidegger, Martin, *The Fundamental Concepts of Metaphysics. World, Finitude, Solitude*, Bloomington, IN: Indiana University Press, 1995.
Kull, Kalevi, 'Jakob von Uexküll: An introduction', *Semiotica*, 134, 1–59, 2001.
Magnus, Riin and Kull, Kalevi 'Roots of culture in the Umwelt', in Valsiner, Jaan (ed.), *The Oxford Handbook of Culture and Psychology*, Oxford: Oxford University Press, 2012, pp. 649–661
Rüting, Torsten, 'History and significance of Jakob von Uexküll and of his institute in Hamburg', *Sign Systems Studies*, 32 (1/2) 35–72, 2004.
Tønnessen, Morten, 'Umwelt Transitions: Uexküll and Environmental Change', *Biosemiotics* 2 (1), 47–64, 2009.
Uexküll, Thure von, 'Introduction: The sign theory of Jakob von Uexküll', *Semiotica* 89 (4), 279–315, 1992.

CARLO BRENTARI

FRANK LLOYD WRIGHT 1867–1959

What, then, is architecture? It is man in possession of his earth. It is the only true record of him ... While he was true to earth his architecture was creative.[1]

Frank Lloyd Wright was an American architect whose early designs were the catalyst for the emergence of modern architecture around 1900, and whose seventy-two-year career has been the single greatest influence on the architecture of the twentieth century. Today, forty years after his death, Wright is the most famous architect in the world, and his designs, including Unity Temple, Fallingwater and the Guggenheim Museum, are among the most well-known works of architecture built in the twentieth century.

Wright was born in Richland Center, Wisconsin, in 1867, and raised in a family where the study of nature, the Unitarian faith and the ideas of American transcendental philosophy were all powerfully present. Aged twenty, without any formal university training, Wright moved to Chicago and entered the practice of architecture. After five years in the office of Louis Sullivan, leader in the development of 'organic' architecture and the skyscraper, Wright opened his own practice in 1893. During the next sixty-six years, Wright designed over six hundred built works, revolutionizing architecture as we understand it in the modern world.

Wright idealized Nature (which he spelled with a capital N) as the absolute reference and evaluative measure for the works of man. Nature was the source of both ethical principles, for the living of life, and formal principles, for the design of architecture. Wright based this interpretation of nature on the writings of the American transcendental thinkers, Walt Whitman, Henry Thoreau, Horatio Greenough and, most importantly, his beloved Ralph Waldo Emerson. The transcendentalists held as fundamental the fact that the material and spiritual worlds were inseparable, being in fact one and the same. Nature was the ideal manifestation of divine order, and Emerson called on his readers to 'esteem nature a perpetual counselor, and her perfections the exact measure of our deviations'.[2] Every physical thing, natural or man-made, was the consequence of, and had consequences for, spiritual thought – all form had moral meaning. 'All form is an effect of character',[3] Emerson said, and Wright believed that a person's character was an effect of the form and construction of the place in which they dwelled:

> Whether people are fully conscious of this or not, they actually derive countenance and sustenance from the 'atmosphere' of the things they live in or with. They are rooted in them just as a plant is in the soil in which it is planted.[4]

Wright's formal principles of architectural design were also drawn from the natural world. The formative experience of working on his uncle's farm during the summers of his childhood established Wright's great love and respect for nature. The Friedrich Froebel kindergarten training Wright received transformed this naïve love of nature into a precise method of making form. Based on learning from nature, Froebel training taught the child to seek the fundamental geometries underlying all natural forms. From this training, reinforced by his later studies of nature-based ornament with Sullivan, Wright would develop his definition of architectural design: discovering the underlying geometric structure of nature and building with it. Wright believed that man does not learn from nature by merely copying its surface effects – the underlying structure and geometry of nature were nature's true gifts for the architect, to be discovered only through close analysis of both natural forms and their determining functions. The ideal of an 'organic' architecture, first proposed by Horatio Greenough in his 1852 essay 'Form and Function', and defined thirty-five years later by Wright's mentor Louis Sullivan as 'form follows function', was redefined by Wright as 'form and function are one'. Wright sought to build an architecture that attained the perfect fusion of geometric form and life-giving function he found in his studies of nature.

Yet for Wright, nature as the ideal source of geometric order for design (Whitman's 'the square deific') was not to be confused with the particular building site or landscape in which he was called upon to work. While idealized Nature was sacred, the inhabited landscape was always in need of the redemptive power of design. Wright believed that no site selected for building was ever untouched by the hand of man. 'Fallingwater', the most famous modern house in the world, was designed by Wright in 1935 on what most visitors today assume was a 'wild, natural' site, but which in fact had been inhabited for more than forty years – this most 'natural' house itself sits in the hill cut of a pre-existing road. Wright believed that humans never built in and inhabited the natural world without fundamentally changing it, but he felt that, if the architect worked with the underlying geometric order of nature, it was possible to make the built landscape as beautiful, in its own way, as wild nature.

The vast majority of Wright's buildings were built in the American suburbs, where the original landscape had been subdivided into lots served by street grids and utilities, and where often most of the original trees and vegetation had been removed before any houses were built – these suburbs were far indeed from being 'natural' places. From the very beginning of his career in Oak Park, a suburb of Chicago where Wright built his house in 1889, Wright conceived of the architect's task in designing houses for the American suburbs to be one of reconstitution of a lost natural balance, a nature now fundamentally changed through the inhabitation of man. For Wright, man was an integral part of nature;

Man takes a positive hand in creation whenever he puts a building on the earth beneath the sun. If he has birthright at all, it must consist in this: that he too is no less a feature of the landscape than the rocks, trees, bears, or bees of that nature to which he owes his being.[5]

Wright thus conceived of architectural design as encompassing both the landscape and the architecture that engaged it. The 'Prairie Houses' of 1900–15, Wright's first important domestic design innovation, also involved an equally innovative (if rarely noted) strategy of relating to the landscape. Wright's Prairie Houses were often located at the edge of their suburban lots, allowing their gardens to occupy the geometric centre of the sites (usually reserved for the house itself), and weaving together interior and exterior spaces so that the house and landscape were inextricably bound to one another. Rather than the free-standing object in the landscape, so typical of much later modern architecture, Wright from the very beginning of his career constructed a remarkable interdependence between house and landscape, such that neither appears complete without the other.

Wright's was a truly 'organic' design ethic, embracing both architecture and landscape, and all that takes place within them:

[B]uildings are the background or framework for the human life within their walls and the natural efflorescence without; and to develop and maintain the harmony of a true chord between them … These ideals take the buildings out of school and marry them to the ground.[6]

Wright began each design by incorporating the formative power of the landscape as the primal place of inhabitation – the building literally began with the ground on which it was to stand: 'It is in the nature of any organic building to grow from its site, come out of the ground into the light – the ground itself held always as a component basic part of the building itself.'[7]

Wright believed that architecture was determined by 'the nature of materials' of which it was constructed. He believed that the way a space was experienced was directly related to the way it was made or constructed. Wright built with both the underlying structures of nature (the cantilevered skyscraper based upon the tree) and the actual materials of nature. Wright employed each material in its natural state, displaying its inherent colours and texture – whether it was stone mined from a nearby quarry, concrete cast into ornamented block, or wood cut in the mill – and exposing the marks of cutting and shaping inevitably involved in taking materials from nature and preparing them for use in construction. Wright employed each material so that it contributed its own unique character to the spatial experience of inhabitation – the 'natural house' was literally made from nature. In this way, Wright believed his buildings were natural places within and without, where man could truly be at home in nature.

For Wright, architecture was literally mankind's place in nature, our particular manner of dwelling on the earth, under the sky. Whatever the commission, Wright always designed for a balanced condition – man in nature and nature in man. In the public urban building, such as Unity Temple, Johnson Wax and the Guggenheim Museum, vertical sunlight fell from above, filtered through the 'natural' geometric forms of skylights, bringing nature deep into the very heart of the city. In the private suburban house, such as the Coonley House, the Robie House and the Jacobs House, horizontal views, sheltered by the brow of the broad overhanging roof, opened to the surrounding landscape, bringing nature all the way into the hearth at the centre of the house. The public urban building was given the arc of the sun, and the private suburban house was given the line of the land – sky and horizon, as respective boundaries of the natural world, brought by Wright into the spaces of daily life.

Wright held that it was essential for daily life to be lived in direct communion with nature, and that architecture should be designed as a place in nature. Wright believed, following Emerson and Thoreau, that because man was a product of nature, he was only able to learn about his own essential nature through regular and intimate contact with the natural landscape. Wright felt that the American democratic experiment would ultimately fail unless all its citizens had the opportunity to live intimately in nature. In 1935, the same year he designed Fallingwater, Wright designed the first of his 'Usonian' Houses, modestly priced prototype homes for the growing American middle class. The Usonian Houses were L-shaped in plan, framing two edges of their suburban sites in such a way that the garden was the centre of both the site and the spatial composition of the house itself. Flooded with light, these gardens became the focus of the house and the life that took place within it; Wright strove to 'make the garden be the building as much as the building will be the garden, the sky as treasured a feature of daily indoor life as the ground itself'.[8]

A life taking place in nature was what Wright sought to make possible through his house designs, and thus his opposition to the flattening of landscape contours or the mechanical control of climate:

> To me air conditioning is a dangerous circumstance … I think it far better to go with the natural climate than to try to fix a special

artificial climate of your own. Climate means something to man. It means something in relation to one's life in it.[9]

The remarkable energy-efficiency and unerring solar orientation of Wright's houses from the very beginning of his career, though unprecedented in architectural practice, is entirely consistent with his vision of architecture's harmony with nature. While often considered 'ahead of his time' in his willingness to embrace and employ technical developments, Wright remained absolutely opposed to the instrumental aspects of the modern industrial era that in any way diminished mankind's experience of being at home in nature. Primary among these were land speculation and speculative building, which Wright believed were inherently evil and unnatural, noting that in the typical American suburb 'architecture and its kindred, as a matter of course, are divorced from nature in order to make [architecture] the merchantable thing ... It is a speculative commodity.'[10]

Wright's designs engaged both the natural land form and the history of human occupation of the site. He believed agriculture (to care for and cultivate) and architecture (to build and to edify) were related human activities on the earth – the tending and transforming of the landscape. Broadacre City, designed in 1935, was Wright's greatest and most comprehensive counter-proposal to the crowding of the traditional city, but also to the isolation of both agrarian life and the developer's speculative suburb. For Wright, culture and cultivation were closely related, and the level of culture of a society was directly indicated in the level of cultivation of its landscape: 'You will find the environment reflecting unerringly the society.'[11] At the most fundamental level, Wright believed that the natural environment should be integrated into daily domestic life: each of his designs was intended 'to be a natural performance, one that is integral to site, integral to environment, integral to the life of the inhabitants'.[12]

Notes

1 'Architecture and Modern Life', *Collected Writings*, vol. 3, p. 222, 1937.
2 Ralph Waldo Emerson, 'Prudence', *Emerson's Essays*, New York: Harper & Row, p. 166, 1926, 1951.
3 Emerson, 'The Poet', ibid., p. 269.
4 Wright, *The Natural House*, New York: Horizon Press, p. 135, 1958.
5 'Architecture and Modern Life', p. 223.
6 'In the Cause of Architecture', *Collected Writings, vol. 1*, p. 95, 1908.
7 *The Natural House*, p. 50.
8 Ibid., p. 53.
9 Ibid., pp. 175–8.
10 'Architecture and Modern Life', p. 237.

11 'Concerning Landscape Architecture', *Collected Writings, vol.* 1, p. 57, 1900.
12 *The Natural House*, p. 134.

See also in this book
Emerson, Ruskin, Thoreau

Frank Lloyd Wright's major built works
Larkin Company Building (demolished), Buffalo, NY, 1903.
Darwin Martin House, Buffalo, NY, 1904.
Unity Temple, Oak Park, IL, 1905.
Avery Coonley House, Riverside, IL, 1907.
Frederick Robie House, Chicago, IL, 1907.
Frank Lloyd Wright House/Studio, 'Taliesin', Spring Green, WI, 1911–25.
Midway Gardens (demolished), Chicago, IL, 1913.
Imperial Hotel (demolished), Tokyo, Japan, 1914–22.
Aline Barnsdall 'Hollyhock' House, Los Angeles, CA, 1919.
Samuel Freeman House, Los Angeles, CA, 1923.
Edgar Kaufmann House, 'Fallingwater', Mill Run, PA, 1935.
Johnson Wax Buildings, Racine, WI, 1935, 1944.
Herbert Jacobs House, Madison, WI, 1936.
Frank Lloyd Wright House/Studio, 'Taliesin West', Scottsdale, AZ, 1937.
Florida Southern College, Lakeland, FL, 1938–59.
Guggenheim Museum, New York, 1943–59.
H.C. Price Company Tower, Bartlesville, OK, 1952.
Beth Sholom Synagogue, Elkins Park, PA, 1954.
Marin County Civic Center, San Rafael, CA, 1957.

Frank Lloyd Wright's major writings
Frank Lloyd Wright: Collected Writings, ed. Bruce Brooks Pfieffer, New York: Rizzoli, 1992–5:
Volume 1: 1894–1930 (1992)
Volume 2: 1930–1932 (1992)
Volume 3: 1937–1939 (1993)
Volume 4: 1939–1949 (1994)
Volume 5: 1949–1959 (1995)

Further reading
Hoffmann, Donald, *Frank Lloyd Wright: Architecture and Nature*, New York: Dover, 1995.
Levine, Neil, *The Architecture of Frank Lloyd Wright*, Princeton, NJ: Princeton University Press, 1996.

McCarter, Robert, *Frank Lloyd Wright*, London: Phaidon Press, 1997.

McCarter, Robert (ed.), *On and By Frank Lloyd Wright: A Primer of Architectural Principles*, London: Phaidon Press, 2005.

Sergeant, John, *Frank Lloyd Wright's Usonian Houses*, New York: Whitney Library of Design, 1976.

ROBERT McCARTER

MAHATMA GANDHI 1869–1948

> The next step should not be destructive agriculture but the planting of plenty of fruit trees and other vegetation.[1]

Although Gandhi has become a household name, the lean, saintly looking bespectacled son of India who took on the British Empire with his sharp wit and prolific pen is better known for his ethics of non-violence and truth-force than for his environmental philosophy. However, just as leaders of non-violent civil rights movements across the globe attribute their inspiration to Gandhi's strategy of making the oppressors confront their own unjust practices, leading environmental theorists and activists in India and other parts of the world defer to Gandhi's insights and practices in the area of ecology as well. While much of what Gandhi said or wrote on ecology is of an anecdotal nature, his criticism of structures antithetical to a healthy ecological life-world ramified into ideas which developed and were put into action in different areas of environmental concern. Gandhi's importance as an environmental thinker may be marked in terms of the strategies and vistas opened up by his pursuits, both public and private, towards a sustained animal and environmental liberation struggle. Looked at another way, Gandhi's environmental thinking is rooted in his larger philosophical and moral thinking.

The Mahatma ('great soul') was born Mohandas Karamchand Gandhi in Porbandar, now in the State of Gujarat, on 2 October 1869. As a child he had learned to appreciate the beauty of the coastal region washed by the Arabian seas and surrounded by temples, churches and mosques. Although by caste the Gandhis were merchants, his family held high office in the sovereign province's court and were devout Hindus. Very early on he came to the realization that morality is an inexorable part of the objective reality he preferred to call Truth rather than God, and that nature was a substance within this reality. Hence, as in traditional wisdom, nature was not there merely for human use or as an appendix to civilization but was a presence, much like one's nourishing nurse, to be respected. Gandhi's Hindu background taught him about the basic elements that constituted the physical and material world – namely, earth, water, fire, ether and space, which he saw ritually invoked in home worship (*puja*) as well as in meditational practices. Indeed, Hindu biocosmology, with its large pantheon of gods and goddesses, appeared to share these elemental constituents in varying measures and permutations.

During his education in England, Gandhi rediscovered the virtues of his family's vegetarianism, albeit on the moral grounding articulated by Henry Salt, and inspired by Shelley, Thoreau, Whitman and Ruskin. At the same time Gandhi sought out theosophists who initiated him into a non-ritual moral reading of the *Bhagavad Gita*; this instilled humanitarian ideals that were to take Gandhi further towards a complete break with Western civilization. In South Africa, where he went to practise as an attorney, Gandhi withdrew from time to time to deepen his understanding of Tolstoy, the Upanishads, Quakerism, the Gospels through contacts with Trappists, Methodists and Jewish acquaintances. He also tried his hand at living in a commune. The influence of Ruskin's *Unto This Last* led Gandhi to write his own treatise on *sarvodaya* ('welfare for all') which became the basis of the movement of the same name which he launched upon his return to India in 1914. It was part of the larger programme he envisioned for India of *swadeshi* or 'self-sufficiency' and had outlined in the 1908 treatise *Hind Sawaraj*. Both socio-ethical directives, as well as that of non-violent resistance (*ahimsa*), were propelled by a common volitional determination he called '*satyagraha*' or 'truth-force'. Gandhi acknowledges the influence of the Jaina ethical precept of non-injury (which Buddhism and Hinduism also heed and which has its parallel in the Golden Rule of 'turning the other cheek' or 'nonresistance', as Tolstoy had christened this practice). Under Gandhi's impetus, however, this basically passive and individual stance becomes a positively empowering and collective experience with enormous potential for unleashing liberative but, at times, also coercive and indignant energies.[2]

From these general articulations and stances also sprang the more practical ideal of minimal or 'reactionary' economy and Luddite manufacturing skills, such as the humble spinning wheel (*charkha*) and weaving of yarns (*khadi*), and small-scale farming. Gandhi also experimented extensively with 'earth treatments' and 'dietetics' as means of healing and rejuvenation that did not depend on chemical-based medicines and toxic pollutants. Personal ecology for him was the basis for social and environmental ecologies as well.[3] Traditional methods of farming, husbandry and irrigation were explored in the Ashrams which Gandhi helped set up in different regions.

Gandhi's overall social and environmental philosophy is based on what human beings need rather than what they want. His early introduction to the teachings of Jains, theosophists, Christian sermons, Ruskin and Tolstoy, and most significantly the *Bhagavad Gita*, were to have profound impact on the development of Gandhi's holistic thinking on humanity, nature and their ecological interrelation. His deep concern for the disadvantaged, the poor and rural population created an ambience for an alternative social thinking that was at once far-sighted, local and immediate. For Gandhi was acutely aware that the demands generated by the need to feed and sustain human life, compounded by the growing industrialization of India, far outstripped the finite resources of nature. This might nowadays appear naïve or commonplace, but such pronouncements were as rare as they were heretical a century ago. Gandhi was also concerned about the destruction, under colonial and modernist designs, of the existing infrastructures which had more potential for keeping a community flourishing within ecologically-sensitive traditional patterns of subsistence, especially in the rural areas, than did the incoming Western alternatives based on nature-blind technology and the enslavement of human spirit and energies.

Perhaps the moral principle for which Gandhi is best known is that of active non-violence, derived from the traditional moral restraint of not injuring another being. The most refined expression of this value is in the great epic of the *Mahabharata*, (*c*.100 BCE to 200 CE), where moral development proceeds through placing constraints on the liberties, desires and acquisitiveness endemic to human life. One's action is judged in terms of consequences and the impact it is likely to have on another. Jainas had generalized this principle to include all sentient creatures and biocommunities alike. Advanced Jaina monks and nuns will sweep their path to avoid harming insects and even bacteria. Non-injury is a non-negotiable universal prescription. Gandhi relates this principle to the value that the *Bhagavad Gita* places on the welfare of all beings:

> The one whose self is disciplined by yoga
> Sees the self abiding in every being
> And sees every being in the self;
> He sees the same in all beings.[4]

The transcendence of the self from constricting human conditions of desire and attachment and the prudential ethic of not causing injury to other beings for fear of attracting more karma into one's soul is turned by Gandhi into a categorical value: one does X because X is right and it is also just from the position of the other.

This principle, more than anything else, becomes the foundation stone for Gandhi's approach to environmental ethics. Much that can be gleaned from Gandhi's own practices, as noted earlier, is of anecdotal value. His obsession with the hygiene of man and animals alike – safer waste disposal systems and cleanliness of both the body and the surrounding environs – have been meticulously noted in the Gandhiana literature and his own writings. Gandhi's weakness, as many writers have pointed out, is that he did not compose a systematic treatise on this subject, nor did he lead a major ecological campaign in the way that he did political campaigns, such as the symbolic 'Salt March', an act of nationalist defiance against the British monopoly over access to sea salt. His impact, nevertheless, has been tremendous, and Gandhi's visions, if not his words, have certainly left traces in the great works on ecological thinking, especially those of Arne Naess and other 'deep ecology' or pan-ecotheistic thinking in recent decades. Gandhi, with his advocacy of *sarvodaya* and radical empowerment of localized or microeconoculture, was a forerunner of the avant-garde movements nowadays associated with 'deep ecology' and the Greens. But Gandhi went further in some respects with his emphasis on the aboluteness of non-violence and *dharma*.

Gandhi was also adamant about the need for a rigorous ethic of non-injury in our treatment of animals.[5] On active environmental renewal projects, Gandhi wrote in 1926 that for India the next step should not be destructive agriculture but the planting of fruit trees and other vegetation as these provide nourishment, stability in the soil, and attract rainfall as well as provide fodder for the insect and animal world. The implications of such simple ecological wisdom have only just begun to dawn on a tech-fested agricultural economics. Likewise, Gandhi's symbolic insistence on *khadi* spinning was instructive for avoidance of factory-emitted pollution, desalination of soil through over-cultivation and dependence on raw materials produced through suffering caused to animals (e.g. silk and wool). Gandhi's advocacy of simple living through the principles of non-violence and holding steadfastly to truth challenge modern-day Hindus to reconsider their lifestyle engendered by pressures of contemporary consumerism. They have had to consider whether social duty can be expanded to include ecological community and whether the Hindu tradition can develop new modalities of caring for the earth.[6] Can *dharma* be re-interpreted in earth-friendly terms to meet the challenges of modern post-industrial 'civilization'?

Gandhian activists have attempted to deal with just these challenges. *Sarvodaya* has increasingly become a basis for a number of *asarkari* or non-governmental organizations across India. Inspired by Gandhi and especially his wife-partner Kasturba's dedicated *sarvodaya seva* or service ideal, these groups regularly travel to remote villages to teach women and youth the virtues and simple practices of hygiene and earth-care. Rural development and alternative technology programmes have been helping villagers to construct *chulas* or smokeless ovens, mudbrick dwellings, and to utilize non-toxic organic fertilizers. Schools and colleges have been established to explore and promote safe ecological practices. Tribal groups have been encouraged to preserve the wild bushland, to curtail excessive use of wood for cooking, and to develop a technology for dealing with local conditions while resisting the technologies and wares brought in by profit-driven urban and corporate enterprises. Gandhians have not been unanimous on a complete biospheric egalitarianism, and most have come to accept small-scale 'soft' technology supplemented heavily with hand-crafting and local cottage industries.[7]

Active in northern regions of the subcontinent is Sunderlal Bahuguna, best known for his spectacular Himalayan campaign, known as the Chipko ('Hug the Trees') Movement, aimed at resisting environmental destruction, particularly by governmental agencies and corporate interests which, in exploiting the hill regions, leads inexorably to irreparable deforestation.[8] Bahuguna is also a great believer in locally renewable 'sustainable economy'; hence, he has been one of the leading critics of India's current policy of economic liberalization which has allowed the influx of multinational companies and unilateral concessions on produce and plant variety rights forced upon India by World Trade Organization treaties.

Another scene which has been drawing world-wide attention, where similar non-violent resistance tactics have been used to raise awareness of environmental concerns, is the Narmada Bacho Andolan in southern Gujarat. Environmentalists led by the veteran Medha Patkar have ceaselessly argued that the 3,200 dams planned on the Narmada and tributary rivers would cause immense damage to surrounding land mass which would also lead to the dislocation of 2,500 families in nearly 60 villages of tribal people who have lived along the river basins and maintained a healthy eco-community for countless generations. The Gandhian spirit lives on. There are numerous other grassroots groups and movements that invoke traditional wisdom and practical ethics in their expression of resistance to and concerns for radical transformations of the local environment. The supply of safer drinking water to rural areas, conserving rain water and utilizing dead water from hydro-electric dams, have become joint initiatives of NGOs, religious leaders and some state governments as well (e.g. the southern taluks around Puttaparthi in Andhra Pradesh).

The Bhopal incident in 1984 where the ill-maintained Union Carbide chemical plant unleashed thousands of tons of poisonous chemical fumes into the atmosphere, killing and disabling thousands of people, perhaps highlighted a particular kind of challenge facing Gandhian environmentalists. The challenges of industrialization, modernity, globalization and a rapidly expanding liberal economy present Gandhians with a very different set of circumstances and contexts from those that Gandhi could have foreseen. These call for quite different sorts of responses on the environmental front, and they can only be forthcoming case by case. Still, there are a number of Gandhian followers who are prepared to 'risk their all' in order to meet these challenges for the sake of non-violent truth and to bring greater welfare to all beings on the planet Earth.

Notes

1   Letter to Kaka Kaleklar, 1926, quoted in I. Harris, *Gandhians in Contemporary India: The Vision and the Visionaries*, Lewiston, NY: Edward Mellon, p. 274, 1998.
2   Joan V. Bondurant, *Conquests of Violence*, Princeton, NJ: Princeton University Press, 1985.
3   *An Autobiography*, p. 271.
4   *The Teaching of the Gita*, VI.29.
5   Gandhi, *My Socialism*, Ahmedabad: Navajivan, pp. 34–5, 1959.
6   Christopher Key Chapple, 'Hinduism, Jainism and Ecology', *Earth Ethics*, Fall, pp. 16–18, 1998.
7   P. Bilimoria, 'Indian Religious Traditions', in D.E. Cooper and J.A. Palmer (eds), *Spirit of the Environment: Religion, Value and Environmental Concern*, London and New York: Routledge, pp. 1–14, p. 12, 1998.
8   Harris, op. cit., p. 265.

See also in this book
Buddha, Naess, Ruskin, Tagore

Gandhi's major writings
An Autobiography: The Story of My Experiments with Truth, Boston, MA: Beacon Press, 1957.
Collected Works of Gandhi, 90 vols, New Delhi: Ministry of Information and Broadcasting, India, 1958–94.
The Essential Gandhi, ed. L. Fischer, New York: Vintage Books, 1962.
The Teaching of the Gita, Bombay, India: Bharatiya Vidya Bhavan, 1962.

Further reading
Bilimoria, P. and McCulloch, J., *Environmental Ethics*, Victoria, Canada: Deakin University, 1992.

Bilimoria, P., Chapple, C.K., and Wong, P., 'Ethical Studies Overview: Eastern Traditions', in L. Kurtz (ed.), *Encyclopedia of Violence, Peace & Conflict*, 2nd edn, Oxford, UK: Elsevier Inc., 2008, pp. 720–39.

Brown, Judith, *Gandhi: Prisoner of Hope*, New Haven, CT: Yale University Press, 1989.

Chapple, C.K., *Nonviolence to Animals, Earth, and Self in Asian Thought*, Albany, NY: State University of New York Press, 1993.

Joshi, Nandini, *Development Without Destruction*, Ahmedabad, India: Navajivan, 1992.

Macy, Joanna, *Dharma and Development: Religion as Resource in the Sarvodaya Self-Help Movement*, West Hartford, CT: Kumarian Press, 1983.

Naess, Arne, *Gandhi and the Nuclear Age*, trans. A. Hannay, Totowa, NJ: Bedminster Press, 1965.

Weber, Thomas, *Hugging the Trees: The Story of the Chipko Movement*, Delhi, India: Viking Press, 1988.

PURUSHOTTAMA BILIMORIA

ALBERT SCHWEITZER 1875–1965

Man has lost the capacity to foresee and to forestall. He will end by destroying the earth.[1]

The publication of Rachel Carson's *Silent Spring* in 1962 is frequently regarded as the beginning of the modern environmental movement. It was to Albert Schweitzer that Carson dedicated the work, and she opened her text using his above words.

'In terms of intellectual achievement and practical morality', Schweitzer has been described as 'probably the noblest figure of the twentieth century.'[2] Born in 1875, he was brought up at Gunsbach in Alsace. His intellectual achievements span four major disciplines. He learnt the organ under Widor in Paris and eventually published *J.S. Bach, le musicien-poète* in 1905. He studied theology and philosophy at Strasbourg, Paris, and Berlin, and published major works of New Testament scholarship, most notably *The Quest of the Historical Jesus* (English translation 1910). In 1896 he made his famous decision to live for science and art until age 30 and then devote his life to serving humanity. Accordingly, despite his international reputation as a musician and theologian, he turned to medicine and qualified as a physician. In 1905 he resigned as principal of the theological college in Strasbourg and founded the hospital at Lambaréné in the heart of what was French Equatorial Africa.

By 1962 Schweitzer had already become a legend in his lifetime. Although his work in Lambaréné captured the public imagination, earning him the Nobel Prize for peace in 1952, Schweitzer considered that his most meaningful contribution, the one for which he most wished to be remembered, was his ethic of 'reverence for life'. Travelling slowly upstream in a tug-steamer – amidst the panorama of the tropical forest – on Gabon's Ogowe River, the 'unforeseen and unsought' phrase, 'reverence for life', 'flashed' into his mind. The phrase, simple as it is profound, unlocked for him the 'iron door' of ethical thought.

Although the concept of 'reverence for life' is now well known, it has been subject to a range of distortions, and it is important that we confront these in order to understand what Schweitzer meant by this term.

The first distorting lens is *legalism*. Contrary to many commentators, Schweitzer does not propound reverence as a new moral law but rather as 'ethical mysticism'. Ethical mysticism emerges out of reflection upon the 'will-to-life' (*Wille zum Leben*). 'The essential thing to realise about ethics', he writes, 'is that it is the very manifestation of our will-to-live.'[3] His use of the term 'will-to-live' is derived from Arthur Schopenhauer, the principal advocate of the German Voluntarist school, who articulated the phrase in *The World as Will and Idea* (1819). Schweitzer follows Schopenhauer's conviction that 'the essence of things in themselves, which is to be accepted as underlying all phenomena', is 'will-to-live'.[4] Whereas Immanuel Kant denied that the 'thing-in-itself' (his term for an 'object considered as it is independently of its cognitive relation to the human mind'[5]) was knowable, Schweitzer believed that the 'thing-in-itself' was the 'will-to-live' and readily ascertainable through the physiological make-up of animate phenomena. That which underlies all life – actually its very essence – is the will-to-live.

Schweitzer's metaphysics begins with the supposition that despite the diversity of individual things in the world, they all manifest the same inner essence. From a comprehension of oneself (the microcosm), one is able to acquire knowledge of the world (the macrocosm); the key to understanding the world is proper self-understanding. Schweitzer's argument largely rests on whether knowledge that originates from the inner experience of the will-to-live is more reliable than knowledge derived from empirical examination of the outer, physical world. The non-empirical quality of the will-to-live as the core self is a presupposition of his work. His view is that all empirical reality must, like himself, have an inner nature (will-to-live), and he uses this notion to offer a new account of the relationship between the self, the natural world and God.

It is from this reflection on the will-to-live that Schweitzer derives the ethic of reverence for life. Though he starts from the personal ('I am life which wills-to-live'), he goes on to assert the radical interdependence of all life. Each life 'wills-to-live' not in isolation, but 'in the midst of other wills-to-live'. This assertion is not an ingenious dogmatic formula but rather a personal revelation:

> Day by day, hour by hour, I live and move in it. At every moment of reflection it stands fresh before me ... A mysticism of ethical union with Being grows out of it.[6]

This immediate, experiential identification of one's individual will-to-live (or life) with other life, and through life with Being, is the foundation of his ethical mysticism. Indeed, the mystical nature of the experience of reverence is implicit in the very word: 'reverence' (*Ehrfurcht*) implies 'awe', 'wonder' and 'mystery'.

The second distorting lens is *inviolability*. Many commentators have assumed that Schweitzer is proposing the moral inviolability of all life of whatever kind. It is true that he sometimes writes in such a way as to invite this misunderstanding. The ethical person, he maintains,

> tears no leaf from a tree, plucks no flower, and takes care to crush no insect. If in the summer he is working by lamplight, he prefers to keep the window shut and breathe a stuffy atmosphere rather than see one insect after another fall with singed wings upon his table.
>
> If he walks on the road after a shower and sees an earthworm which has strayed on to it, he bethinks himself that it must get dried up in the sun, if it does not return soon enough to ground into which it can burrow, so he lifts it from the deadly stone surface, and puts it on grass. If he comes across an insect which has fallen into a puddle, he stops a moment in order to hold out a leaf or a stalk on which it can save itself.[7]

At first sight the sheer practical impossibility of these injunctions presents itself. But what Schweitzer offers here are not *rules* but rather *examples* of what reverence for life may require in a given situation. Schweitzer's basic definition of the moral is that 'it is good to maintain and to encourage life, it is bad to destroy life or obstruct it'.[8] Beyond this statement, he affords the reader only instances of the kind of action expected from one who upholds this ethic.

The third distorting lens is *inconsistency*. Since Schweitzer defines reverence as an 'absolute' ethic which enjoins 'responsibility without limit towards all that lives',[9] it is perhaps not surprising that reverence is judged to entail inconsistency in practice. And Schweitzer himself has not escaped this charge. He notoriously captured fish to feed his sick pet pelican, engaged in a pre-emptive strike against poisonous spiders, and did not fully embrace vegetarianism until later in life. These apparent inconsistencies are made more glaring by his rejection of any moral hierarchy:

> The ethics of reverence for life makes no distinction between higher and lower, more precious and less precious lives. It has good reasons for this omission. For what are we doing, when we establish hard and fast gradations in value between living organisms, but judging them in relation to ourselves, by whether they seem to stand closer to us or farther from us? This is a wholly subjective standard. How can we know the importance other living organisms have in themselves and in terms of the universe?[10]

Some commentators have interpreted Schweitzer at this point as suggesting that no form of life should ever be destroyed and that all creatures from human beings to microbes should have the same moral standing. It is doubtful whether this was Schweitzer's intention. Rather what he is doing is rejecting here the long tradition of moral hierarchy which places humanity at the top of the pyramid of descending moral worth. Schweitzer would have admitted (as his personal examples demonstrate) that it is sometimes necessary to make choices between one form of life and another, but what he wanted to emphasize was the essentially *subjective* and *arbitrary* nature of these declarations.

Any time life is sacrificed or injured, either 'for the sake of maintaining [one's] own existence or welfare' or 'for the sake of maintaining a greater number of other existences or their welfare', one is no longer wholly 'within the sphere of the ethical'.[11] In other words, killing may be '*necessary*' but it can never be '*ethical*' as such. When one is constrained by 'necessity', one must bear the 'responsibility' and 'guilt' of having injured life. 'Whenever I injure life of any sort', wrote Schweitzer, 'I must be quite clear whether it is necessary. Beyond the unavoidable, I must never go, not even with what seems insignificant.'[12]

Having clarified aspects of Schweitzer's thought, it is now possible to indicate some of his main contributions to the development of ecological consciousness.

The first and most important contribution concerns *the mystical apprehension of the value of life*. At the heart of many environmental controversies is the issue of value: whether beings outside of ourselves have value, of what kind, and why. What Schweitzer emphasizes is that the recognition and appreciation of the value of life is actually a mystical apprehension. This apprehension is 'primary' because all subsequent decisions and choices depend upon it. To understand Schweitzer at this point we do best perhaps to make a comparison with Plato. Plato describes philosophers in a democratic state as those who 'wrangle over notions of right in the minds of men who have never beheld *Justice* itself'.[13] Likewise, Schweitzer would maintain that one can have no proper sense of oneself and others in the world unless, first and foremost, one has a sufficient sense of the value of *Life* itself. Everything depends practically upon this prior recognition of value.

The second contribution concerns *service to life as practical mysticism*. In contrast to most mystics, Schweitzer maintains that the goal of union with the Divine is achieved not through contemplation, but primarily through service to other life:

> Ethics alone can put me in [a] true relationship with the universe by my serving it, co-operating with it; not by trying to understand it ... Only by serving every kind of life do I enter the service of that Creative Will whence all life emanates ... It is through the community of life, not community of thought, that I abide in harmony with that Will. This is the mystical experience of ethics.[14]

The phenomenon we call 'life', in short, is not something put here for our use or pleasure; we are part of 'life' (or as Schweitzer would say 'the will-to-live') and our role is to enhance and serve each and every manifestation of it.

The third contribution concerns the recognition of *the tragedy of life in conflict with itself*. Schweitzer is not a pantheist – that is, someone who thinks that the world is God or coterminous with God. Indeed he is sharply critical of those who seek to deify the natural world as it is instead of recognizing its essentially tragic and incomplete nature. Schweitzer writes movingly of the world as 'the ghastly drama of the will-to-live divided against itself'.[15] To affirm life and the value of life is not to affirm the parasitical and predatory aspect of nature itself. Schweitzer's own preaching is clearly eschatological – that is, he looks forward to a time when creation will be renewed and redeemed. His pioneering work in the field of New Testament scholarship – especially on the teaching of Jesus and Paul – emphasizes Jesus as an eschatological figure who will inaugurate 'the Kingdom', understood as the liberation of all creation from its present predation and suffering. Reverence for life was for Schweitzer 'practical eschatology'.

The fourth contribution concerns *non-injury to life as the central ethical imperative*. 'A man is truly ethical', Schweitzer writes, 'only when he obeys the compulsion to help all life which he is able to assist, and shrinks from injuring anything that lives.'[16] 'The time is coming ... when people will be astonished that humankind need so long a time to learn to regard thoughtless injury to life as incompatible with ethics.'[17]

Schweitzer regarded traditional philosophy which restricted ethics to human-to-human relations as spiritually impoverished. He was deeply critical of animal experimentation, opposed hunting for sport and eventually embraced a vegetarian diet. His hospital at Lambaréné was a model of ecological responsibility: he went out of his way to preserve trees and flora, re-used every piece of wood, string and glass, and rejected modern technological developments which would have resulted in environmental degradation.

It is unsurprising then that Rachel Carson, and others, have found in Schweitzer an inspiration for a wider ecological ethic. When Carson received the Schweitzer Medal from the Animal Welfare Institute in 1963, she summed up her work in Schweitzerian-like terms: 'What is important is the relation of man to all life'.[18] An inscribed photograph of Schweitzer (together with a letter of thanks for the dedication of Silent Spring) were encased, centre-stage, in her study. According to Carson's housekeeper, Ida Sprow, it was 'her most cherished possession'.[19]

Notes

1 Rachel Carson, *Silent Spring*, New York: Houghton-Mifflin Company, p. v, 1962.
2 Magnus Magnusson and Rosemary Goring (eds), *Chambers Biographical Dictionary*, London: Chambers Harrap, p. 1314, 1990.
3 'The Ethics of Reverence for Life', p. 229.
4 *The Philosophy of Civilisation*, p. 236.
5 Ted Honderich (ed.), *Oxford Companion to Philosophy*, Oxford: Oxford University Press, p. 871, 1995.
6 *The Philosophy of Civilisation*, p. 310.
7 Ibid.
8 Ibid., p. 309.
9 Ibid., p. 311.
10 *The Teaching of Reverence for Life*, p. 47, 1965.
11 *The Philosophy of Civilisation*, p. 325, reprint of New York: Macmillan, 1949.
12 Ibid., p. 318.
13 Plato, *The Republic*, trans. F.M. Cornford, Oxford: Oxford University Press, VII.518, p. 232, 1969; emphasis added.
14 'The Ethics of Reverence for Life', p. 239.
15 *The Philosophy of Civilisation*, p. 312.
16 Ibid., p. 310.
17 Ibid., pp. 310–11.
18 Linda Lear, *Rachel Carson: Witness for Nature*, London: Allen Lane, p. 440, 1997.
19 Ibid., p. 438.

See also in this book
Carson

Schweitzer's major writings

Kulturphilosophie I: Verfall und Wiederaufbau and *Kulturphilosophie II: Kultur und Ethik*, Bern: Paul Haupt, 1923. Both volumes published together as *The Philosophy of Civilisation*, trans C.T. Campion, New York: Prometheus Books, 1987.

Aus Meinem Leben und Denken, Leipzig: Felix Meiner, 1931; *Out of My Life and Thought: An Autobiography*, trans. A.B. Lemke, New York: Henry & Company, 1990.

'The Ethics of Reverence for Life', *Christendom*, I (winter), 1936.

The Teaching of Reverence for Life, trans. Richard and Clara Winston, New York: Holt, Rinehart & Winston, 1965.

A Place for Revelation: Sermons on Reverence for Life, trans. David Larrimore Holland, ed. Martin Strege and Lothar Stiehm, New York: Macmillan Press, 1988.

Further reading

Barsam, Ara, *Reverence for Life: Albert Schweitzer's Great Contribution to Ethical Thought*, Oxford: Oxford University Press, 2008.

Brabazon, James, *Albert Schweitzer: A Biography*, London: Gollancz, 1976.

Cicovacki, Predrag, *The Restoration of Albert Schweitzer's Ethical Vision*, New York: Continuum International Publishing Group, 2012.

Clark, Henry, *The Ethical Mysticism of Albert Schweitzer*, Boston, MA: Beacon Press, 1962.

Joy, Charles, *Schweitzer: An Anthology*, Boston, MA: Beacon Press, 1956.

Linzey, Andrew, *Animal Theology*, London: SCM Press, 1994.

Meyer, Marvin and Kurt Bergel (eds), *Reverence for Life: The Ethics of Albert Schweitzer for the Twenty-First Century*, Syracuse, NY: Syracuse University Press, 2002.

Seaver, George, *Albert Schweitzer: The Man and His Mind*, London: A. & C. Black, 1947.

ARA BARSAM and ANDREW LINZEY

ALDO LEOPOLD 1887–1948

If the individual has a warm personal understanding of the land, he will perceive of his own accord that it is something other than a breadbasket. He will see land as a community of which he is only a member ... He will see the beauty as well as the utility of the whole, and know that the two cannot be separated. We love (and make intelligent use of) what we have learned to understand.[1]

Aldo Leopold was born in Burlington, Iowa, in 1887, the eldest of Carl and Clara's four children, an American family of German ancestry. Aldo had two brothers and one sister. His father was fond of hunting and introduced his sons to it at an early age; and before the advent of game laws, he imposed upon himself and his sons a sporting ethic, including closed seasons and bag limits. Leopold dedicated his first published book, *Game Management*, to his father, 'a pioneer in sportsmanship'.[2] His mother was interested in music, especially opera; from her he acquired a keen aesthetic sensibility. Leopold was educated in Burlington public schools, the Lawrenceville Preparatory School in New Jersey, and the Sheffield Scientific School and Forest School of Yale University, from which he was graduated in 1909 with a Master's degree. He immediately joined the U.S. Forest Service and was posted to District 3, the Arizona and New Mexico territories. After a shaky start, Leopold advanced through the ranks to become Assistant District Forester in Charge of Operations, the second highest position in the unit. He married Estella Bergere in 1912, and together they reared five children, Starker, Luna, Nina, Carl and Estella – all of whom have gone on to distinguished careers in the geo-biological sciences or conservation. In 1924, Leopold accepted a transfer to a comparable position at the Forest Products Laboratory in Madison, Wisconsin. During his fifteen-year tenure in the southwest, he was primarily interested in 'secondary' forest uses, especially recreational hunting. After four years in his job as Assistant Director of the Forest Products lab, Leopold resigned to pursue his vocation full time. Supported by the Sporting Arms and Ammunition Manufacturers' Institute, he conducted game surveys of eight midwestern states and worked on a textbook of game management. In 1933, after several months of unemployment during the depths of the Great Depression, Leopold joined the University of Wisconsin as the nation's first professor of game management. He spent the rest of his life conducting research, teaching, writing and shaping conservation policy in this capacity.

Leopold died suddenly of a heart attack in 1948, just a week after learning that his new book manuscript, which would become *A Sand County Almanac*, had been accepted for publication by Oxford University Press.

Leopold was an innovator and a visionary. He was indeed a founder of a number of environmental fields. First, and most obviously, Leopold was a founder of game management, which became wildlife management in his own lifetime, then wildlife ecology, and finally now conservation biology. This, Leopold's central and eventually professional interest, grew directly out of his lifelong passion for hunting, which he acquired in boyhood. The original idea was simple: for a variety of reasons American game was growing scarce in the first quarter of the twentieth century, and game management was essential for 'producing something to shoot'.[3] At the end of his life, Leopold envisioned the desire to 'seek, find, capture, and carry away' game animals becoming transformed into the desire to seek, find, capture and carry away knowledge about animals of all kinds; that is, the transformation of the consumptive sport of hunting into the non-consumptive 'sport' of wildlife research.[4]

At the end of the intellectual trail that Leopold blazed, which began with a passion for hunting, lies a most ambitious project – to which he devoted his masterpiece, *A Sand County Almanac* – worldview remediation.[5] Addressing his colleagues in 1940, Leopold said, 'We find that we cannot produce much to shoot until the landowner changes his ways of using his land, and he in turn cannot change his ways until his teachers, bankers, customers, editors, governors, and tresspassers change their ideas about what is land is for. To change ideas about what land is for is to change ideas about what anything is for. Thus we started to move a straw and end up with the job of moving a mountain.'[6] With *A Sand County Almanac*, Leopold hoped to replace the toxic mix of human exceptionalism and consumerism with an evolutionary-ecological worldview: 'Conservation,' he wrote,

> is getting nowhere because it is incompatible with our Abrahamic [i.e., biblical] concept of land. We regard land as a commodity belonging to us. When we see land as a community to which we belong, we may begin to use it with love and respect. ... That land is a community is the basic concept of ecology.[7]

And ecology, historically, was an offshoot of evolutionary biology.

Leopold was a founder of the North American Wilderness Movement. In the 1920s he argued passionately and voluminously for a system of wilderness reserves in the national forests, primarily for purposes of primitive and virile kinds of recreation. The first National Forest Wilderness Area, surrounding the headwaters of the Gila River in Arizona, was designated the year he left the region. Leopold helped form the Wilderness Society in 1935. His understanding of the importance of wilderness shifted, during his university years, from an emphasis on recreation to biological conservation. Designated wilderness areas were important to conservation for two reasons. First, they afforded a vital habitat for some 'threatened species' – those that, for whatever reason, do not co-exist well with human beings, our cities, suburbs, factories, dwellings, farms, ranches and mines.[8]

Second, wilderness areas provide 'a base-datum of normality, a picture of how healthy land maintains itself as an organism'.[9] By reference to such base-data in 'each biotic province' we can measure the health of similar areas (which should also be conserved to the extent possible) that are used for timber extraction, grazing and farming.[10]

Leopold was a founder of ecological restoration, another very recently emerged formal conservation discipline. In his view, the main purpose of the University of Wisconsin Arboretum and Wildlife Refuge in Madison was 'to construct a sample of original Wisconsin, a sample of what Dane County looked like when our ancestors arrived here in the 1840s'.[11] In addition to his restoration work at the Arboretum, Leopold spent leisure hours during the last thirteen years of his life restoring a property that he bought in 1935 on the banks of the Wisconsin River. The first part of his chief work, *A Sand County Almanac*, is devoted to literary sketches of this place. In the Foreword, Leopold describes this section of the book in terms of ecological restoration.

> Part I tells what my family sees and does at its week-end refuge from too much modernity: 'the shack.' On this sand farm in Wisconsin, first worn out and then abandoned by our bigger and better society, we try to rebuild, with shovel and axe, what we are losing elsewhere.[12]

Leopold was a founder of ecosystem-management forestry, to which the U.S. Forest Service has been converting since 1992. For most of the twentieth century, the Forest Service was devoted to an agronomic model of forestry, the purpose of which was, in Leopold's words, 'to grow trees like cabbages, with cellulose as the basic forest commodity'.[13] The alternative that Leopold envisioned 'sees forestry as fundamentally different from agronomy because it employs natural species and manages a natural environment rather than an artificial one'.[14] The current policy of ecosystem management adopts another conservation concept that Leopold formulated – 'land health' – as its norm. The basic idea is to manage forest ecosystems with the primary goal of restoring or maintaining their health, with commodity extraction an ancillary or subordinate goal. Land health is a concept that Leopold struggled to articulate during the last years of his life. He most frequently characterized it as 'the capacity for self-renewal in the biota'.[15]

Leopold contributed foundationally to conservation philosophy. He himself closely associated land health with land integrity, the full complement of the native species of a biotic province in their characteristic numbers. He believed that preserving its integrity was a necessary and sufficient condition for preserving a particular piece of land's health. Today, ecosystem health and biological integrity are not so tightly coupled. The biological integrity of an area is a sufficient, but not a necessary, condition for its ecosystem health.

Certainly, that is, an ecosystem containing the full complement of its native species populations in their characteristic numbers will be healthy, but an ecosystem with a simplified biota, including nonnative species, may also exhibit normal processes and functions. Leopold himself was especially interested in promoting land health as the conservation norm for the extensively modified farmscapes of southern Wisconsin. To this central conservation concern of his latter years, he linked both his concern for wilderness preservation and ecological restoration. Wilderness provided 'the most perfect norm' of ecosystem health; less perfect, but still useful is 'a reconstructed sample of old Wisconsin to serve as a benchmark ... in the long and laborious job of building a permanent and mutually beneficial relationship between civilized men and a civilized landscape'.[16] Leopold frequently characterized this relationship as 'a state of harmony between men and land'.[17]

Contemporary practitioners in many other environmental fields can (and do) legitimately claim Leopold as an important figure in its development. Take range management: while with the Forest Service, Leopold discovered connections between over-grazing, fire suppression and the disastrous shift from grassy forage to unpalatable brush in the southwest, and recommended management strategies for range recovery in the region. Take erosion control: Leopold was alarmed by the extensive grazing-related erosion he encountered in the southwest, and continued to be concerned about it in the midwest; and he worked in both regions to stanch it. So great, indeed, was his concern about erosion that it may give a more literal sense to his 'land ethic'. Take sustainable agriculture: Leopold's Chair of Game Management at the University of Wisconsin was at first located in the Department of Agricultural Economics and much of his work on the ground was with farmers, first to encourage them to 'grow' a 'crop' of wild game, and later to practise methods of farming that are more accommodating to wildlife of every kind. Shortly after assuming his academic duties, Leopold began monthly broadcasts over the university's extension radio station, addressed to farmers; the next year he began offering a Farmer's Short Course in Game Management; and, between 1938 and 1942, he published a series of thirty-four short 'how-to' pieces in the *Wisconsin Agriculturist and Farmer*. Leopold was among the first to observe and decry 'the tremendous momentum of industrialization ... spread to farm life' and to conceive an alternative 'new vision of "biotic farming"'.[18] Take environmental history: Leopold's essay 'Good Oak' in *Sand County* is a pioneering contribution to the field. Take environmental policy and law: Leopold chaired a blue-ribbon American Game Association Committee on Game Policy and was the senior author of its influential 1930 *Report*. He was offered the job of Chief of the United States Bureau of Biological Survey (forerunner of the present U.S. Fish and Wildlife Service) in 1934, but turned it down. He was also appointed to the Wisconsin Conservation Commission in 1943 and was embroiled for the rest of his life in bitter controversy over his recommended state deer management policy. His conservation policy advice was sought on every scale, from the local and private to the public and national. Take environmental education: in addition to training graduate students for careers in wildlife management, Leopold offered an undergraduate Wildlife-Ecology course open to any University of Wisconsin student. He published a paper addressed to fellow academics in the field titled 'The Role of Wildlife in a Liberal Education', the most important advice of which was 'to use wildlife ecology to teach the student how to put the sciences together' – because 'all the sciences and arts are [conventionally] taught as if they were separate', but 'they are separate only in the classroom'; all one need do is 'step out on the campus and they are immediately fused'.[19] Take nature writing: *A Sand County Almanac* has become more than a classic in the field; it is a genre exemplar.

Of all the environmental fields that Leopold either founded or that his genius shaped, none is of more lasting significance than environmental ethics. The climactic essay of the *Almanac*, 'The Land Ethic', is the seminal text in this new field of philosophy. After all of his years working for a *public* conservation agency, the U.S. Forest Service, and helping to formulate policy and law for such newer agencies as the National Park Service and the Bureau of Biological Survey, Leopold came to believe that conservation would never succeed without a land ethic on the part of individual, *private* landowners. Government alone could not do the job.

Leopold situated his proposed land ethic in the broader evolutionary-ecological worldview that he gradually exposed and inculcated in the preceding essays of the *Almanac*. From Charles Darwin he borrowed an account of ethics as a necessary condition for human social organization. 'All ethics so far evolved', Leopold wrote,

> rest upon a single premise: that the individual is a member of a community of interdependent parts. His instincts prompt him to compete for his place in that community, but his ethics prompt him also to co-operate (perhaps in order that there may be a place to compete for).[20]

From Charles Elton, he borrowed the concept of a 'biotic community', a social model of the inter-relationships of plants and animals studied in ecology. Ecology, Leopold wrote, 'simply enlarges the boundaries of the community to include soils and waters, plants, and animals, or collectively:

the land'.[21] Putting these two elements together, he formulated 'a land ethic', which 'changes the role of Homo sapiens from conqueror of the land community to plain member and citizen of it. It implies respect for his fellow-members and also respect for the community as such.'[22] The golden rule of the land ethic is this: 'A thing is right when it tends to preserve the integrity, stability, and beauty of the biotic community. It is wrong when it tends otherwise.'[23] Leopold, of course, intended for the land ethic to supplement our human-to-human ethics, not replace them. And in light of subsequent developments in ecology, in which 'stability' is down-played, the golden rule of the land ethic may have to be revised. Nevertheless, the very idea of a land or environmental ethic, and Leopold's sketch of its contours, has taken the contemporary environmental movement out of the domain of mere utility and into that of morality. If for no other reason, then for this one Leopold would deserve the frequently conferred metonym of 'prophet' and his masterpiece that of 'the bible of the contemporary environmental movement'.

Notes

1 'Wherefore Wildlife Ecology?', in *The River of the Mother of God*, pp. 336–7, p. 337.
2 *Game Management*, p. v.
3 'The State of the Profession', in *The River of the Mother of God*, pp. 276–80, p. 280.
4 *A Sand County Almanac and Sketches Here and There*, p. 168.
5 See J. Baird Callicott, *Thinking Like a Planet: The Land Ethic and the Earth Ethic*, for a full account.
6 'The State of the Profession', in *The River of the Mother of God*, p. 280.
7 *A Sand County Almanac and Sketches Here and There*, p. viii.
8 'Threatened Species', in *The River of the Mother of God*, pp. 230–4.
9 'Wilderness as a Land Laboratory', in *The River of the Mother of God*, pp. 287–9, p. 288.
10 Ibid., p. 289.
11 J. Baird Callicott, '"The Arboretum and the University: The Speech and the Essay", Apendix A: The Speech "What Is the University of Wisconsin Arboretum and Wild Life Refuge, and Forest Experiment Preserve?" by Aldo Leopold', *Transactions of the Wisconsin Academy of Sciences, Arts and Letters*, 87: p. 15, 1999.
12 *Sand County*, pp. vii–viii.
13 Ibid., p. 221.
14 Ibid.
15 'The Land-Health Concept and Conservation', in *For the Health of the Land*, pp. 218–26, p. 219.
16 'Wilderness as a Land Laboratory', p. 288; J. Baird Callicott, '"The Arboretum and the University: The Speech and the Essay", Appendix A', p. 17.
17 *Sand County*, p. 207.

18 'The Outlook for Farm Wildlife', in *The River of the Mother of God*, pp. 323–6, p. 326; *Sand County*, p. 222.
19 'The Role of Wildlife in a Liberal Education', in *The River of the Mother of God*, pp. 301–5, p. 302.
20 *Sand County*, pp. 203–4.
21 Ibid., p. 204.
22 Ibid.
23 Ibid., pp. 224–5.

See also in this book
Callicott, Darwin

Leopold's major writings

Game Management, New York: Charles Scribner's Sons, 1933.

A Sand County Almanac and Sketches Here and There, New York: Oxford University Press, 1949.

Round River: From the Journals of Aldo Leopold, ed. Luna B. Leopold, New York: Oxford University Press, 1953.

The River of the Mother of God and Other Essays by Aldo Leopold, ed. Susan L. Flader and J. Baird Callicott, Madison, WI: University of Wisconsin Press, 1991.

For the Health of the Land: Previously Unpublished Essays and Other Writings by Aldo Leopold, ed. J. Baird Callicott and Eric T. Freyfogle, Washington, DC: Island Press, 1999.

Further reading

Callicott, J. Baird (ed.), *Companion to A Sand County Almanac: Interpretive and Critical Essays*, Madison, WI: University of Wisconsin Press, 1987.

Callicott, J. Baird, *In Defense of the Land Ethic: Essays in Environmental Philosophy*, Albany, NY: State University of New York Press, 1989.

Callicott, J. Baird, *Beyond the Land Ethic: More Essays in Environmental Philosophy*, Albany, NY: State University of New York Press, 1999.

Callicott, J. Baird, *Thinking Like a Planet: The Land Ethic and the Earth Ethic*, New York: Oxford University Press, 2013.

Flader, Susan L., *Thinking Like a Mountain: Aldo Leopold and the Evolution of an Ecological Attitude Toward Deer, Wolves, and Forests*, Columbia, MO: University of Missouri Press, 1974.

Meine, Curt, *Aldo Leopold: His Life and Work*, Madison, WI: University of Wisconsin Press, 1988.

J. BAIRD CALLICOTT

ROBINSON JEFFERS 1887–1962

Robinson Jeffers' statement in 1928 that 'I'd sooner, except for the penalties, kill a man than a hawk' startled the reading public with a different understanding of human significance in the world.[1] Twenty years later, Jeffers labelled his philosophy 'inhumanism', which he defined as 'a shifting of emphasis and significance from man to not-man; the rejection of human solipsism and recognition of the transhuman magnificence'.[2]

After stunning initial success, Jeffers found that his high regard for the natural world and low regard for humans combined with his isolationist political stance had earned him the scorn of public taste-makers, especially during the Great Depression and the Second World War. Since his death in 1962, however, he has been hailed as the foremost American poet of environmental politics and a philosopher-poet who, in giving voice to the coastal landscape of the Big Sur area, has set a pattern followed by other writers. He has also taught his readers a different understanding of human relationships to the natural world.

Jeffers was born in 1887 in a suburb of Pittsburgh, Pennsylvania. His father, a stern and scholarly professor of Old Testament literature and history, began his son's study of Greek at age 5. When he was 9, Jeffers said, 'my father began to slap Latin into me, literally, with his hands.'[3] For much of his childhood, Jeffers attended schools in Europe. His father brought him back to US to start college as a sophomore, and he graduated from Occidental College at 18.

Jeffers entered graduate school in literature at the University of Southern California. A year later he returned to Europe and at the University of Zurich began studying philosophy, Old English, French literary history, Dante, Spanish literature, and the history of the Roman Empire. He returned to Los Angeles to enter the USC Medical School, where he ranked highest in his class. He briefly taught physiology at the USC Dental College. In 1910, deciding that medicine would leave him too little time to write, he entered the University of Washington, where he studied forestry until 1913.

While a graduate student at USC, Jeffers met Una Call Kuster, who was to become, along with the Big Sur landscape, the greatest influence on his life. Married to a well-to-do lawyer, Una carried on a love affair with Jeffers until, after seven years of rumour and scandal, she divorced her husband and married Jeffers. The two planned to move to Europe where Jeffers could write, but the outbreak of the First World War led them to move up the coast to Carmel instead. As soon as they arrived, Jeffers felt he had found the place about which he would write.

On their land overlooking the Pacific Ocean at Carmel Point, Jeffers began a daily pattern which would hardly vary for many years: he wrote every morning, quarried stone in the afternoon for their granite house and forty-foot stone tower, took walks with Una and their twin boys in the late afternoon, and read Shakespeare and other literature aloud to the family in the evening. Jeffers lived with no telephone, no electricity until the 1940s, and no heat beyond what was produced by a Franklin stove and fireplaces.

Jeffers' first volumes to gain widespread attention were *Tamar and Other Poems* in 1924 and *Roan Stallion* in 1925. The *New York Herald Tribune* enthusiastically reviewed Jeffers' work; critic Mark Van Doren praised it in the *Nation*, calling Jeffers 'a major poet'; and critic Babette Deutsch compared reading *Tamar* to Keats looking into Chapman's Homer.[4] Leonard and Virginia Woolf's Hogarth Press published editions of Jeffers' work in Great Britain, and a French edition went through five printings. Within three years, his work appeared in eight anthologies. Jeffers' work was widely discussed, his books were bestsellers, and he was well paid. Soon Jeffers was even on the cover of *Time* magazine.

Jeffers' time at the pinnacle of American letters did not last long. His relegating of humanity and human consciousness to the importance of basalt and lichen offended socially progressive sensibilities of the 1930s. Jeffers' opposition to American involvement in the Second World War (or in any war) and his portrayal of all the leaders – Stalin, Roosevelt, Churchill, Hitler, Mussolini – as equally evil in leading their people to war offended many. Religious conservatives condemned Jeffers' portrayal of Jesus and his anti-Christian stance; for Jeffers, Jesus and Christianity turn people away from the beauty of the physical world, which is wrong because the beauty of the physical world is God, or at least the manifestation of God, and deserves our worship. Moralists condemned Jeffers' acceptance of violence as an essential aspect of life, and they condemned Jeffers' use of sexual acts, especially his use of incest to illustrate the human obsession with humans and human things to the exclusion of the beauty of the greater outside world.

Yet Jeffers has always had fervent admirers and a general readership. Selections of Jeffers' poetry have sold continuously over the decades. Jeffers articulates ideas that readers outside the academy and outside literary circles know to be vitally important: the non-human world is complex, interactive, conscious, a whole; every aspect of the non-human world is beautiful, and can lead people to a greater understanding of God and of our temporary and insignificant position in the cosmos; our scientific and our religious ways of knowing have serious flaws. Jeffers presents his ideas in memorable narratives, characterizations, images and metaphors which anyone, not just experienced readers of poetry, can understand.

What is the right relationship between nature and humans? Deriving ideas from Lucretius, Herodotus, Nietzsche and Schopenhauer, his four main philosophical pillars, Jeffers offers a series of answers in direct opposition to the prevailing Western belief that 'no man is an island, entire of itself': we should turn from our 'incestuous' involvement with each other in our corrupt 'communal' life, and pay attention to nature. The problem with humanity, Jeffers says, is our self-absorption. He describes how we might look to future ages:

> We shall seem a race of cheap Fausts, vulgar magicians.
> What men have we to show them? but inventions and appliances.
> Not men but populations, mass-men; not life
> But amusements; not health but medicines.[5]

The solution for Jeffers is to turn to permanent, natural things. It doesn't matter where we turn to nature for instruction: 'this ocean will show us / The inhuman road',[6] and 'there are left the mountains'.[7] In 'A Little Scraping', Jeffers says to 'Shake the dust from your hair', and he lists various elements of the landscape that are 'real' and more worthy of observation: a mountain sea-coast, lean cows which 'drift high up the bronze hill', a 'heavy-necked plough team', gulls, rock, 'two riders of tired horses' on a cloudy ridge, topaz-eyed hawks, and more.[8]

Though city dwellers can't very easily lean on rocks and contemplate hawks soaring, they still partake of the permanent reality because we all have bodies, and we eat:

> Broad wagons before sunrise bring food into the city from the open farms, and the people are fed.
> They import and they consume reality. Before sunrise a hawk in the desert made them their thoughts.[9]

The landscape in Jeffers, then, is bodily consumed and afterwards influences everyone's thoughts; we must pay attention to the values the landscape might bring us. We should even attempt to imitate the landscape, for it provides examples of how humans should live:

> The beauty of things is the face of God: worship it;
> Give your hearts to it; labor to be like it.[10]

If we labour enough to be like the beauty of the natural world, we might experience the feeling of union with the landscape of the sort one of Jeffers' characters describes: 'I entered the life of the brown forest ... / ...

and I was the stream / Draining the mountain wood; and I the stag drinking; and I was the stars, / Boiling with light ... / ... I was mankind also, a moving lichen / On the cheek of the round stone'.[11] The speaker is one with the universe, experiencing a feeling of union with all creation more common to mystics than to heroes of Euro-American narratives.

This religious feeling in his poetry, Jeffers says, 'is the feeling – I will say the certainty – that the universe is one being, a single organism, one great life that includes all life and all things; and is so beautiful that it must be loved and reverenced; and in moments of mystical vision we identify ourselves with it'.[12] Jeffers explains that

> 'This is, in a way, the exact opposite of Oriental pantheism. The Hindu mystic finds God in his own soul, and the outer world is illusion. To this other way of feeling, the outer world is real and divine; one's own soul might be called an illusion, it is so slight and so transitory.[13]

The importance of a holistic understanding of reality permeates Jeffers' writing. In a response to a request for a comment on his 'religious attitudes', Jeffers says,

> I believe that the universe is one being, all its parts are different expressions of the same energy, and they are all in communication with each other, influencing each other, therefore parts of one organic whole. (This is physics, I believe, as well as religion.)[14]

This is dialogism, we might add, as well as physics, religion and ecology: all entities are in communication with each other, creating each other by their interaction. In what has become perhaps one of Jeffers' most quoted passages, he summarizes his holistic view and its benefits:

> a severed hand
> Is an ugly thing, and man dissevered from the earth and stars and his
>   history [...]
> Often appears atrociously ugly. Integrity is wholeness, the greatest
>   beauty is
> Organic wholeness, the wholeness of life and things, the divine
>   beauty of the universe. Love that, not man
> Apart from that, or else you will share man's pitiful confusions, or
>   drown in despair when his days darken.[15]

The search for truth is 'foredoomed and frustrate', Jeffers says, 'until the mind has turned its love from itself and man, from parts to the whole'.[16]

Jeffers has been called a prophet in the Old Testament pattern, a mystic, a seer, a religious teacher. Jeffers, ever the son of a Calvinist minister, does not come to know the conventional Christian God through creation; rather, Jeffers asserts that 'Things are the God', and he gives a formula for arriving at this understanding: 'Lean on the silent rock until you feel its divinity'.[17] Other routes to God tend to obscure reality, and throughout his poetry Jeffers aims great invective at saviours of all kinds. Religion apart from the landscape can only lead to disaster.

Jeffers emphasizes the importance of understanding science, too, in relating to the natural world: 'The happiest and freest man is the scientist investigating nature, or the artist admiring it', he tells us.[18] Jeffers considers 'a scientific basis' to be 'an essential condition' for the thinker. Jeffers says, 'We cannot take any philosophy seriously if it ignores or garbles the knowledge and view-points that determine the intellectual life of our time.' While an artist need not know science well, Jeffers says that if an artist has no familiarity with modern science, 'his range and significance would be limited accordingly'.[19]

Jeffers' own background in science, including medical school and forestry school, seems to have given him not only a will and an ability to incorporate a scientific outlook into his poetry, but also a sense of the limits of science as only a symbolic parallel to reality. Science 'has fallen from hope to confusion at her own business / Of understanding the nature of things', Jeffers says.[20] The echo of *De rerum natura* in the expression 'the nature of things' suggests that Lucretius, for all his pessimism, was on the right track. The methods of science have also gone awry. 'Man, introverted man ... cannot manage his hybrids', Jeffers writes in 'Science'; 'Now he's bred knives on nature turns them also inward'.[21] Science itself is admirable, Jeffers implies throughout his poetry; it is a means to knowledge and hence to truth. But as it is practised in the twentieth century, science needs severe critiquing.

Jeffers achieves a relative complacency about environmental destruction by taking a longer view of reality, in geological and astronomical time rather than human time:

> Man's world puffs up his mind, as a toad
> Puffs himself up; the billion light-years cause a serene and wholesome deflation.[22]

Jeffers has a clear idea of the extent of the destruction, as in the depiction of the death of a canyon of redwoods or an abandoned mine, where 'The sweat of men laboring has poisoned the earth'.[23] Jeffers is particularly affected by such abuse of the landscape when it occurs close to home: 'This beautiful place defaced with a crop of suburban houses'.[24] Simply looking harms a landscape as well. In an application of Heisenberg's Uncertainty Principle to everyday life, Jeffers says, 'Whatever we do to a landscape – even to look – damages it.'[25]

Yet change is inevitable, and beautiful places especially call for tragedy involving violence and pain. Jeffers is remarkable in part because he can so easily think beyond the greatest tragedy for the human race – our extinction. The world, he says, will think, 'It was only a moment's accident, / The race that plagued us', and then resume 'the old lonely immortal splendor'.[26]

Notes

1 'Hurt hawks', *Collected Poetry*, vol. 1, p. 377.
2 'Preface', *The Double Axe*, p. xxi.
3 *Selected Letters of Robinson Jeffers, 1897–1962*, ed. Ann N. Ridgeway, Baltimore, MD: Johns Hopkins University Press, 1968, p. 353.
4 Alex A. Vardamis, *The Critical Reputation of Robinson Jeffers: A Bibliographical Study*, Hamden, CT: Archon, 1972 p. 9.
5 'Decaying Lambskins', *Collected Poetry*, vol. 2, p. 604.
6 'The Torch-bearers' Race', ibid., vol. 1, p. 99.
7 'Shine, Perishing Republic', ibid., p. 15.
8 'A Little Scraping', ibid., vol. 2, p. 282.
9 'Meditation on Saviors', ibid., vol. 1, p. 399.
10 'The Inhumanist', ibid., vol. 3, p. 304.
11 'The Tower Beyond Tragedy', ibid., vol. 1, pp. 177–8.
12 *Robinson Jeffers: Themes in My Poems*, San Francisco, CA: Book Club of California, 1956; repr. in M.B. Bennett, *The Stone Mason of Tor House: The Life and Work of Robinson Jeffers*, Los Angeles: Ward Ritchie, *1966*, p. 182.
13 Ibid.
14 *Selected Letters*, p. 221.
15 'The Answer', *Collected Poetry*, vol. 2, p. 536.
16 'Theory of Truth', ibid., p. 610.
17 'Sign-post', ibid., p. 418.
18 *Themes*, p. 184.
19 *Selected Letters*, p. 254.
20 'Triad', *Collected Poetry*, vol. 2, p. 309.
21 'Science', ibid., vol. 1, p. 113.
22 'Animula', ibid., vol. 3, p. 420.
23 'Metamorphosis', ibid., p. 417.
24 'Carmel Point', ibid., p. 399.
25 'An Extinct Vertebrate', ibid., p. 438.
26 'The Broken Balance', ibid., vol. 1, p. 375.

See also in this book
Emerson, Ruskin, Thoreau

Jeffers' major writings

Tamar and Other Poems, New York: Peter Boyle, 1924.

Roan Stallion, Tamar and Other Poems, New York: Boni & Liveright, 1925.

The Women at Point Sur, New York: Liveright, 1927.

Cawdor and Other Poems, New York: Liveright, 1928.

Dear Judas and Other Poems, New York: Liveright, 1929.

Thurso's Landing and Other Poems, New York: Liveright, 1932.

Give Your Heart to the Hawks and Other Poems, New York: Random, 1933.

Solstice and Other Poems, New York: Random, 1935.

Such Counsels You Gave Me and Other Poems, New York: Random, 1937.

The Double Axe and Other Poems, New York: Random, 1948; repr. New York: Liveright, 1977.

The Collected Poetry of Robinson Jeffers, ed. Tim Hunt, 5 vols, Stanford, CA: Stanford University Press, 1988–2001.

The Collected Letters of Robinson Jeffers, with Selected Letters of Una Jeffers, ed. James Karman, 3 vols, Stanford, CA: Stanford University Press, 2009–2011.

Further reading

Brophy, R. (ed.), *Robinson Jeffers: Dimensions of a Poet*, New York: Fordham University Press, 1995.

Karman, J., *Robinson Jeffers: Poet and Prophet*, Stanford, CA: Stanford University Press, 2015.

Karman, J. (ed.), *Stones of the Sur: Poetry by Robinson Jeffers, Photographs by Morley Baer*. Stanford, CA: Stanford University Press, 2002.

Zaller, R. (ed.), *Centennial Essays for Robinson Jeffers*, Newark, DE: University of Delaware Press, 1991.

Zaller, R., *Robinson Jeffers and the American Sublime*, Stanford, CA: Stanford University Press, 2012.

MICHAEL McDOWELL

MARTIN HEIDEGGER 1889–1976

Man is not the lord of beings. Man is the shepherd of Being.[1]

Martin Heidegger was born on 26 September 1889 in the village of Messkirch in southern Germany. After a brief period spent training for the Roman Catholic priesthood, he studied philosophy at Freiburg

University – from 1919 as assistant to the renowned philosopher Edmund Husserl. His reputation as an incisive and radical thinker was sealed in 1927 with the publication of his magnum opus, *Being and Time*. His reputation as a man, on the other hand, was later sullied by his support of Nazism during the early 1930s. In these early years of the Reich, Heidegger saw in Nazism a means to combat the rise of technologism and globalization and to thereby recover the rootedness of the German people in their homeland. After the mid-1930s, however, he became both increasingly disillusioned with Nazism[2] and increasingly dissatisfied with the philosophical approach he had taken in *Being and Time*. Now the 'later' Heidegger came to see his earlier work as being infused with the anthropocentrism or 'humanism' of the Western philosophical tradition. Accordingly, in his later years Heidegger concerned himself with the possibility of recovering a non-anthropocentric way of 'dwelling' in harmony with nature. Heidegger died on 26 May 1976 and was buried in the churchyard of his beloved Messkirch.

Heidegger's thought has had repercussions throughout the intellectual world, having influenced fields as diverse as literary theory, theology, architecture, political theory and cognitive science. In the 'Continental' philosophical tradition, the movements of existentialism, hermeneutics and deconstruction all take their cue from his work, philosophers of the prominence of Sartre, Foucault and Derrida all admitting their indebtedness to him. Although Anglo-American philosophers have traditionally dismissed Heidegger's work, in recent years many have come to recognize his status as a (post-)modern thinker of a stature comparable perhaps only to Wittgenstein.

Before considering Heidegger's relevance to environmental thought, one must first get to grips with some basic features of his analysis of the human condition. Heidegger maintains that to be human is not fundamentally to be a particular sort of thing, but to be a space or 'clearing' in which things show up *as* things. In *Being and Time* he expresses this point by claiming that Being (the process whereby things 'reveal' themselves as things) occurs only within the space provided by human being (referred to in these earlier works as *Dasein*). Later, after the so-called 'Turn' (*die Kehre*) in the direction of his thought, he came to reject this 'existentialist' position in favour of a less anthropocentric conception of humans as participants in a wider clearing of Being. For the later Heidegger, to realize one's essence as a human is to keep watch over the revealing of things, to act as a humble 'shepherd of Being'.[3]

Heidegger claims that Being is essentially historical in the sense that different things reveal themselves in different historical epochs (and to different cultures). A witch, for instance, might reveal herself to a medieval but not to a post-Enlightenment European; an individual citizen might reveal themselves to a modern-day American but not to a fourth-century Chinese. (Note the language here: for Heidegger, these are not changes in perspective or world-view but the results of Being 'granting' different things in different epochs.) Heidegger contends that in the modern world we are increasingly finding that things come to reveal themselves 'technologically'.[4] He associates this sort of revealing with a 'setting upon' or 'challenging' of nature: the 'unreasonable demand' that all things reveal themselves as 'standing reserve' (*Bestand*), as mere resources for our use.

Heidegger's account jars with common sense. For surely – one might protest – technology consists of certain artefacts, such as food blenders, dynamos and microwave ovens, which, while they can be put to good or bad uses, are in themselves neither good nor bad. Heidegger, however, would contend that this is precisely how things would seem to someone who was fully inculcated into the technological way of revealing. For to suppose that technology is nothing more than a neutral means to certain ends is, he would suggest, to offer a technological account of technology and hence to remain blind to its essence as a way of revealing.

Heidegger attributes to technology the peculiar and 'dangerous' power to 'drive out' all other ways of revealing. As it encroaches into all areas of life, non-technological understandings find themselves levelled down and destroyed: poetry, for instance, becomes nothing more than clever wordplay; great artworks, divested of their intrinsic power, become mere decorations, or, perhaps worse, investments. Were he still alive, Heidegger would no doubt complain that the authentic appreciation of wild nature has become levelled down to a pitiable concern with the proper *management* of natural *resources*. He sees the greatest danger in the possibility that technology might eventually come to extinguish all other modes of revealing. In such a nightmarish future, all would have been sacrificed to the modern technological idols of efficiency and management. The world would have become a featureless expanse of standing reserve, a domesticated world shorn of 'otherness' and mystery, and impoverished as a result.

How, then, can we resist this insidious spread of technologism? For Heidegger, the question is inappropriate: we will not, he claims, be able to halt the encroachment of the technological understanding through an act of will, for our will counts for nothing compared to the remorseless 'destining' of history. We and our technological world are the powerless products of the blind dictates of the history of Being. To contend otherwise, he maintains, is to exhibit a characteristically technological arrogance. Yet our situation is not entirely without hope. Heidegger affirms the possibility of salvation, not indeed through stubborn resistance to technological developments, but through 'questioning' or meditating on the essence of technology itself. For, he suggests, deep questioning reveals that technology is itself a mode of revealing, and this realization invites us to discover our essential nature as clearings wherein things reveal themselves. Accordingly, Heidegger calls for us to recognize the flip side of our historical destiny, namely, the *contingency* of the technological mode of revealing. We can, for instance, contemplate the fact that other peoples in other eras – the Ancient Greeks, say – were free from the urge to 'technologize' the world. Reflecting on technology also reveals that the world is not entirely technological, that other modes of revealing still persist, at least for the moment. Thus Heidegger calls for us to remain open to those facets of life which have so far resisted being subsumed in the technological understanding and have been marginalized as a result. We must cherish art and beauty, for instance, as well as simple pleasures such as hiking, fishing or enjoying the company of friends.

Elsewhere, Heidegger offers a series of meditations on the nature of a non-technological way of life he calls 'dwelling'.[5] To describe what such a life would involve, he develops a quasi-mythic account of a world consisting of a 'fourfold' of 'earth, sky, mortals and gods'. Dwelling, he writes, involves a way of being which allows things to reveal themselves in such a way that they come to unite or 'gather' these four dimensions. In this manner, even a lowly and unremarkable thing such as an earthenware jug can become resplendent with world, coming to gather the 'dark slumber' of the earth, the cool radiance of the sky, the nobility of authentic mortal life and the promise of divine deliverance. In these meditations, Heidegger seems to be articulating what might nowadays be called a 'deep ecological' holistic vision of nature. Yet it must be noted that Heidegger's holism does not involve the dissolution of things into some idealized whole – the Environment, Nature or whatever. Rather, the experience of Heidegger's dweller combines a realization of wholeness with an appreciation of the inherent worth of individual things. Accordingly, dwelling involves, not reverence of some nebulous idealization of nature, but a 'poetic' sensitivity to particular things, a sensitivity of the sort one might associate with a Zen *haiku* poet, for instance.[6] Heidegger maintains that to live one's life in this way is to be 'at home' in the world. (In this respect, the deep affection he retained throughout his life for the land of his birth is surely significant.)

Not all writers welcome the prospect of a Heideggerian environmental philosophy. Some argue that the man's thought cannot be abstracted from his disturbing commitment to what he once called the 'inner truth and greatness of National Socialism'.[7] In fact, Heidegger himself told Karl Löwith, one of his former students, that his 'political engagement' was based on his philosophical concept of historicity.[8] Such considerations have led some critics to see in Heidegger's appropriation by deep ecologists a cause to fear the rise of so-called 'eco-fascistic' elements in radical ecological thought.[9] Yet even if such concerns are justified, the correct response, surely, is not to dismiss Heidegger's work without reading it, but to determine more precisely the relations between his political views and the relevant aspects of his thought.[10] Such efforts would be worthwhile, for Heidegger's work has much to offer environmental thinkers: not just eyes to see the spread of technology, but a vision of a more wholesome way of dwelling on earth.

Notes

1 *Basic Writings*, p. 245.
2 Nonetheless, the fact that he never offered a satisfactory apology for his involvement has struck many as appalling. See, for instance, Emmanuel Levinas, 'As If Consenting to Horror', reprinted in *Critical Inquiry*, 15, 485–8, 1989.
3 Heidegger explains this change of tack in his 'Letter on Humanism', *Basic Writings*, pp. 217–65.
4 See especially 'The Question Concerning Technology' in either *The Question Concerning Technology and Other Essays*, trans. William Lovitt, New York: Harper & Row, 1977, or *Basic Writings*.
5 See Heidegger's essays 'The Thing' in *Poetry, Language, Thought* and 'Building Dwelling Thinking' in *Basic Writings*.
6 Similarities with Zen may not be accidental: Heidegger took a keen interest in East Asian philosophical traditions, Daoist and Buddhist ones in particular. See Graham Parkes (ed.), *Heidegger and Asian Thought*, Honolulu: University of Hawaii Press, 1987.
7 *An Introduction to Metaphysics*, trans. Ralph Manheim, New Haven, CT: Yale University Press, p. 199, 1959.
8 Cited in Richard Wolin, *The Heidegger Controversy: A Critical Reader*, Cambridge, MA: MIT Press, p. 142, 1993.
9 See further, Michael E. Zimmerman, 'Martin Heidegger: Antinaturalist Critic of Technological Modernity', in David Macauley (ed.), *Minding Nature: The Philosophers of Ecology*, New York: Guilford Press, 1996.
10 For a good introduction to the question of Heidegger's politics (and much else in Heidegger's philosophy) see Richard Polt, *Heidegger: An Introduction*, London: UCL Press, pp. 152–64, 1999.

See also in this book
Marx, Merleau-Ponty, Uexküll, Zhuangzi

Heidegger's major writings

Basic Writings, ed. David Farrell Krell, New York: Harper & Row, 1996.

Being and Time, trans. John Macquarrie and Edward Robinson, Oxford: Blackwell, 1996.

Poetry, Language, Thought, trans. Albert Hofstadter, New York: Harper & Row, 1971.

Further reading

Dreyfus, Hubert L., 'Heidegger on the Connection Between Nihilism, Technology and Politics', in Charles Guignon (ed.), *The Cambridge Companion to Heidegger*, New York: Cambridge University Press, 1993.

Foltz, Bruce, *Inhabiting the Earth: Heidegger, Environmental Ethics, and the Metaphysics of Nature*, New Jersey: Humanities Press, 1995.

Malpas, Jeff, *Heidegger's Topology: Being, Place, World*, Cambridge, MA: MIT Press, 2006.

Thomson, Iain, 'Environmental Philosophy', in Hubert L. Dreyfus and Mark A. Wrathall (eds), *A Companion to Phenomenology and Existentialism*, Oxford: Blackwell, 2009.

SIMON P. JAMES

EVE BALFOUR 1898–1990

The health of man, beast, plant and soil is one indivisible whole.[1]

Eve Balfour was the most powerful and sustained voice in Britain – and many of its former colonies – to argue in favour of organic food and farming for most of the second half of the twentieth century. During the Second World War, Eve emerged as the leader of a new, international network of people concerned about the impacts of chemical-based, industrial agriculture. Never one to claim to be an original thinker, Eve's contribution was to articulate powerfully the links between a series of previously disparate arguments about what would become known as organic farming. These ideas had been published by a scattered and eclectic group of thinkers in continental Europe, Britain and India during the 1920s and 1930s. What they had in common was a vision of agricultural production aimed at making farming's primary goal that of maximising the nutritional value of crops and animal products by paying close attention to soil quality. This was a vision of farming as a force to replenish soil, crops, livestock and, in turn, humans – rather than deplete them. Eve's 1943 bestselling book, *The Living Soil*, promoted what was then often termed as humus or compost farming, and raised important questions about the potential risks posed by new approaches to farming, reliant on chemical inputs rather than on sound soil management.[2] Shortly after the book's publication, Eve became the primary force behind the creation of a new British-based but international organic food and farming organisation, the Soil Association. Founded in 1946, Eve would serve as its leader and/or chief spokesperson for almost 30 years.

But it wasn't until Eve was about 40 years old that she began to dedicate herself to the organic cause. Until then she was a conventional farmer, educated in 'modern' scientific techniques and not one to question the benefits of nitrogen fertilisers. She had grown up a confident and outspoken child within one of Britain's most important political families of the late nineteenth and early twentieth centuries. With more than one prime minister amongst her beloved uncles, Eve's aunts and mother were also politically active, with several committed to women's suffrage and access to education. Family legend states that Eve declared her plan to become a farmer at the age of 12, inspired by the beauty of the well-managed farms she saw on her family's estate in the Scottish Borders. She attended agricultural college during the First World War and soon after graduating she and one of her sisters became the proud owners of a 64 hectare farm in eastern England, later buying an adjacent farm as well. For two decades Eve owned and ran a conventional, mixed farm, rarely managing to stay in profit for long. Always one to add an extra string to her bow, Eve's youthful experiments included playing in a jazz band and writing detective novels, but as time passed these light-hearted enterprises gave way to increasingly political extra-curricular activities. During the mid-1930s, Eve's concern about the negative impacts on farmers of Britain's commitment to free trade led her to champion a wave of protests against the payment of archaic tithes. This role in the tithe protest movement brought Eve national – even a bit of international – fame and developed her skills as a public speaker and author of political pamphlets.[3]

As the Second World War loomed, Eve encountered ideas about humus or compost farming. She was particularly drawn to claims about humus farming's power to transform the health of those whose diet was based on food grown and raised using these natural, non-chemical techniques. She read the works of authors such as Gerard Wallop, Albert Howard, Robert McCarrison and G.T. Wrench, who in different ways all argued that any type of farming reliant on inorganic fertilisers and reduced applications of farmyard manure, compost or other natural wastes would undermine soil fertility. A less fertile soil would lead to less healthy crops and livestock, food of lower nutritional value and, eventually, poorer human health. Eve was captivated by these arguments and soon embarked on a plan to turn her farm into a scientific research centre. She would prove the central argument about the nutritional superiority of organic farming methods. Eve's efforts to attract support for this farm-based research project prompted a leading publisher to commission her to write a book. Published in 1943, *The Living Soil* proved a huge international success, with Eve using the same title for a ten-part radio series on the BBC's Africa service, which aired in 1944. Both the book and the radio series raised her profile and spread her message widely and rapidly.

Today, it is often ecological arguments in favour of organic farming that are influential, particularly when they are presented by powerful nature conservation organisations. Such bodies strengthen their case by referencing scientific evidence indicating that organic farming can be used to achieve higher levels of farmland biodiversity. Arguments linked to nature conservation were largely absent from *The Living Soil* and were used only infrequently by the Soil Association during its first decade. In Eve's ground-breaking book it is not nature that will benefit from organic agriculture, but people. Eve argued persuasively in *The Living Soil* that unless crops are grown in fertile soil that has been managed in ways that support it as a habitat for a myriad of health-promoting microorganisms, then the plants growing out of that soil will lack 'vitality'. This in turn will mean that the animals and people eating these plants will be less 'vital' themselves and therefore weaker and more prone to disease. It was the goal of achieving and maintaining optimum health – rather than protecting biodiversity – that Eve articulated in *The Living Soil* and that prompted the creation of the Soil Association. 'The health of man, beast, plant and soil is one indivisible whole,' wrote Eve.

> The health of the soil depends on maintaining its biological balance, and ... starting with a truly fertile soil, the crops grown in it, the livestock fed on those crops and the humans fed on both, have a standard of health and a power of resisting disease and infection, from whatever cause, greatly in advance of anything ordinarily found in this country; such health as we have almost forgotten should be our natural state, so used have we become to subnormal physical fitness.[4]

It is a matter of record that Britain's post-war agricultural revolution proved to be the antithesis of what Eve and the Soil Association had hoped post-war reconstruction would bring. After the war, the British government asked farmers to deliver substantial increases in domestic food production by accelerating the industrialisation and 'chemical-isation' of agricultural practices, a process that was already underway and which war had facilitated.[5] Eve and the Soil Association accepted that an increase in food production was a priority, but insisted that problems with poor food quality were just as urgent. Instead of looking to chemistry and mechanisation to transform the agricultural sector, the newly-emergent organic movement argued that the overriding priority of both public health and agricultural policy should be to protect and maintain soil fertility by 'natural' methods. Large-scale composting by all farmers, supported by further scientific research aimed at perfecting composting techniques, was the way forward. During the 1950s, Eve's preoccupation with food quality was shared by a small, health-focused minority within the British public, but not by the farming sector, the British government or the majority of consumers. For most people, the issue was simple: more food, available at low prices.

Public concern about industrial agriculture – if not food quality – began to grow in the 1960s and 70s, however the organic movement in Britain remained on the fringes and did not succeed in broadening its support base until the 1980s, once a younger group of organic supporters had taken over. Nevertheless, publication of Rachel Carson's international bestseller *Silent Spring* in 1962,[6] the success in Britain of Ruth Harrison's book *Animal Machines*[7] and the creation of the Nature Conservancy drew the public's attention to the damage being done in the name of modern agriculture. The mood of the public and of some elements within the political class began to shift subtly. The Soil Association had no formal connection to Rachel Carson, but the latter had been assisted in her research for *Silent Spring* by American organic supporters with links to Eve Balfour. Carson maintained a scrupulous public distance between her work and the organic movement,[8] nevertheless the Soil Association was correct in identifying similarities between its organic vision for agriculture and the types of changes to farming practices that Carson called for. It would take years, but by the mid-late 1980s the Soil Association would succeed in leveraging the ecological arguments of the nature conservation community – which increasingly acknowledged the need for an organic-style of agriculture in order to protect biodiversity. But by this time Eve Balfour had retreated from active campaigning and was the elderly figurehead of the movement.

An assessment of Eve Balfour's contribution to the organic food and farming would not be complete without acknowledgment of religion and the role this played in her campaigning and in shaping the resistance the Soil Association faced during the decades when industrial agriculture was firmly in ascendance. Eve held profound and unconventional religious beliefs throughout her adult life, about which she was somewhat secretive. She had grown up in a family that was not only politically powerful, but also devoted to the practice of Spiritualism. As an adult, Eve's religious worldview evolved into a fluid mix of 'New Age' ideas that allowed her and others to ascribe deep spiritual significance to her earthly actions.[9] Given this perspective, Eve's leadership of the Social Association can be seen as a contribution to the 'directed energy of Creative Mind'.[10] Eve was also heavily influenced by the core New Age idea that ill-health in an individual is connected to deeper, spiritual malaise.

Many of the Soil Association's members held similar, dissident religious worldviews and although the organisation never openly declared itself a spiritual enterprise, by the mid-1950s it and its leader were well on their way to earning a reputation as purveyors of 'muck and mystery'. The muck was compost, of course, while the mystery was a form of esotericism now generally referred to as New Age. This charge of muck and mystery was not meant as a compliment. Throughout the 1960s and 1970s, opponents of the organic movement, including agricultural scientists and representatives of agri-chemical manufacturers, were pointed in their attacks on Eve and the Soil Association. Over time, they succeeded in generating suspicion about the organisation and in undermining Eve's reputation as a campaigner capable of mounting scientifically credible arguments about the dangers of industrial agriculture. Whether those who attacked her were aware that Eve and other Soil Association members had connections to the New Age community in Findhorn, Scotland as well as the esoteric network led by Sir George Trevelyan, is not clear. Regardless, those uncomfortable with Eve's criticism of chemical-based farming mounted a recurring accusation: that the Soil Association and its leader were opposed to science, that they were 'anti-science' or at the very least that they did not understand scientific research.

Eve died in 1990, a time when public support for organic food and farming was quietly growing. The Soil Association would go on to make significant gains over the following 15 years. Eve's contribution to environmental activism during the immediate post-war period – a time when 'green' campaigning was at a particularly low ebb – helped to link late nineteenth-century, Edwardian and interwar ecological ideas with the flowering of environmentalism in the 1970s. *The Living Soil* articulated organic ideas to a wider group, and more powerfully than had been achieved to date. Her participation in BBC radio and print media debates about farming throughout the 'wilderness years' of the 1950s, combined with her speaking tours of the UK, Europe, the USA, Australia and New Zealand were instrumental in building a community of concern that would eventually capitalise on the emergence of a new, youth-based environmental movement in the 1980s. A series of food safety scares throughout the 1990s further strengthened the organic cause. Today, the Soil Association continues to represent the organic food and farming movement in the UK, and *The Living Soil* is viewed as its 'classic' text.

Notes
1 Eve Balfour, *The Living Soil*, London: Faber & Faber, 1943, p. 21.
2 Erin Gill, 'Chapter 2: The Second World War, Eve Balfour's organic conversion and *The Living Soil*,' in *Lady Eve Balfour and the British Organic Food and Farming Movement*, 2011, pp. 88–109.
3 Michael Brander, *Eve Balfour: The Founder of the Soil Association & Voice of the Organic Movement*, Haddington, UK: The Gleneil Press, 2003.
4 *Living Soil*, p. 21.
5 Short, Watkins and Martin eds, *The Front Line of Freedom: British Farming in the Second World War*, Exeter, UK: British Agricultural History Society, 2007.
6 Rachel Carson, *Silent Spring*, Boston, MA: Houghton Mifflin, 1962.
7 Ruth Harrison, *Animal Machines*, London: Vincent Stuart, 1964.
8 Linda Lear. *Rachel Carson: The Life of the Author of 'Silent Spring'*, London: Penguin, 1997.
9 Gill, 'Chapter 4: Eve Balfour, the Soil Association, science and religion,' in *Lady Eve Balfour and the British Organic Food and Farming Movement*, 2011, pp. 169–227.
10 Balfour, *9,600 Miles Through the USA in the Station Wagon*, London: The Soil Association, 1954 or 1955, p. 107.

See also in this book
Berry, Carson

Balfour's major writings
The Living Soil, London: Faber & Faber, 1943.

Further reading
Conford, Philip. *The Origins of the Organic Movement*. Edinburgh, UK: Floris Books, 2001.
Brander, Michael. *Eve Balfour: The founder of the Soil Association & Voice of the Organic Movement*. Haddington, UK: The Gleneil Press, 2003.
Gill, Erin. *Lady Eve Balfour and the British Organic Food and Farming Movement*. Unpublished PhD: Aberystwyth University, 2011. Downloaded from: www. ladyevebalfour.org.

ERIN GILL

RACHEL CARSON 1907–1964

Only within the moment of time represented by the present century has one species – man – acquired significant power to alter the nature of this world …

The most alarming of all man's assaults upon the environment is the contamination of air, earth, rivers, and sea with dangerous and even lethal materials. This pollution is for the most part irrecoverable; the chain of evil it initiates not only in the world that must support life but in living tissues is for the most part irreversible. In this now universal contamination of the environment, chemicals are the sinister and little-recognized partners of radiation in changing the very nature of the world – the very nature of its life.[1]

Silent Spring (1962) embodies a connectedness with nature, a kinship with other species, a feeling of the responsibility to take personal action. From what wellsprings had such a mature environmental philosophy flowed? What influences led to such profound insights, personal courage and ultimate heroism in face of vilification by the American corporate and scientific establishment? What experiences led Carson to such deep understanding of nature awareness and how it might be conveyed, to her surpassing understanding of human ecology, and to her realization of the end of nature as humans had known it throughout time?

Carson was born in 1907 in Springdale, Pennsylvania, and had positive formative experiences in nature and in literature under the tutelage of her mother, Maria. She was strongly influenced in a romantic view of nature by several children's magazines. She published her first story in *St Nicholas* at age 11. She attended the private Pennsylvania College for Women, now Chatham College, in Pittsburgh just a few miles from home. Carson was educated at great financial sacrifice to her family and with the aid of several scholarships. She continued her close relationship with her mother throughout college. She had faculty mentors – in writing, Grace Croff, and in biology, Mary Scott Skinker. She followed the latter into science as her academic major, to advanced study at Johns Hopkins University, and into work as a government scientist.

Rachel Carson's writing about the sea fulfilled a childhood inland dream. Her favourite line of poetry while studying English and science in Pennsylvania, proved prescient: 'For the mighty wind arises, roaring seaward, and I go.'[2] She wrote, 'I can still remember my intense emotional response as that line spoke to something within me, seeming to tell me that my own path led to the sea – which then I had never seen – and that my own destiny was somehow linked with the sea.'[3]

She became a biologist at the U.S. Fish and Wildlife Service, eventually working her way to the position of editor-in-chief. She ultimately became, in writing a trilogy of popular books, *Under the Sea-Wind* (1941), *The Sea Around Us* (1951) and *The Edge of the Sea* (1955), what she called a biographer of the sea. She was captivated by the eternal mysteriousness of the sea. She wrote in such a way as to express the facts as well as the beauty of nature – the knowledge as well as the poetry. According to Paul Brooks, her editor and biographer:

> Though she had the broad view of the ecologist who studies the infinitely complex web of relationships between living things and their environment, she did not concern herself exclusively with the great impersonal forces of nature. She felt a spiritual as well as physical closeness to the individual creatures about whom she wrote: a sense of identification that is an essential element in her literary style.[4]

This writing was in the spirit of John Muir, William Burroughs, Anna Botsford Comstock and others in the nature-study era of North American environmental writing. Through her nature-study writing, voice was given both to nature and to the unexpressed sensibilities of readers. The depth and power of her insights and the authority of her research educated readers to a world they did not know. Carson shared a subject she believed was vital. Through effective, powerful writing, she vivified the sea. Carson contextualized her scientific writing within this vastness of time and space. Her writing conveys a sense of proportion – a soul aware of sitting at the edge of the continent, at not only the edge of the sea, but the edge of the sea of stars. These vast expanses of time and the cycles of recurring natural events situate her insights of the human place in science. She understood the limits of science – even when she enriched its definition to include humans as feeling and socially responsible participants in its study. She saw the power of science to reveal knowledge of natural processes and to raise questions of the human relationship to such processes and to human knowledge of them. Finally, Carson saw science as needing to evoke the sister of identification and knowledge – personal responsibility.

In addition to scientific knowledge of the 'nearly eternal', she would have us feel the poetic essence of our response to nature – and of reverence for it. Rachel Carson's combined knowledge and love of nature has been compared in a feminist critique of her writing to Barbara McClintock's 'feeling for the organism'.[5]

The scholar Vera L. Norwood has plumbed the subtext of Carson's work and its epistemology. Carson's thinking and feeling lead her to question how we know what we know. She has no Godlike perspective apart from nature and human nature, rather she struggles to locate herself. In 'The Nature of Knowing: Rachel Carson and the American Environment', Norwood writes:

The occasions when the economic metaphor shatters against the unwillingness of the natural world to 'produce' meaning provide her most telling critiques of human limitations and lead her to doubt all context, [sic] Carson becomes more than a nature writer; she raises fundamental questions about how human knowledge is constructed, questions that reveal the epistemological hubris underlying much human understanding. These questions prompt her later normative work in *Silent Spring* and *The Sense of Wonder*.[6]

Carson's nature writing has been celebrated for sensitivity, complexity and depth. She taught about life in the sea but also to stand in reverence of how little is known. She educated towards another way to know – to *feel* nature. And finally, she raised questions not only about nature, but about the nature of the knowledge by which we know nature.

A qualitatively different stage of Carson's writing began in 1956 when she wrote an article for *Woman's Home Companion* entitled 'Help Your Child to Wonder'. This was the first time in seven years she had not had a book in production. She wanted to leave the sea for a time; she wrote to her editor 'like that old scorpionlike thin in the Silurian, I have come out on land'.[7] She had hopes to develop the article as a book, but soon she was to start her research on pesticides, and she never did. The article was published posthumously in 1965 as a book, *The Sense of Wonder*.

It is in this work that Carson is explicitly an environmental educator and can be best critiqued for her philosophical and pedagogical contributions to the field. She asks:

What is the value of preserving and strengthening this sense of awe and wonder, this recognition of something beyond the boundaries of human existence? Is the exploration of the natural world just a pleasant way to pass the golden hours of childhood or is there something deeper?

I am sure there is something much deeper, something lasting and significant. Those who dwell, as scientists or laymen, among the beauties and mysteries of the earth are never alone or weary of life.[8]

In the most direct statement of Carson's rationale for her kind of environmental education, she assures us of a deeper meaning, a hidden soul, that lies just beyond our experience in the natural world. Much of education teaches not to trust wonder, intuition and the ineffable sources of human strength. Yet these are part of our knowledge of nature, Carson says. She offers a validation of the power and authority of childhood experience and an invitation to reconsider its depths. The reader is enticed to wonderment in the sensual experience of nature. This work, with its explicit inclusion of affect and questions of value, foreshadows the raising of these questions by educators in the 1970s. She gives permission to explore the actual and perceived landscapes of childhood. Rachel Carson's philosophy of environmental education speaks of sensory creatures in a sensory world, humble citizens of a mysterious universe, and people free to place themselves 'under the influences of earth, sea, and sky and their amazing life'.[9]

In 1958, Carson decided to write a brief article on the impact of DDT spraying upon bird life – her next four and a half years were spent researching and writing one of the most influential books of the age. She told her long-time friend Dorothy Freeman she had proposed an article about it in 1945. The 1945 article became her magnum opus in 1962. *Silent Spring* has demonstrated remarkable vitality. It has been translated into 'nearly every language on the planet'.[10] Thirty-eight years after US publication, it has never been out of print and continues to sell. It is given great credit for changing the way we see our world. According to H. Patricia Hynes, '*Silent Spring* crystallized an "ethic of the environment" which inspired grassroots environmentalism, the "deep ecology" movement and the creation of the Environmental Protection Agency (EPA) and its state counterparts; it influenced the ecofeminist movement and feminist scientists.'[11]

Through her research on pesticides, Rachel Carson saw the vast destruction of which humans are capable. Hynes, in a chapter entitled '*Silent Spring*: A Feminist Reading', writes:

> Rachel Carson told students of Scripps College in 1962 that 'in the days before Hiroshima,' she thought that there were powerful and inviolate realms of nature, like the sea and vast water cycles, which were beyond man's destructive power. 'But I was wrong,' she continued. 'Even these things, that seemed to belong to the eternal verities, are not only threatened but have already felt the destroying hand of man.'[12]

Her dedication of *Silent Spring* is instructive of her environmental world-view. She quoted Albert Schweitzer, 'man has lost the capacity to foresee and to forestall. He will end by destroying the earth.'[13] She was among the very first to appreciate the gravity of the human impact on nature and her writing in this period precedes the concern to follow in the years leading up to Earth Day 1970 and the popular recognition of the seriousness of the environmental crisis. Rebecca Raglon gives Carson a new place in the context of the tradition of women's writing in nature-study and of nature writing by both women and men. She writes:

> *Silent Spring* marks the origin of a new kind of nature writing: a dark new genre that deals with the horrific consequences of human actions upon the earth ... Carson's legacy has insured that such innocent nature appreciation will now have to occur within a much darker context.[14]

Destined to be considered a seminal work in environmentalism, and perhaps one of the most important books of the twentieth century, its writing is meticulously chronicled by Carson's editor Paul Brooks in his biography *The House of Life: Rachel Carson at Work*:

> The storm aroused in certain quarters by the publication of *Silent Spring*, the attempts to brand the author as a 'hysterical woman,' cannot be explained by the concern of special interest groups for their power or profits. The reasons lie deeper than that. Rachel Carson's detractors were well aware of the real danger to themselves in the stance she had taken. She was not only questioning the indiscriminate use of poisons but declaring the basic responsibility of an industrialized, technological society toward the natural world. This was her heresy. In eloquent and specific terms she set forth the philosophy of life that has given rise to today's environmental movement.[15]

Carson's environmental philosophy raises questions about the nature of nature and human knowledge of it; it invites the reader to stand in wonder at the depth of nature's influence upon values and attitudes; and it calls a people to their responsibility to halt its destruction. Indeed, the recent intellectual history of environmental thought owes much to the wisdom of this remarkable scientist, writer, educator, elder and lover of nature.

Notes

1 *Silent Spring*, pp. 5–6.
2 Alfred Lord Tennyson, 'Locksley Hall'.
3 Paul Brooks, *The House of Life: Rachel Carson at Work*, Boston, MA: Houghton-Mifflin Company, p. 18, 1972.
4 Ibid., pp. 7–8.
5 H. Patricia Hynes, *The Recurring Silent Spring*, New York: Pergamon Press, p. 57, 1989.

6 Vera L. Norwood, 'The Nature of Knowing: Rachel Carson and the American Environment', *Journal of Women in Culture and Society*, 12, 747–52, 1987.

7 Brooks, op. cit., p. 201.

8 *The Sense of Wonder*, p. 88.

9 Ibid., p. 95.

10 Michael Brosnan quoted in Caskie Stinnett, 'The Legacy of Rachel Carson', *Down East*, June, p. 43, 1992.

11 Hynes, op. cit., p. 9.

12 Ibid., p. 181.

13 *Silent Spring*, p. v. This is, fascinatingly, a slight misquoting of Schweitzer, who said, 'Modern man no longer knows how to foresee or to forestall. He will end by destroying the earth from which he and other living creatures draw their food. Poor bees, poor birds, poor men …'

14 Rebecca Raglon, 'Rachel Carson and her Legacy', in Barbara T. Gates and Ann B. Shteir, *Natural Eloquence: Women Reinscribe Science*, Madison, WI: University of Wisconsin Press, pp. 198, 207, 1997.

15 Brooks, op. cit., p. 284.

See also in this book
Comstock, Griffin, McKibben, Muir, Schweitzer

Carson's major writings

Under the Sea-Wind, New York: Oxford University Press, 1941.

The Sea Around Us, New York: Oxford University Press, 1951.

The Edge of the Sea, Boston, MA: Houghton-Mifflin Company, 1955.

Silent Spring, Boston, MA: Houghton-Mifflin Company, 1962.

The Sense of Wonder, New York: Harper & Row, 1965. This edition is now out of print. The Nature Company has published a new version (1994) as has HarperCollins (1998).

Lost Woods: The Discovered Writing of Rachel Carson, ed. with an Introduction by Linda Lear, Boston, MA: Beacon Press, 1998.

Further reading

Freeman, Martha (ed.), *Always, Rachel: The Letters of Rachel Carson and Dorothy Freeman, 1952–1965*, Boston, MA: Beacon Press, 1995.

Graham, Frank, Jr, *Since Silent Spring*, Boston, MA: Houghton-Mifflin Company, 1970.

Lear, Linda, *Rachel Carson: Witness for Nature*, New York: Henry Holt & Company, 1997.

Quartiello, Arlene Rodda, *Rachel Carson: A Biography*, Westport, CT: Greenwood Press, 2004.

Souder, William, *On a Farther Shore: The Life and Legacy of Rachel Carson, Author of Silent Spring*, New York, NY: Broadway Books, 2012.

Further readings and information may be found at the following websites:

www.rachelcarson.org
www.chatham.edu/rachelcarson/rachelcarson.cfm
www.rachelcarsoncouncil.org/

PETER BLAZE CORCORAN

LYNN WHITE, JR 1907–1987

What people do about their ecology depends on what they think about themselves in relation to things around them.[1]

Lynn Townsend White, Jr was born in San Francisco, California, on 29 April 1907. After his academic training at elite schools in the USA,[2] his first academic post was at Princeton University from 1933 to 1937. In 1938 he joined the faculty of his alma mater, Stanford, and remained there until 1943. From 1944 until 1957 Lynn White served as President of Mills College, a women's college in Oakland, California. In the midst of his stint at Mills, White penned a provocative book entitled *Educating Our Daughters*, which spoke to the problems women faced in higher education in the USA at the time. White clearly made his mark at Mills College: a residence hall and an endowed chair still carry his name. In 1958 White joined the history faculty at the University of California – Los Angeles where he remained until retiring from academic life in 1974. Lynn White, Jr is widely and most notably recognized as the 'founder of all serious modern study' of the history of technology in medieval Europe. His most famous and still classic, *Medieval Technology and Social Change*, was once declared by Joseph Needham to be 'the most stimulating book of the century on the history of technology'. On 30 March 1987 Lynn White, Jr died of heart failure; he was 79. In his lifetime, White was known for both his scholarly as well as his more popular writings, for his timely and controversial intellectual boldness, and for his insistence that scholarly parochialism was antithetical to the life of the mind. It is said that throughout his life White remained a Christian, a fact that might seem curious to some given the nature of his most obvious contribution to environmental thought.

In the cold of the Washington, DC winter of 1966, Lynn White, Jr presented a ground-breaking and controversial paper at the annual meeting of the American Association for the Advancement of Science. In the paper, which was published the following year in the prestigious journal *Science*,[3] White laid much of the blame for our current environmental predicament upon the doorstep of Christianity. It is, therefore, something of an irony that 'The Historical Roots of Our Ecologic Crisis', a paper considered to be so critical of the Christian tradition, was presented by a life-long Christian thinker on the day after Christmas.

Within just a few short years of its publication, the article had already been dubbed a 'classic'. The essay provoked both immediate and long-term reactions: literally dozens of responses to White's essay have since been published,[4] the essay remains a staple for university 'environmental' courses, and it continues to be reprinted in a wide variety of anthologies and textbooks. As a scholar who attended carefully and knowingly to the connection between our ideas and our actions, White often seemed surprised by the response to his essay. However, 'as the tide of protest from churchmen flowed across his desk in a growing stream of letters and articles', White apparently kept his sense of humour, joking with colleagues that he 'should have blamed the scientists'.[5] In an essay written by White in 1973 in response to his critics, he comments that as criticisms poured forth he 'was denounced, not only in print but also on scraps of brown paper thrust anonymously into envelopes, as a junior Anti-Christ, probably in the Kremlin's pay, bent on destroying the true faith'.[6]

It is, of course, not ordinary to consider someone a key environmental thinker on the basis of essentially a single essay. But this is no ordinary essay. Seldom has the splash of a single work created such enduring ripples. There are two noteworthy contributions made by White's essay.

Understandably the point most people immediately fixated on was White's attack on Christianity. White begins by pointing out that although 'all forms of life modify their context' current anthropogenic environmental impact 'has so increased in force that it has changed in essence'. Whereas past environmental impact was local and point-source impact, currently we are witness to environmental impact that is not just a difference in the degree but, in reality, a different kind of impact all together. We now possess and exercise an ability to affect the globe as a whole: to change the planet's climate, to be the agents of the planet's sixth great biodiversity extinction episode. As White put it, 'the impact of our race upon the environment has so increased in force that it has changed in essence'. In fact, the bulk of his scholarly work was an attempt to show how even quite minute alterations in technology – such as the use of horse power and the resulting heavy plough – can and did eventuate in a radical escalation in the ability of humans to alter nature.[7] Hence, as an historian, White provides us with an explanation of how it is that humans have impacted and altered the environment so extensively. However, one of the most unique and important intellectual contributions White makes is his denial that our current rate of environmental change, resulting in our environmental crisis, is merely a result of an increase in our ability to manipulate our context with the tools of modern science and technology. Instead, White argues that because the fusion of science and technology during the seventeenth-century Scientific Revolution occurred within a larger context – a Christian conceptual framework – and because Christianity had been interpreted as dictating an essentially despotic relationship between humans and the rest of nature – a relationship where 'technological advance was seen as superlatively virtuous' – Christianity is ultimately responsible for our contemporary environmental crisis. As White puts it:

> The artifacts of a society, including its political, social and economic patterns, are shaped primarily by what the mass of individuals in that society believe, at the sub-verbal level, about who they are, about their relation to other people and to the natural environment, and about their destiny. Every culture, whether it is overly religious or not, is shaped primarily by its religion.[8]

According to White, the message we have gleaned from Christianity is that we humans are uniquely created in the image of God, a quality that cuts us out from the rest of creation, making us not only separate but special, and that therefore our role on this earth with regard to the rest of God's creation is to dominate and subdue it. Because of these background assumptions, humans believe 'we are superior to nature, contemptuous of it, willing to use it for our slightest whim'. Where 'formerly man has been part of nature, now he was the exploiter of nature'. Hence, Christianity not only allows for the anthropogenic exploitation of nature that has resulted in our environmental crisis, it also can be seen as sanctioning and even encouraging it.

What, then, is the solution to our environmental crisis? White dismisses a focus on a simple increase in science and technology. Since 'our science and technology have grown out of Christian attitudes toward man's relation to nature', it would be unwise to believe that merely tinkering around the edges of our problems, applying the same old approach in a slightly different way, would be an appropriate way to respond to our environmental problems. In White's words, since

> what we do about our ecology depends on our ideas of the man–nature relationship ... more science and more technology are not

going to get us out of the present ecologic crisis until we find a new religion, *or rethink our old one*.[9]

Since 'Christianity in absolute contrast to ancient paganism and Asia's religions ... not only established a dualism of man and nature but also insisted that it is God's will that [humans] exploit nature for his proper ends' – and since it is, at least in part, the message taken from the establishment of this dualism – the goal, for White, is to not jettison Christianity but, rather, it 'is to find a viable equivalent to animism'.[10]

A central feature of animistic traditions is the belief that, not just humans, but natural entities and nature itself are inspirited, possessive of those qualities that endow it with personhood. Personhood is a morally relevant quality. Therefore, to say we must find a 'viable equivalent to animism' is to say that we must find a way to reinterpret Christianity as non-anthropocentric, as a rational and ethical structure imbuing both the human and the non-human world with intrinsic value. White is adamant about this point: 'We shall continue to have a worsening ecologic crisis until we reject the Christian axiom that nature has no reason for existence save to serve man.'[11]

White's proposal, therefore, is to rethink our interpretation of Christianity, to focus on the possibility of an alternative message about the human–nature relationship, to focus on a message of stewardship. To this end, White concludes his essay with a tribute to St Francis of Assisi who, White claims, was not only 'the greatest radical in Christian history since Christ' but who delivered the required nature-sympathetic message of stewardship. In fact, White even goes so far as to propose St Francis as the environmental 'patron saint'. Interestingly, since the publication of 'Historical Roots' in 1967, Christians seem to have taken up the task White lays out. In fact, the advent of a Christian environmental stewardship is arguably the most powerful thing to happen to the environmental movement since Rachel Carson's *Silent Spring*.

Apart from attributing the West's brazenly opprobrious environmental behaviour to our narrowly focused and anthropocentric interpretation of Christianity, there is a subtle yet powerful subtext flowing throughout White's essay. White asserts that to solve our environmental crisis we must 'clarify our thinking', 'think about fundamentals', and 'rethink our axioms'. In other words, we must philosophize. White's entire essay, therefore, is a call for, and stamp of approval on, the new field of environmental ethics; a subdiscipline of philosophy in its infancy when White's challenge broke. In fact, because of this essentially philosophical subtext, environmental philosopher J. Baird Callicott has even gone so far as to dub White's essay 'the seminal paper in environmental ethics'.[12]

Although often misunderstood, Lynn White, Jr's contribution to environmental thought was, and continues to be, both important and profound. He boldly challenged us to think deeply about the roots of our environmental problems and to exercise bravery and intellectual honesty in the fact of reconsidering those fundamental anthropocentric axioms asserting our human superiority. White lays before us a formidable task: we must learn humility. We must learn to care for ourselves as well as for God's creation, and not to assume that if we are doing the former we are in turn doing the latter.

Notes
1 'The Historical Roots of Our Ecologic Crisis', *Science*, 155, p. 1205, 1967.
2 BA, Stanford, 1928; MA, Union Theological Seminary, 1929; MA, Harvard, 1930; PhD, Harvard, 1934.
3 'The Historical Roots of Our Ecologic Crisis', pp. 1203–7.
4 One of the best is James Barr, 'Man and Nature: The Ecological Controversy and the Old Testament', *Bulletin of the John Rylands Library*, 55, 9–32, 1972.
5 Bert S. Hall, 'Obituary of Lynn White, Jr', *ISIS*, 79, p. 480, 1988.
6 'Continuing the Conversation', in Ian Barbour (ed.), *Western Man and Environmental Ethics*, p. 60.
7 See especially *Medieval Technology and Social Change*, p. 1203.
8 White in Barbour, op. cit., pp. 59, 57.
9 Ibid., pp. 1205–6, emphasis added.
10 White in Barbour, op. cit., p. 63.
11 'The Historical Roots of Our Ecologic Crisis', *Science*, 155, p. 1207, 1967.
12 J. Baird Callicott, 'Environmental Philosophy is Environmental Activism: The Most Radical and Effective Kind', in D.E. Marietta and L. Embree (eds), *Environmental Philosophy and Environmental Activism*, Lanham, MD: Rowman & Littlefield, p. 30, 1995.

See also in this book
Bartholomew, Callicott, Carson, Nasr, Saint Francis of Assisi

White's major writings
Latin Monasticism in Norman Sicily, Cambridge, MA: Medieval Academy of America, 1938.
Educating Our Daughters: A Challenge to the Colleges, New York: Harper & Brothers, 1950.
Medieval Technology and Social Change. London: Oxford University Press, 1962.
Machina ex deo: Essays in the Dynamism of Western Culture, Cambridge, MA: MIT Press, 1968.
Medieval Religion and Technology: Collected Essays, Berkeley, CA: University of California Press, 1978.

Further reading

Barbour, Ian G. (ed.), *Western Man and Environmental Ethics: Attitudes Toward Nature and Technology*, Reading, MA: Addison-Wesley, 1973.

Callicott, J. Baird, *Earth's Insights: A Multicultural Survey of Ecological Ethics from the Mediterranean Basin to the Australian Outback*, Berkeley, CA: University of California Press, 1994.

Callicott, J. Baird, 'Genesis and John Muir', in *Beyond the Land Ethic: More Essays in Environmental Philosophy*, Albany, NY: State University of New York Press, 1999.

Gimpel, Jean, *The Medieval Machine: The Industrial Revolution of the Middle Ages*, New York: Holmes & Meier, 1974.

Gottlieb, Roger (ed.), *This Sacred Earth: Religion, Nature, Environment*, London: Routledge, 1996.

Hill, Brennan R., *Christian Faith and the Environment: Making Vital Connections*, Maryknoll, NY: Orbis, 1998.

Kinsley, David, *Ecology and Religion: Ecological Spirituality in Cross-cultural Perspective*, Englewood Cliffs, NJ: Prentice-Hall, 1995.

Sauer, Thomas J. and Nelson, Michael Paul, "Science, ethics, and the historical roots of our ecological crises – was White right?" in T.J. Sauer, J.M. Norman, and M.V.K. Sivakumar (ed.) *Sustaining Soil Productivity in Response to Global Climate Change – Science, Policy and Ethics*. Wiley-Blackwell, Chichester, UK, 2011, pp. 3–16.

MICHAEL PAUL NELSON

MAURICE MERLEAU-PONTY 1908–1961

Because perception gives us faith in a world, in a system of natural facts rigorously bound together and continuous, we have believed that this system could incorporate all things into itself, even the perception that has initiated us into it. Today we no longer believe nature to be a continuous system of this kind ... We have then imposed upon us the task of understanding whether, and in what sense, what is not nature forms a 'world', and first what a 'world' is, and finally, if world there is, what can be the relations between the visible world and the invisible world.[1]

Nature is one of the most important concepts in the thought of Maurice Merleau-Ponty, a French philosopher who, it could be argued, devoted his entire philosophical career to developing a better understanding of humanity's relationship to the natural world. Educated at the École Normale Supérieure along with such luminaries as Simone de Beauvoir and Jean-Paul Sartre, Merleau-Ponty pursued a more academic path, teaching first in the lycée system before holding posts at the University of Lyon (1945–48), the Sorbonne (1949–52), and ultimately the Collège de France (1952 until his death), the most prestigious academic chair in the country. Merleau-Ponty was not only an academic, however: along with Sartre and de Beauvoir, he was on the editorial board of *Les Temps Modernes*, an important cultural and political journal, serving as the political editor. While contemporary environmentalists might not recognize Merleau-Ponty's philosophy as being explicitly ecological or concerned with the kinds of political questions we deal with today, his work is nevertheless an interesting touchstone for the contemporary environmental movement due to how his philosophy entwines how nature is conceptualized, how we regard ourselves as human beings, and how both of those philosophical commitments shape our political engagement with the surrounding world.

From the environmental perspective, one of the most interesting aspects of Merleau-Ponty's philosophy is the centrality of relations. What Merleau-Ponty offers is an ecological philosophy in the sense that he attempts to understand the nature of the surrounding world, living beings, and human society in terms of the types and qualities of relations that exist. What is novel, however, is that he offers this ecological perspective without appealing to a holistic vision such as is found in the work of Arne Naess or Fritjof Capra. On the contrary, Merleau-Ponty seeks to understand the specificity of relations in terms of how both larger structures emerge on the basis of smaller scale relations, and individuals emerge as such through relations of differentiation.

Merleau-Ponty's first book, *The Structure of Behavior* (1942), lays out the problematic that he will continue to refine and revise throughout his philosophical career: how might we best conceive of the relationship between nature, humanity, and other forms of animal life. The two dominant ways of thinking about nature—that of a causal system held together by laws and that of a system constructed in the human mind—are found to be lacking, both because neither approach can adequately explain the phenomenon of life and because both approaches rely upon something external to nature, laws or the mind, in order to explain nature's coherence. Both of these perspectives are still present within scientific and philosophical thinking today, though there are signs that their limitations are becoming increasingly clear.

In his early work, the key to understanding nature lies in the notion of "form." Form is an order that we perceive within the natural world, an organization of processes that depends upon the actual behavior of those processes in order to appear coherent. So, in the perception of natural order, there is always an element that is contributed by the perceiver and an element that is contributed by the world itself. Forms are *real*: they exist as a perceived system of relations, a perceived orientation of processes toward some set of future processes. As perceived, however, forms are also relative in two ways: on the one side, relative to the consciousness perceiving them and, on the other, relative to the processes perceived. Nature, in this view, refers to the world of processes prior to their organization into forms through their relationship with consciousness.

It is important, however, not to hear Merleau-Ponty as arguing that nature is some underlying world of process that is somehow tainted, corrupted, or otherwise inaccessible as a result of its engagement with a conscious mind. Rather, his idea is similar to that of independent co-arising within Buddhist discourses: both beings become what they are through their relation to the other. But, for Merleau-Ponty, there is never a simple two-term relation: there will always be mediating terms within relations. Consider a scientist observing an amoeba under a microscope or stellar phenomena by means of a radio telescope: in both cases the observations made and interpretations of them offered are not direct, but mediated through technological apparatus, a network of scientific institutions, a history of theories, etc. In this way, nature becomes an extremely complex phenomenon: the forms we perceive in the natural world are not only a product of the behavior of natural phenomena, but also the product of the complex histories and institutions that accompany our perception.

These ideas launched Merleau-Ponty's inquiries into the two areas of philosophy for which he is most famous: that of the philosophy of perception and that of artistic expression. Very simply, Merleau-Ponty believes that perception is best understood as a mutable relation between a body and its surroundings such that objects within the world guide our perception of them, shape the body—in some cases to the extent that the body assimilates objects into its normal functioning, e.g., eyeglasses, prosthetic limbs, or, arguably, smart-phones—and call for specific responses. Because our bodies are only susceptible to a given range of stimuli, we are *not* related to everything that exists as is sometimes claimed, and the body develops behavioral patterns within a delimited perceptual field. Thus, as we engage with, alter, and otherwise live within our environments, the environment exerts a reciprocal influence upon us.

The philosophy of expression picks up at this point: confronted by the world, familiar or not, there is always the possibility for a novel response. In human terms, this relation is readily apparent: faced with a familiar context in which one possesses a habitual set of behaviors, we view ourselves as capable of taking up that situation and operating in a manner independent of our habits. The interesting question is how and why this is possible. For Merleau-Ponty, this ability to alter our habitual patterns is a result of the malleability of relations internal to a given form resulting from the mediation of those relations. Because bodies change and are capable of interacting with their environment, it is possible to *transform* the world and ourselves within the limits of our bodies, which are themselves mutable. Hence, all knowledge of the world will be perspectival, emerging from a bodily situation and grappling with the perceived forms of nature.

To put that point in concrete terms, we can utilize Merleau-Ponty's example of Paul Cézanne. Merleau-Ponty situates the value of Cézanne's painting in a paradox that his works embody: "investigate reality without departing from sensations, with no other guide that the immediate impression of nature, without following the contours, with no outline to enclose the color, with no pictorial composition."[2] In other words, Cézanne attempted an objective realism by means of subjective impressions, to express inhuman nature through a human vision and a human expressive medium. But the lesson of these works is not the difficulty of representing reality given that the world is always experienced through certain subjective categories. Rather, the lesson of Cézanne is that the thing itself comes into being through its relation to otherness: what something *is* is a result of the kinds of relations it is capable of entering into, and what these relations are is not given in advance. Our habitual means of perceiving the world are inherited, and so the value of attending to the process of expression is in the subsequent ability to take up the relations in the world actively and to perceive them in a different manner. In this way we are able to alter our behavior, change our inherited culture, and develop new means of expression.

These concepts serve as the foundation for Merleau-Ponty's understandings of the sciences and of nature. Science is a human activity; it is the taking up of a specific perspective upon the world in an attempt to render nature as it is. For this reason, Merleau-Ponty shares the commonly held view that science remains our surest guide in understanding nature. As he puts the point, "If Nature is an all-encompassing something, we cannot think starting from concepts, let alone deductions, but we must rather think it starting from experience, and in particular, experience in its most regulated form—that is, science."[3] We must be cautious in how we utilize science to understand nature, however, as we frequently forget its embeddedness within experience and the consequences of that embeddedness for our interpretations of scientific results. Consider how the power of science comes from its objectivity, where its objectivity is derived from its purported ability to transcend individual situations in order to discover something universal about nature or an object. This standpoint, this view from nowhere, is itself a relationship we adopt toward objects of our experience; it is, as he says, a highly regulated experience but an experience nevertheless.

So what is the experience of nature like within the sciences? To answer this question, the first thing to note is where the mediating terms lie within the sciences: in specific *procedures* or *operations*, in the *processes* and *institutions* that govern the practice of science, and in the *instruments* and *tools* (especially computers, more contemporarily) scientists rely upon to observe and interpret the world. Far from a direct experience of nature, scientific experience is highly mediated. This mediation, however, does not destroy the objectivity of science; it is its guarantor. Following the operations, systems of translation from instrumental results into meaningful answers, noting the distinct permutations of reality that obtain in response to specific alterations of defined conditions—all of these practices aid us in understanding our openness to the world and the openness of bodies to other bodies. Although it is not the objectivity sought after in the modern period, it is objectivity nevertheless: a description of the experience of the processes that are nature, but understood as the set of relations that create the determinations of what is and what is not possible. In this way, the sciences provide us with facts, but note that the term "experience" is a loaded one here: while the processes described are not contested, they are subject to the vagaries of experience and as such are open to being expressed in different ways. What we find, then, in the sciences is not a conception of nature in itself, but rather, "we find in it what we need to eliminate false conceptions of nature."[4]

But where does this leave us with the question of how to think of nature? Nature consists in a set of relations provisionally organized into systems that achieve contingent and mutable stabilities over a period of time as a result of the relative stability of the bodies that inhabit a given place at a given time. Thus, nature is a continually evolving form, and, following Alfred North Whitehead, Merleau-Ponty calls nature "the memory of the world."[5] Just as our memories determine our possibilities without making any one necessary, the past of a natural system determine its possibilities without there being any one healthy state of a natural system. As with a human body, there is room for significant variation and new stabilities. We are a part of this nature, and throughout human existence we have played a role in shaping how nature is.

As stated above, Merleau-Ponty has been most influential in the fields of aesthetics and cognitive psychology, but he has also inspired contemporary thinkers of interest to environmentalists such as Tim Ingold and David Abram. He has done so, above all, by discrediting the environmentally pernicious idea of a separation and opposition between the natural world and human culture. "It is impossible," he wrote, "to say that here nature ends and the human being or expression begins."[6] This inseparability requires us to rethink many of our most common standards for what constitutes an environmental harm and opens up the possibility of imagining new ways of living within the natural world where our presence adds to the beauty of the earth rather than blighting it. While it remains to be seen if Merleau-Ponty's ideas will play a part in uprooting the conceptions of nature and culture that have been so devastating to the earth, the manner in which he integrates science, an ecological vision, and a role for human participation within the natural world should at least give us reason to hope.

Notes
1   *The Visible and the Invisible*, pp. 26–7.
2   "Cézanne's Doubt," p. 72.
3   *Nature*, p. 87.
4   *Nature*, p. 86.
5   *Nature*, p. 120.
6   "Eye and Mind," p. 376.

See also in this book
Heidegger, Naess

Merleau-Ponty's major writings
"Cézanne's Doubt." Translated by Michael B. Smith, revised by Leonard Lawlor. In *The Merleau-Ponty Reader*, edited by Ted Toadvine and Leonard Lawlor, pp. 69–87. Evanston, IL: Northwestern University Press, 2007 (1945).
"Eye and Mind." Translated by Michael B. Smith, revised by Leonard Lawlor. In *The Merleau-Ponty Reader*, edited by Ted Toadvine and Leonard Lawlor, pp. 351–78. Evanston, IL: Northwestern University Press, 2007 (1964).
Nature: Course Note from the Collège de France. Translated by Robert Vallier. Evanston, IL: Northwestern University Press, 2003 (1994).
Phenomenology of Perception. Translated by Donald Landes. London: Routledge, 2013 (1945).
The Structure of Behavior. Translated by Alden Fisher. Pittsburg, PA: Duquesne University Press, 1963 (1943).
The Visible and the Invisible. Translated by Alphonso Lingis. Evanston, IL: Northwestern University Press, 1968 (1964).

Further reading
Abram, David. *The Spell of the Sensuous*. New York: Vintage, 1996.

Abram, David. "Merleau-Ponty and the Voice of the Earth." *Environmental Ethics* 10 (1988): 101–20.

Bannon, Bryan E. *From Mastery to Mystery*. Athens, OH: Ohio University Press, 2014.

Bannon, Bryan E. "Flesh and Nature: Understanding Merleau-Ponty's Relational Ontology." *Research in Phenomenology* 41 (2011): 327–357.

Cataldi, Sue and William Hamrick. *Merleau-Ponty and Environmental Philosophy*. Albany, NY: SUNY Press, 2007.

Fischer, Sally. "Ecology of the Flesh: Gestalt Ontology in Naess and Merleau-Ponty." *International Studies in Philosophy* 34 (2002): 53–67.

Hamrick, William. *Nature and Logos*. Albany, NY: SUNY Press, 2011.

James, Simon P. "Merleau-Ponty, Metaphysical Realism, and the Natural World." *International Journal of Philosophical Studies* 15 (2007): 501–19.

Russon, John. "Embodiment and responsibility: Merleau-Ponty and the Ontology of Nature." *Man and World* 27 (1994): 291–308.

Toadvine, Ted. *Merleau-Ponty's Philosophy of Nature*. Evanston, IL: Northwestern University Press, 2009.

BRYAN E. BANNON

E.F. SCHUMACHER 1911–1977

> The fight against pollution [cannot] be successful if the patterns of production and consumption continue to be of a scale, a complexity, and a degree of violence which, as is becoming more and more apparent, do not fit into the laws of the universe, to which man is just as much subject as the rest of creation.[1]

Schumacher was working on his ideas at a time when the dominant ideology was 'the bigger the better'. Large institutions, multinational corporations, industrial mergers, unlimited economic growth and ever-increasing consumption, were considered symbols of progress. Schumacher said, 'We suffer from an almost universal idolatry of giantism.'[2]

In response to such idolatry, Schumacher encapsulated an alternative world-view in his seminal collection of essays, *Small is Beautiful* (1973), which became one of the most popular books amongst members of the British Parliament. The suggestion that many of the environmental and social problems facing the world were the result of idolatry to giantism intrigued Jimmy Carter, President of the USA, and consequently Schumacher was invited to the White House to advise the president in 1977. The Governor of California at that time, Jerry Brown, became so convinced by Schumacher's analysis that he initiated a number of measures embodying the 'small is beautiful' approach.

Ernst Fritz Schumacher was born in Bonn, Germany, in 1911. He came to England in 1930 as a Rhodes scholar to read Economics at New College, Oxford. After a short spell of teaching Economics at Columbia University, New York, followed by dabbling in business, farming and journalism, he became an economic advisor to the British Control Commission in Germany (1946–50), followed by a long career in the National Coal Board in Britain.

It was Schumacher's involvement in the economics of developing countries that challenged and changed his economic philosophy. He realized that the Western pursuit of unlimited economic growth on a gigantic scale is neither desirable nor practicable for the rest of the world. If anything, the West itself needs to learn the simplicity, spirituality and good sense of other cultures which are not yet in the grip of technological imperatives. 'In the excitement of the unfolding of his scientific and technological powers, modern man has built a system of production that ravishes nature, and a type of society that mutilates man.'[3]

The turning point came in 1955 when he was sent as Economic Development Advisor to the government of Burma. He was supposed to introduce the Western model of economic growth in order to raise the living standards of the Burmese people. But he discovered that the Burmese needed no economic development along Western lines, as they themselves had an indigenous economic system well suited to their conditions, culture and climate. As a result of his encounter with this profound and practical Buddhist civilization, he wrote his well-known essay, 'Buddhist Economics' (1966). Schumacher was perhaps the only Western economist to dare to put these two words, Buddhism and economics, together. The essay was printed and reprinted in numerous journals and anthologies.

Recalling his time in Burma he told me that the Burmese needed little advice from him. In fact, Western economists could learn a thing or two from the Burmese. They had a perfectly good economic system, which supported a highly developed religion and culture and produced not only enough rice for their own people but also a surplus for the markets of India. He further commented that when he had published his findings under the title of 'Buddhist Economics', a number of his economist colleagues had asked, 'Mr Schumacher, what does economics have to do with Buddhism?' His answer was simply that 'Economics without Buddhism is like sex without love.' Economics without spiritual values can only give temporary and physical gratification; it cannot provide lasting fulfilment. Buddhist economics includes service to fellow human beings and compassion for all life as well as making a profit and working efficiently. We need both economics and spirituality and we need them simultaneously.

During his time in Burma, Schumacher encountered the Buddhist concept of the Middle Way. He wanted to apply it to technology. He saw that people are either stuck with the sickle or they seek a combine harvester, thus he developed the Schumacher Principle of the Disappearing Middle, referring to the way that when a new, advanced technology is developed it displaces its immediate predecessor. Consequently, what is left is either expensive, sophisticated, state-of-the-art technology or very simple hand tools, whereas what small farmers and manual workers require is a technology between these two extremes.

In 1970, after many years of gestation, Schumacher founded the Intermediate Technology Development Group (ITDG) with an article in the pages of the *Observer* newspaper. He received an overwhelming response from the general public. ITDG became the practical expression of respect for cultural diversity. It pursued economic development within people's cultural context, rather than looking at the non-industrialized world as 'under-developed'. Intermediate Technology was envisioned to be environment-friendly, non-polluting and non-exploitative of people or nature. Therefore, it also became known as 'appropriate technology' or 'alternative technology'. The concept was initially applied to non-industrialized countries, but technologies of renewable energy, of recycling and of ecological restoration in the West, became part of the same movement of a technology for a sustainable future.

Complementary to Intermediate Technology was his involvement with sustainable agriculture; he spent much time on his organic garden and became president of the Soil Association. He believed that 'in the simple question of how we treat the land ... our entire way of life is involved'.[4] He had no doubt that 'a callous attitude to the land and to the animals thereon is connected with, and symptomatic of, a great many other attitudes, such as those producing heedless urbanization, needless industrialization, and a kind of fanaticism which insists on playing about with novelties – technical, chemical, biological and so forth – long before their long term consequences are even remotely understood'.[5]

For Schumacher, care for the land and for the soil was fundamental to caring for the whole of the natural world, as well as a way of creating a just and equitable society. In the 1960s and 1970s attention to 'Mind, Body, Spirit' was becoming popular amongst alternative circles. Schumacher found this too narrow, human-centred and individualistic. It was all about the human mind, human body and human spirit. It left out the issues of social justice and caring for the earth. The spiritual dimension for

Schumacher was paramount: individual development and personal growth were necessary, but only in the context of social wellbeing and the wellbeing of the Earth. Therefore, Schumacher's philosophy led away from the personal focus of 'mind, body and spirit' to the broader and more inclusive concerns of what I have called, 'soil, soul and society'.

Schumacher was a holistic and ecological economist. Modern economics looks at the world as a resource for ever-increasing profit, and at human beings as units of labour for the profitability and continuity of the economic system. Schumacher saw it the other way round. That is why he subtitled *Small is Beautiful* 'a study of economics as if people mattered'. Furthermore, economics must be a way of sustaining, restoring and maintaining the immense diversity and complexity of the biosphere in addition to nourishing, nurturing and fulfilling appropriate human needs. Economics is to serve people *and* planet. In order to achieve this kind of economic system, it must remain under local control and within a human scale, not becoming subservient to the so-called 'economy of scale'.

The importance of small-scale and local production became crystal clear to him when Schumacher saw a lorry full of biscuits being brought from Manchester to London, and minutes later another lorry full of biscuits being taken from London to Manchester. Schumacher gasped: what could be the economic rationale of this activity? Having failed to see any good reason for this transportation which caused air pollution and wasted fossil fuels and human labour, Schumacher said in frustration: 'As I am not a nutritionist, I wonder if the nutritional value of the biscuits is increased by this transaction?! Otherwise, if Manchester has a special kind of biscuit it could simply send the recipe to London on a postcard, and vice-versa.'

To Schumacher it was logical and natural to produce, consume and organize as locally as possible, which inevitably meant on a smaller scale. Therefore, to him the question of size was an overriding and over-arching principle. He refused to accept that largeness was necessary for prosperity: 'Small units are highly prosperous and provide society with most of the really fruitful new developments.'[6] Again, he wrote: 'The question of scale is extremely crucial today, in political, social and economic affairs just as in almost everything else.'[7]

Beyond a certain scale the people involved are disempowered and a bureaucratic machine takes over. For example, in a school of 1,000 children, parents do not know the teachers, teachers cannot know all the children, the children cannot know each other, and the surrounding community is overwhelmed by the influx of pupils who do not belong to that community. In this situation children become numbers, and the aim of education becomes meeting the requirements of the system and the league tables rather than the development of the whole child.

Similarly, large hospitals, large factories and large businesses lose the purpose of enriching human wellbeing and become obsessed with maintaining and perpetuating the organization for its own sake. Therefore, it could be said almost invariably that if there is something wrong, there is something too big. Also, big organizations will have big problems, and small organizations will have small problems, which can be solved more easily.

As in economics, so in politics. Schumacher was greatly influenced by the Austrian philosopher Leopold Kohr, whom he considered his mentor. In the book, *Breakdown of Nations*, Kohr outlines the case against giantism and against big nations. In countries such as Sweden or Switzerland, there is much more political participation and flexibility. When people in these countries want to bring about change, they can do so with greater ease than in countries like China, India or the USA. So, Schumacher believed in small nations, small communities and small organizations. Small, simple and non-violent were his three philosophical precepts.[8] These were to determine all relationships – economic, political and cultural – within human societies, as well as between humans and the natural world.

Schumacher died in September 1977, in Switzerland. He wrote only two books, *Small is Beautiful* and *A Guide for the Perplexed*, the latter published posthumously. A collection of his speeches was later published under the title of *Good Work*. Yet his influence was vastly greater than the volume of his published work might suggest. He was more than an economist, he was also a very practical man. He inspired many people through his busy schedule of lectures, private meetings and through his support of grassroots projects. 'Pollution must be brought under control and mankind's population and consumption of resources must be steered towards a permanent and sustainable equilibrium'[9] was his advice to the groups with whom he worked.

His legacy continues to be felt. Immediately after his death the Schumacher Society was established in Britain, which continues to promote the ideas of ecological economics. The Society holds annual lectures in Bristol, Liverpool and Manchester. Some of these lectures have now been published. Schumacher Societies have also sprung up in the USA, Germany and India.

His writings have inspired people in different disciplines. In education a number of Small Schools have been established, where the emphasis is on 'education as if children matter'. A college named after him has also been established at Dartington, Devon, exploring an ecological world-view from many different perspectives, while students practise a lifestyle built around the precepts of small, simple, local and non-violent. In economics, the New Economics Foundation encourages ideas of local economies, local currencies and local trading. In the field of development, ITDG continues to promote indigenous and small-scale projects. In the field of energy, the National Centre of Alternative Technology, Wales, attracts thousands of visitors keen to see methods of renewable energy. *Resurgence* magazine, for which Schumacher wrote regularly, continues to examine and expound the 'small is beautiful' ethos.

Notes
1 *Small is Beautiful*, p. 247.
2 Ibid., p. 54.
3 Ibid., p. 246.
4 *This I Believe*, p. 181.
5 Ibid.
6 *Small is Beautiful*, p. 53.
7 Ibid., p. 55.
8 *Resurgence*, January/February 1974.
9 Ibid., p. 248.

See also in this book
Balfour, Buddha, Sukhdev

Schumacher's major writings
Small is Beautiful; A Study of Economics as if People Mattered, 1973, London: Abacus, Sphere Books, 1988. Includes the essay 'Buddhist Economics'.
A Guide for the Perplexed, 1977, London: Abacus, Sphere Books, 1989.
Good Work, London: Jonathan Cape, 1979. A collection of speeches.
This I Believe, Totnes, UK: Green Books, 1997. A collection of twenty-one articles published in *Resurgence;* includes 'Buddhist Economics'.

Further reading
Button, John, *The Green Fuse*, London and New York: Quartet Books, 1990. A collection of Schumacher Lectures.
Kohr, Leopold, *The Breakdown of Nations*, London and New York: Routledge & Kegan Paul, 1986 (1957).
Kumar, Satish (ed.), *The Schumacher Lectures Vol. I and II*, London: Blond & Briggs, 1980 and 1984 respectively.
McRobie, George, *Small is Possible*, London: Jonathan Cape, 1981.
Wood, Barbara, *Alias Papa: A Biography*, London: Jonathan Cape, 1984.

SATISH KUMAR

JOHN CAGE 1912–1992

> We must consider ecology even more than the individual. It is not simply by observing the individual, but by reintegrating individuals into nature, by opening the world to the individual, that we will get ourselves out of this mess.
>
> *Cage 1981, p. 56*

John Cage was born in 1912 in Los Angeles, California. He was a highly influential composer of avant-garde music, as well as a celebrated writer, visual artist, and counter-cultural figure. He was also a noted mycologist, an environmentalist, and one of the first Americans to actively engage with Zen Buddhism. Cage met the Buddhist philosopher D.T Suzuki at Colombia University in the late 1940s, and Zen influenced his work and his thinking about ecological issues profoundly. Cage lived with his partner and frequent collaborator, the dancer and choreographer Merce Cunningham, until his death in New York in 1992.

Cage began writing music in the 12-tone form pioneered by his teacher Arnold Schoenberg. In 1939 he began writing music for the 'prepared piano'—a piano that he had modified by adding bits of rubber, wood, metal, and other objects to the strings. This innovation eventually led to further experiments in sound with found objects, tape recorders, radios, and early electronic instruments. Crucial to his practice as a composer was the discovery of chance operations, a compositional procedure which allowed him to reduce personal intention. Cage began using chance operations in 1951, when he was given a pocket version of the *I Ching* by his student, the composer Christian Wolff. This ancient Taoist book of oracles accords with the Buddhist principle that the universe is impermanent, existing as it does in a continual state of flux. Cage used the text to determine such compositional factors as the number and duration of sounds in a musical composition, or the choice and placement of words in writing.

Cage composed his most noted (and most infamous) piece of music, *4'33"*, using chance operations. In this composition, performers simply remain present at their instruments for four minutes and thirty-three seconds, without playing a single note. Cage derived the seemingly random duration of this piece by throwing the *I Ching*, thereby removing a high degree of his own subjectivity as composer. *4'33"* presents a significant challenge to the Western music tradition by offering its audience a set period of time to listen to sounds in the surrounding environment, rather than to sounds performed by musicians on instruments. After the first performance of *4'33"* in 1952 by pianist David

Tudor, Cage remarked that the audience did not seem to hear the sound of the wind and the raindrops falling outside the concert hall during the recital—an indication of his turn towards noise as music, and in this instance, of his embrace of the ambient sounds of nature. Moreover, his use of the *I Ching* as a compositional tool to limit his own subjectivity illustrates his desire to step back from dominance over the environment, and to frame the environment as a co-creator of the work. This elision of the boundary between human intention and the actions of the natural world remained crucial to his compositional practice, as well as to his environmentalist philosophy, throughout the rest of his career.

Originally Cage threw coins in the traditional manner prescribed by the *I Ching* to determine the formal features of a composition. During the late 1960s, however, he made his first experiments with computer-generated compositional methods based on the *I Ching*, and in the early 1980s he began using IC ("*I Ching*"), a computer program developed by his assistant Andrew Culver to speed up the generating process. This approach "revolutionized" Cage's working practices, because all he had to do was refer on the computer printout to the hexagram, use that hexagram as numerical data for determining compositional form, and then to cross out the data and move on to the next hexagram.[1] It is important to note that Cage was "less interested in the *I Ching* as a book of wisdom than as a mechanism of chance operations that produces random numbers from 1 to 64."[2] The point of relying so heavily on the oracle was not to receive arcane solutions or esoteric knowledge, but to compose intricate scores based on numerical values, and to reconfigure subjective intention by limiting personal choice. This limitation was central to Cage's writing and music in relation to his non-dualistic breakdown of the distinction between human intention and the operation of nature.

Although Cage was predominantly interested in Zen Buddhism as it was taught by Suzuki, he frequently quoted the South-Asian Hindu philosopher Ananda Coomaraswamy's definition of art as "the imitation of Nature in her manner of operation." Coomaraswamy's view of nature owes much to Thomas Aquinas, who applied the term "Nature" to the broadest context, "indicating the universal, natural order through which individual phenomena are created."[3] Cage's cites Coomaraswamy in his 1961 essay "On Robert Rauschenberg, Artist and his Work" when he points out that Rauschenberg uses an "over-all" compositional method, "where each small part is a sample of what you find elsewhere."[4]

Cage situates his discussion of Rauschenberg's compositional practices immediately before his citation of Coomaraswamy, perhaps suggesting that the painter's work exemplifies Coomaraswamy's definition of art as an imitation of natural process, in which art and nature function analogously in their manner of operation. Note the emphasis on process, a theme and practice that remained crucial to Cage's thinking about nature throughout his career. Cage further writes in the Rauschenberg essay that art is the imitation of nature "or a net."[5] Here Cage sutures Coomaraswamy's thinking on nature with Zen Buddhism. "Indra's Net," writes Buddhist scholar Stephen Batchelor, is "a vast grid of interconnected mirroring spheres, each one reflecting all the others."[6] The third century CE *Avatamsaka Sutra*, or the "Flower Adornment" sutra, employs the "Jewelled Net of Indra" as a metaphor for interdependence. This teaching was more fully developed in China by the *Hua-yen* school during the sixth century, and then later transmitted to Japan under the Japanese name *Kegon*. Suzuki had the highest respect for the teachings of the *Avatamsaka*, and taught it to his classes at Columbia. And it is precisely this form of holographic, infinite interdependency that Cage finds in Rauschenberg's "all-over" compositional method. Where Tu Shun (a patriarch of the *Hua-yen* / *Kegon* school) writes, "[i]f you sit in one jewel, then you are sitting in all jewels in every direction, multiplied over and over,"[7] Cage writes, "each small part is a sample of what you find elsewhere."[8] This holographic, over-all compositional practice stems from Cage's Zen-influenced understanding of a non-dualistic relationship between the human and natural world.

Cage's most sustained engagements with the language of nature occur in his various intertextual rewritings of nineteenth-century American philosopher and writer Henry David Thoreau, whose work on the relationship between social conditions and nature anticipated many contemporary environmentalist concerns. Along with his representation of nature, Thoreau was also interested in Buddhist Philosophy: "I know that some will have hard thoughts of me, when they hear their Christ named beside my Buddha," he writes in *A Week on the Concord and Merrimack Rivers* (1849) "yet I am sure that I am willing they should love their Christ more than my Buddha, for the love is the main thing."[9] In 1967 Wendell Berry introduced Cage to Thoreau's *Journal* (1858), and Cage writes that at that time he was "starved for Thoreau"[10] and that in reading Thoreau's *Journal* he discovered "any idea I've ever had worth its salt."[11] Cage produced several texts based on Thoreau's writing, including *Mureau* (1970), the title of which comes from condensing the words *music* and *Thoreau*, and "Lecture on the Weather" (1975), a musical composition composed for the American Bicentennial which also samples passages drawn from Thoreau's writing.

Mureau recycles letters, syllables, phrases, and complete sentences drawn from Thoreau's *Journal*; Cage wrote it by "subjecting all the remarks of Henry David Thoreau about music, silence, and sounds he heard that are indexed in the Dover publication of the *Journal* to a series of *I Ching* chance operations."[12] Thoreau in Cage's hands becomes a metonymy for natural processes, in which the treated text is a material thing, existing in a contiguous relationship with nature. Free from the dualistic reflection on nature which characterizes so much traditional "nature poetry" and which separates the reader from the text and the text from the world, *Mureau* offers readers an experience of language as material thing—as a non-representational object:

> sparrowssitA gROsbeak betrays *itself* by that peculiar squeakariEFFECT OF SLIGHTEst tinkling measures soundness ingpleasa **We hear!** Does it not rather hear us? **sWhen he hears** the telegraph, he thinksthose bugs have issued forthThe Owl wakes touches the stops, wakes reverberations *d gwalky* In verse there is no inherent *music eof*sttakestakes a man to make a room silent […].[13]

The various fonts, word fragments, unorthodox spacings and non-uniform type-faces here call attention to linguistic materiality, and muddle the clear transmission of information. One could say that his texts' foregrounding of semantic loss serves as a linguistic analogy to the Second Law of Thermodynamics, in which energy (or meaning) is more easily lost than obtained. Moreover, instead of referencing an object in the natural world or a topological scene, *Mureau* self-consciously foregrounds language itself as a natural phenomenon. By juxtaposing various type fonts and foregrounding a cacophony of linguistic excess, Cage complicates any straightforward reading or understanding, thereby calling into question the transparent norms of communication so typical of more traditional forms of nature writing, and presenting instead information as a site for the eruption of non-sense.

Ultimately, Cage may not have presented a well-constructed blueprint for an alternative to the social order and the related environmental degradation of the late twentieth century. What his music, writing, and art continue to provide, however, is a site which foregrounds a lived experience of nature in the manner of its operation. This experience differs from conceptual notions of environmentalism. Cage's poetics offer us a plenitude of experience, a transformation of our sense of what constitutes nature—or rather, a sense of how nature cannot be constituted, conceptualized, or represented.

Notes

1 D. Nicholls. *John Cage*. Urbana and Chicago, IL: University of Illinois Press, p. 99, 2007.
2 Lewallen, Constance in D.W. Bernstein and C. Hatch (eds). *Writings Through John Cage's Music, Poetry, and Art*. Chicago, IL and London: University of Chicago Press, p. 235, 2001.
3 Patterson, David W. 'The Picture That is Not in the Colors: Cage, Coomaraswamy, and the Impact of India', in D. W. Patterson (ed.), *John Cage: Music, Philosophy, and Intention, 1933–1950*, London: Routledge, p. 195, 2002.
4 J. Cage, *Silence*. Middletown: Wesleyan University Press, p. 100, 1961.
5 Ibid., p. 100.
6 Batchelor, Stephen. 'The Sands of the Ganges: Notes Towards a Buddhist Ecological Philosophy', in M. Batchelor and K. Brown (eds), *Buddhism and Ecology*, London: Cassel, p. 35, 1992.
7 Tu Shun. 'Cessation and Contemplation in the Five Teachings of the Hua-yen', in T. Cleary (ed.), *Entry into the Inconceivable: An Introduction to Hua-yen Buddhism*, Honolulu: University of Hawaii Press, p. 66, 1983.
8 Cage, *Silence*, p. 100.
9 H.D. Thoreau, *A Week on the Concord and Merrimack Rivers* (1849), Princeton, NJ: Princeton University Press, p. 67, 1980.
10 J. Cage, *Empty Words: Writing '73–'78*. Middletown, CT: Wesleyan University Press, p. 11, 1978.
11 Ibid., p. 4.
12 J. Cage, *M: Writings '67–'72*. Middletown, CT: Wesleyan University Press, p. ix, 1973.
13 Ibid., p. 35.

See also in this book
Buddha, Nhat Hanh, Thoreau

Cage's major writings

Anarchy. Middletown, CT: Wesleyan University Press, 1988.

A Year From Monday: New Lectures and Writings. Middletown, CT: Wesleyan University Press, 1968.

Empty Words: Writing '73–'78. Middletown, CT: Wesleyan University Press, 1978.

For the Birds: in Conversation with Daniel Charles. London: Marion Boyars, 1981.

M: Writings '67–'72. Middletown, CT: Wesleyan University Press , 1973.

Silence. Middletown, CT: Wesleyan University Press , 1961.

Writings Through Finnegan's Wake. New York: Printed Editions, 1978.

X: Writings '79–'82. Middletown, CT: Wesleyan University Press, 1987.

Further reading

Gann, Kyle. *No Such Thing as Silence: John Cage's 4'33"*. New Haven, CT: Yale University Press, 2010.

Jaeger, Peter. *John Cage and Buddhist Ecopoetics*. London: Bloomsbury, 2013.

Nicholls, David. *John Cage*. Urbana and Chicago, IL: University of Illinois Press, 2007.

Perloff, Marjorie. *The Poetics of Indeterminacy: Rimbaud to Cage*. Princeton, NJ: Princeton University Press , 1981.

Suzuki, Daisetz T. *The Essentials of Zen Buddhism: An Anthology of the Writings of Daisetz T. Suzuki*. ed. Bernard Philips. London: Rider & Co., 1963.

PETER JAEGER

ARNE NAESS 1912–2009

If you hear a phrase like 'all life is fundamentally one!', you must be open to *tasting* this, before asking immediately 'what does this mean?'. Being more precise does not necessarily create something that is more inspiring'.[1]

From the early 1970s, when he first introduced the expression 'deep ecology', Arne Naess became one of the most influential environmental philosophers in the world, his voice heard well beyond the confines of academic discussion. Born in Norway in 1912, Naess was Professor of Philosophy at the University of Oslo from 1939 to 1970. During the war he was active in the Norwegian resistance and after it became recognized as his country's leading philosopher. He was founding editor of the journal *Inquiry* and the central figure in the Oslo school of philosophy. In 1970 he resigned from his Chair in order, partly, to play a more active role in the environmental movement: for example, by engaging in protests against the building of a dam, becoming the Chairman of Greenpeace Norway, and standing in a parliamentary election for the Green Party. He spent much of his later life, however, in mountain retreats where he wrote, skied and enjoyed renown as a climber whose prowess belied his years. The recipient of prestigious honours, in Sweden as well as Norway, Naess remained, in his eighties and nineties, an instantly recognizable and admired figure in his country's intellectual life and in radical environmentalist circles all over the world. Arne Naess died in 2009, aged 96.

The younger Naess's writings on the philosophy of science and 'empirical semantics' provided little indication of the interest in environmental philosophy which was later to dominate his work – an interest inspired, he remarked, by Rachel Carson's *Silent Spring*. Indeed, his earlier enthusiasm for the natural sciences and logical positivism appears to be at odds with his later metaphysical views, influenced more by Spinoza, Romanticism and Eastern thought than by any empiricist tradition. By 1960, however, Naess's attention was turning more towards the history of philosophy and the comparative study of 'total views' of the world and humankind. The difficulty or impossibility of deciding among such views encouraged a respect, recorded in the 1968 book *Scepticism*, for the undogmatic sceptical stance associated with the Hellenistic philosopher, Pyrrho. The Pyrrhonians had drawn from their sceptical premises certain lessons for the conduct of life, including the adoption of a non-aggressive and tolerant attitude not dissimilar from the one later recommended by Gandhi, on whom Naess had written a book, *Gandhi and the Nuclear Age* (1960). By the early 1970s, Naess was already reflecting on the relevance to environmental issues of the views of Gandhi and other thinkers outside the orthodox Western traditions of science and philosophy.

The initial result of those reflections was a short, staccato and seminal paper, 'The Shallow and the Deep, Long-range Ecological Movement', published in 1973. Naess characterized the shallow movement as primarily engaged in a 'fight against pollution and resource depletion', its 'central objective' being 'the health and affluence of people in the developed countries'.[2] As subsequent writings show, what he meant, more widely, by 'shallow ecology' is an 'anthropocentric' position which argues for responsible treatment of the environment solely on the basis of the broadly material benefits which will accrue to human beings – a position, as Naess perceived, which was adopted in the 1980 *World Conservation Strategy* and is still apparent in the goal of 'sustainable development'. Deep ecology is not given a similarly concise characterization. To understand what Naess intends by this label, we might begin with the two words which comprise it. Deep ecology is *deep* because it explores the 'fundamental presuppositions' of our values and experience of the world. It is deep *ecology*, not because it is the empirical science of ecosystems, but because the attitudes it endorses, though inspired by several sources, receive 'rational justification' from the ecologists' demonstration of 'the intimate dependency of humanity upon decent behaviour toward the natural environment'.[3]

Deep ecology is best represented, perhaps, as a set of practical environmental policies underpinned by a set of normative principles which in turn are supported by a scientifically informed, but ultimately philosophical, view of reality and humankind. Among the policies advocated by Naess are radical reduction of the world's population;

abandonment of the goal of economic growth in the developed world; conservation of biotic diversity; living in small, simple and self-reliant communities; and – less specifically – a commitment 'to touch the Earth lightly'. The immediate justification for these policies is to be found in normative principles such as 'Natural diversity has its own intrinsic value' and that of 'bio-spherical egalitarianism', which enjoins respect for 'the equal right' of life forms to 'live and blossom'. The failure to recognize these principles reveals 'racial prejudice' against non-human life.[4] (Egalitarianism has its limits, however. Parents in India have a right, says Naess, to rid the playground of cobras – though he adds that they should have taken care, for the snakes' sake, over the siting of the playground.)

The deep ecologist's case, however, cannot rest with these moral principles. For one thing, 'ethics *follow from* how we experience the world', so that an adequate set of moral principles must be grounded in a proper articulation of experience of the kind that only a philosophy or religion can provide.[5] Second, while a principle like 'Human beings must respect the rights of non-human life!' is fine as a rallying-call, it can also reinforce an assumption to which Naess is resolutely opposed – one which he thinks, moreover, has been largely responsible for our appalling treatment of the natural environment. This is the assumption that humans and non-humans – indeed, beings of any kind – exist independently of one another. Naess is a 'holist', arguing that, at a fundamental level, all organisms are 'intrinsically related' in a 'biospherical net or field'. To distinguish man *from* his environment is to think, therefore, at a 'superficial' and artificial level.[6]

It is this holistic vision which, for Naess, grounds the normative principles and policies of deep ecology – or, as he prefers to call it in later writings, 'Ecosophy T', in order to distinguish his particular position from neighbouring ones. An increasingly central component in this vision is Naess's conception of Self, inspiration for which comes partly from Spinoza's thesis of a single substance, describable as God *or* Nature, but more especially from the Hindu notion of *Atman* (Self). Naess approvingly cites VI.29 of the *Bhagavad Gita*: 'He whose self is disciplined by yoga sees the Self abiding in all beings and all beings in Self'. He does not think, however, that we require 'mystical union' in order to conclude that individual selves are, so to speak, artificial abstractions from a 'comprehensive Self' in which all beings are integrally bound. It is sufficient to reflect on how the identity of each of us is utterly dependent on relations with others and with the world at large, and properly to attend to natural feelings of empathy and sympathy which presuppose that 'one experiences something [as] similar or identical with oneself'.[7]

This conception of a 'comprehensive Self' is used to support the moral imperatives of deep ecology in two ways. First, someone who genuinely internalizes it will be naturally drawn to a universal 'altruism', since he or she no longer recognizes what is presupposed by 'egoism' – the existence, at a basic level, of independent individual selves. Second, it follows from this conception that 'self-realization' requires sympathetic identification with the good of the whole. 'We seek what is best for ourselves, but through the extension of the self, our "own" best is also that of others', and 'when we harm others, we also harm ourselves'.[8] Deep ecologists are sometimes criticized for elevating the good of environment over human interests: but, for Naess, at least, appreciation of the 'comprehensive Self' implies that this contrast is illusory.

Naess's critics have come from several directions. For the most radical Greens and spokespersons for 'animal rights', he does not go far enough, since he accepts that human beings, in virtue of their 'nearness' to one another, are sometimes justified in lending greater moral weight to human wellbeing than to that of non-human life. For more traditional thinkers, his principle of 'biospherical egalitarianism' goes too far, and indeed is belied by his demanding of human beings a degree of self-sacrifice and altruism which it would be absurd to demand of animals.[9] For yet others, Naess's notion of 'self-realization', with its Indian roots, is far too romantic and 'mystical' to provide a foundation for hard-headed environmental policy. This is a charge which Naess, in my opening citation, is rejecting: that a notion cannot be made precise does not mean that we are unable to 'taste' it and be inspired to action by it. The demand for sharp definitions of concepts is, for Naess, a mark of the 'scientistic', technical mindset that is responsible for many of our environmental ills.

Despite his many critics, Naess's influence has been immense. As a successor to his Chair of Philosophy at the University of Oslo states: 'philosophy's place in Norwegian academic life, as in the society at large, is due in large measure to Naess'.[10] Not the least of his contributions to society at large was to environmental education. The Norwegian 'core curriculum' and the Norwegian–Latvian Project in Environmental Education, with their emphasis on, for example, self-awareness and the environment, bear the unmistakable stamp of Naess's ideas.[11] On a broader front, Naess's legacy to the deep ecological tendency in contemporary environmental thought and activism is not simply the name of that tendency. As its most distinguished spokesman among professional philosophers, Naess provided it with a theoretical foundation at which earlier writers of similar sympathies, such as Aldo Leopold, only hinted.

Notes

1 *Ecology, Community and Lifestyle*, p. 8.
2 Naess, 'The Shallow and the Deep, Long-range Ecological Movement', in L. Pojman (ed.), *Environmental Ethics: Readings in Theory and Application*, Boston, MA: Jones & Bartlett, p. 102, 1994.
3 Naess, 'Ecosophy T: Deep *versus* Shallow Ecology', in Pojman, op. cit., pp. 105–10.
4 Ibid., p. 106.
5 *Ecology, Community and Lifestyle*, p. 20, original emphasis.
6 'The Shallow and the Deep, Long-range Ecological Movement', p. 103.
7 'Ecosophy T: Deep *versus* Shallow Ecology', p. 108.
8 *Ecology, Community and Lifestyle*, pp. 174–5.
9 See R. Watson, 'A Critique of Anti-anthropocentric Biocentrism', in Pojman, op. cit., pp. 117–22.
10 A. Hannay, 'Norwegian Philosophy', in Ted Honderich (ed.), *The Oxford Companion to Philosophy*, Oxford, UK: Oxford University Press, p. 627, 1995.
11 See J.A. Palmer, *Environmental Education in the 21st Century*, London: Routledge, pp. 159–63, 244–8, 1998.

See also in this book
Gandhi, Leopold, Spinoza

Naess's major writings
Interpretation and Preciseness: A Contribution to the Theory of Communication, Oslo: Det Norske Videnskapsakademi, 1953.
Scepticism, London: Routledge & Kegan Paul, 1968.
'The Shallow and the Deep, Long-range Ecological Movement', *Inquiry*, 16, 95–100, 1973.
Ecology, Community and Lifestyle, 1976, trans. and ed. D. Rothenberg, Cambridge: Cambridge University Press, 1989.
'The Deep Ecological Movement: Some Philosophical Aspects', in S. Armstrong and R. Botzler (eds), *Environmental Ethics: Divergence and Convergence*, New York: McGraw-Hill, 1993.
Life's Philosophy: Reason and Feeling in a Deeper World, trans. R. Huntford, Athens, GA: University of Georgia Press, 2008.

Further reading
Devall, B. and Sessions, G., *Deep Ecology: Living as if Nature Mattered*, Salt Lake City, UT: Peregrine Smith Books, 1985.
Drengson, A. and Devall, B. (eds), Introduction to *The Ecology of Wisdom: Writings by Arne Naess*, Berkeley, CA: Counterpoint, 2008.
Mathews, F., *The Ecological Self*, London: Routledge, 1991.

Witoszak, N. and Brennan, A. (eds), *Philosophical Dialogues: Arne Naess and the Progress of Ecophilosophy*, Savage, MD: Rowman & Littlefield, 1999.

DAVID E. COOPER

JOHN PASSMORE 1914–2004

> [T]he title of this book [*Man's Responsibility for Nature*] is often misquoted, as man's responsibility *to*, rather than *for*, nature. The difference is fundamental. 'Nature' is not a pseudo-person, to whom human beings are responsible ... Human beings are responsible [only] *for* nature.[1]

The Australian philosopher and historian of ideas, John Passmore, published the pioneering book referred to in the above passage in 1974. A decade later it could still be described as 'the one authoritative treatment of environmental ethics so far produced',[2] and four decades after its appearance its arguments and conclusions remain ones with which all serious environmental philosophers feel obliged to engage.

Passmore was born in 1914, in Manley, New South Wales. A graduate of Sydney University, he taught philosophy at his alma mater until 1949, when he went to teach in New Zealand. After returning to his own country in 1955, he became Professor of Philosophy at the Australian National University, where he remained until retiring in 1969. Ten years after being made a Companion of the Order of Australia – his country's highest civilian honour – John Passmore died in 2004.

Passmore wrote widely in many fields of philosophy, but it was with his magisterial history of the subject since the mid-nineteenth century, *A Hundred Years of Philosophy* (1957), that he first attracted a wide readership. (Thirty years later, a supplement appeared, *Recent Philosophers*, in which Passmore brought the story up to date, and expressed his well-known antipathy to fashionable trends in 'Continental' philosophy, such as deconstruction.) Passmore's skill as a historian was again displayed in *The Perfectibility of Man* (1970). It was to be *Man's Responsibility for Nature* (MRN) and some associated papers, however, which made his name known well beyond the confines of philosophy. A second edition of MRN appeared in 1980, now boosted by a useful new Preface and an Appendix based on a lecture entitled 'Attitudes to Nature', the best succinct introduction to Passmore's position. In later years, his interests did not focus on environmental questions, but he continued to write the occasional piece in this area, including an incisive article on the political aspects of environmentalism in an edited volume on contemporary political philosophy. One of his last books, *Serious Art* (1991), reflected a knowledge of and feeling for a dimension of human life which readers of MRN would already have discerned.

MRN was inspired by two convictions: first, that 'men cannot go on living as they have been living, as predators on the biosphere', but second, that irrationalist tendencies in the burgeoning environmentalist movement are threatening to make matters worse (xiii). The book has three fairly specific aims: to examine historically the religious and other ideas which have shaped current attitudes and behaviour towards the natural world; to argue for a number of solutions to our most pressing environmental problems; and to 'remove the rubbish' of fashionable, obfuscating ecological views which hinder solutions to these problems. These aims are connected by an 'over-arching intention: to consider whether the solution of ecological problems demands a moral or metaphysical revolution' (xiv). Passmore's conclusion is that we require neither. A balanced history of ideas will show that at least the 'seeds' of appropriate environmental action are to be found in the Western tradition. The best way to tackle the pressing problems is to call upon a tradition of scientific reason and upon moral convictions with a long pedigree. Finally, in 'removing the rubbish', one demonstrates the bankruptcy, dangers and sometimes hypocrisy in the calls – by 'deep ecologists', 'nature mystics', 'eco-feminists' and others – for a 'new' morality and metaphysics. What we need, writes Passmore, is 'not so much a "new ethic" as a more general adherence to a perfectly familiar ethic' (187).

A main purpose in the historical chapters is to counter the familiar accusation that the Judaeo-Christian legacy is responsible for our 'predatory' treatment of nature. Passmore concedes that there is 'a strong Western tradition that man is free to deal with nature as he pleases' (27). First, however, the roots of this idea are not Jewish, but Greek, for it was the Stoics who bequeathed to Christianity the teaching that the world was created for the sake of human beings. Second, that teaching cannot, by itself, inspire pernicious treatment of the environment, since it is more likely to encourage the 'quietist' belief that God's world is fine as it is, without our intervention. In order for 'anthropocentrism' to become pernicious, it required the much more recent idea – which emerges in Francis Bacon, and reaches its zenith in Marxist images of nature as wax in man's hands – that the proper life for human beings is one of active transformation of the world about them. Finally, although there indeed exists this 'predatory' tradition, there have also been countervailing ones, emphasizing people's prudent responsibilities towards nature and duties to 'perfect' the world in which they live (39).

It is those 'minority traditions' to which we should turn in addressing the most pressing modern problems – those of pollution, conservation of resources, preservation of relatively untouched areas and overpopulation. In each case, Passmore strives to instil a sense of realism and to strike a balance between extremes. It is a waste of time to propose solutions which for political reasons, say, are totally unworkable. He is especially critical of bland calls to reduce the human population which would require gross violation of the democratic process. In this instance, as in others, environmentalists too readily ignore the question 'How are we to get from here to there?'[3] Workable solutions, he argues, must steer between 'primitivism' and 'despotism' (39): between wholesale rejection of a concern for economic progress or material welfare and the unconstrained, short-sighted pursuit of such goals. Such solutions require the application of scientifically and technologically informed cost–benefit analysis of our present practices and the alternatives to them, together with judgement on the political viability and moral acceptability of these alternatives (71). In keeping with his 'over-arching intention', Passmore argues that there is no need to introduce 'new' moral considerations, such as the 'absurd' idea that nature has 'rights'. Instead, we may justify conserving resources for future generations as an extension of a natural, 'loving' concern for children and grandchildren, just as we can condemn the destruction of wildernesses as a 'vandalism' of the kind always censured by Western morality (125).

The 'rubbish' which Passmore wants removed from recent environmental debate is a mixed pile. To begin with, there is '"mystical rubbish", the view that mysticism can save us, where technology cannot', and the related view that 'nature is sacred' (173–5). Such views, Passmore argues, not only rest on an implausible metaphysics but, unless supplemented by other considerations, have no 'environmentally friendly' implications. The idea that nature is sacred, for example, can also encourage Emerson's confidence that, as 'part and parcel of God', nature cannot really be harmed *whatever* we do to it (176). Second, he is critical of any 'primitivist' rejection of modernity in favour of forms of human life which leave nature untouched. Aside from belonging to the realm of fantasy, such proposals often smack of hypocrisy, since the few who might 'return to nature' will be parasitic on the many who do not. 'The Jain priest can walk abroad only because there are other, less spiritual, men ... to sweep the paths for him' (126). Relatedly, Passmore is dismissive of those who regard man as a 'planetary disease' or 'obscene defiler' of 'flower-sweet Earth', purveyors of 'masochistic nonsense' who are blind both to the achievements of civilization and to the legitimate interests of human beings (181).[4]

In the Preface to the second edition of MRN, Passmore wryly observes that 'it is more than a little disconcerting to be cited both as one of the more virulent critics of economic growth and as an uncritical defender of the status quo' (vii). Certainly, his critics have come from opposite directions. However, although early on he was attacked by economists and planners who resented the intrusion of environmental considerations into the pursuit of economic growth, the bulk of the critics have been fellow environmental thinkers. The most common charge is one of excessive 'conservatism'. At its mildest, the complaint has been that Passmore's cost–benefit approach to the solution of environmental problems, such as pollution, allows for insufficiently radical revaluation of the policies whose costs and benefits are to be assessed.[5] Less mildly, it has been argued that Passmore scores a hollow victory in showing that traditional values suffice for moral appraisal of proposed solutions to our problems, since he wrongly refuses to recognize any moral problems except those which concern the interests of human beings.[6]

Passmore's most hostile critics, unsurprisingly, came from the ranks of those writers whom he has accused of purveying the 'rubbish' discussed above. Deep ecologists, eco-feminists and others convicted Passmore of a complacent and speciesist 'human chauvinism'. His way with such critics was, on the one hand, to charge them with misunderstanding or distorting his position. He points out, for example, that in denying that non-human life can enjoy 'rights', he is in no way denying that we can and do act in morally wrong ways towards animals and environments. Moreover, he might add, his hostile critics overlook the genuinely radical shifts in human attitudes for which he is calling. In some of the most interesting passages of MRN, Passmore argues that it is not just economic greed which has been responsible for our ecological problems. So, ironically, have a 'puritanism' and 'asceticism' which make it difficult for people simply to *enjoy* the world around them 'as itself an object of absorbing interest, not … a resource' (126). A more 'sensuous society' than our own, in which people are 'ready to enjoy the present moment for itself', would never have endured 'the desolate towns … the slag-heaps [and] the filthy rivers' which now surround them (188–9). Passmore's other response to his radical critics was simply and unapologetically to accept their labelling him a 'chauvinist', 'speciesist' and 'shallow'. If the pejorative point of such labels is to condemn anyone who treats human interests as paramount, then Passmore was content to stand condemned (187).

John Passmore's book *Man's Responsibility for Nature* remains the most authoritative statement of a main tendency in environmental ethics, constantly cited by both adherents and opponents of that tendency. Within philosophical circles, it may have been the 'deeper' ecological tendency represented by, for example, Arne Naess which attracted more attention for a number of years. But it is surely the 'shallower' approach of

Passmore which has done more to inform the environmental policies of governments and other organizations for which, ultimately, the interests of human beings must be of paramount concern. The influential efforts in recent years of Pavel Sukhdev and others to guide environmental policy by a sound economic assessment of 'ecosystem goods' or 'nature's services' is a continuation of the cost–benefit approach for which Passmore argued.[7]

Notes
1   *Man's Responsibility for Nature*, p. xii. All page references in the text are to the second edition of *Man's Responsibility for Nature*.
2   R. Attfield, *The Ethics of Environmental Concern*, p. ix.
3   Passmore, 'Environmentalism', in R. Goodin and P. Pettit (eds), *A Companion to Contemporary Political Philosophy*, Oxford: Blackwell, p. 479, 1993.
4   Passmore's targets here are Ian McHarg and W.S. Blunt ('the ecologist's poet-laureate' (p. 180)).
5   See, e.g., C.A. Hooker, 'On Deep versus Shallow Theories of Environmental Pollution', in R. Eliot and A. Gere (eds), *Environmental Philosophy*, Milton Keynes: Open University, pp. 58–84, 1983.
6   See R. Attfield, *The Ethics of Environmental Concern*, pp. 4ff, and Val Routley, 'Critical Notice of John Passmore's *Man's Responsibility for Nature*', pp. 171–85.
7   See P. Kumar (ed.), *The Economics of Ecosystems and Biodiversity*.

See also in this book
Bacon, Emerson, Marx, Naess, Plumwood, Sukhdev

Passmore's major writings
A Hundred Years of Philosophy, London: Duckworth, 1957; republished by Penguin, Harmondsworth, 1968.
The Perfectibility of Man, London: Duckworth, 1970.
Man's Responsibility for Nature: Ecological Problems and Western Traditions, London: Duckworth, 1974; second and enlarged edition, 1980.
Serious Art, London: Duckworth, 1991.

Further reading
Attfield, R., *The Ethics of Environmental Concern*, New York: Columbia University Press, 1983.
Hooker, C.A., 'Responsibility, Ethics and Nature', in D.E. Cooper and J.A. Palmer (eds), *The Environment in Question*, London: Routledge, pp. 147–64, 1992.

Kumar, P., *The Economics of Ecosystems and Biodiversity: Economic and Ecological Foundations*, Abingdon, UK: Routledge, 2012.

Routley, V. (later V. Plumwood), 'Critical Notice of John Passmore's *Man's Responsibility for Nature*', *Australasian Journal of Philosophy*, 53, 171–85, 1975.

DAVID E. COOPER

JUDITH WRIGHT 1915–2000

> South of my days' circle, part of my blood's country,
> rises that tableland, high delicate outline
> of bony slopes wincing under the winter.[1]

Judith Wright was Australia's greatest environmental poet as well as the thinker who most clearly linked ecological concerns to the dispossession of the Indigenous people. From her moving lyrics to her advocacy of other poets such as Oodgeroo Noonuccal to her environmental and antinuclear activism, she was pivotal in reorienting Australian attitudes towards the land as a venue for economic extraction to a setting for ecological and social justice.

Judith Wright came, for what were, in Australian terms, privileged affiliations. Her father, Phillip Wright, was from the "squattocracy"—those Australian landowners who had amassed large estates for grazing cattle and who ran their land in such a way as to amass wealth beyond self-sustenance.[2] Phillip Wright owned a large station (grazing farm) east of Armidale on the tablelands of northern New South Wales, and eventually became the second chancellor of the University of New England in Armidale. This was Australia's earliest and most prestigious country university, and held a good academic reputation throughout the twentieth century.

Judith Wright was born in 1915 and, aside from her undergraduate years, spent her first quarter-century of life in the region of her origins, even helping her father run the station once wartime conscription meant that most of the extant workers were called off to battle. By war's end, though, she was working at the University of Queensland near Brisbane as a statistical researcher. Brisbane is north and east of Armidale and very warm: the disjunction between her then-current state of life and her girlhood brought forth her first great poem, "South Of My Days", with its evocation of

> low trees, blue-leaved and olive, outcropping granite-
> clean, lean, hungry country. The creek's leaf-silenced,

willow choked, the slope a tangle of medlar and crabapple
branching over and under, blotched with a green lichen;
and the old cottage lurches in for shelter.[3]

The mixture of nostalgia and specificity here is galvanized by a willingness
to embrace as beautiful and even more, necessary, that landscape that
does not fall into conventional European aesthetic categories with
"granite-lean, lean, hungry" used positively. Vocabulary that is
unconventional to express admiration of a place in English language
poetry—"silenced", 'choked', "tangled", "lurched"—demonstrates an
adamant yearning to admire a landscape that is above all present and felt.

Wright was a member of a poetic generation that was above all male
and nationalist, and tended to write about iconic male figures, such as
explorers, war heroes, and even opera singers, as a way of attaining cultural
independence from Britain. Like these poets, Wright wrote of a
distinctively Australian landscape. But her imperatives were not
nationalistic, and she wrote necessarily from the perspective of a woman
and later a wife and a mother. Her iconic poem "Woman to Man" speaks
of the embryo in the mother's womb as "the selfless, shapeless seed I
hold", language that makes a direct link between the fertility of maternity
and that of the plant and animal worlds.[4] Importantly, though, she wrote
evocatively of male labour in "South of My Days", drovers who made the
strenuous trek "from Charleville to the Hunter", from remote country to
settled farmland, or the farmer in "Bullocky" whose years of accumulated
toil threatens to run "widdershins in his brain".[5] "Bora Ring" was notable
both in its registering of the European dispossession of the Indigenous
people and its acknowledgement that whites had little capacity to
appropriate or comprehend what remained. Soon after she wrote these
iconic poems, Wright was married, to the appreciably older maverick
philosopher Jack McKinney, and had a daughter, Meredith. In 1950, the
McKinney family settled in the Mount Tamborine area of southern
Queensland, in the mountains behind the Gold Coast, on a homestead
they called "Colanthe". Although "South of My Days" was written from
Queensland as a position of exile, in fact Wright's paternal grandparents
owned a station called Nulalbin in another area of southern Queensland.[6]

In the 1950s, Wright's poetry moved from being landscape poetry to
environmental poetry. In addition, while the 1940s poems had explored
certain traumas and pathologies, including violence, racism, and misogyny,
the 1950s poems took on more of a tone of political protest, such as "The
Two Fires", which gave its name to her 1955 volume of poems and
concerned the two cities in Japan that had suffered nuclear attacks,
Hiroshima and Nagasaki. The poem, though, also resonated with the

Australian landscape, "the ancient kingdom of the fire", where fire can be both nurturer and threat.[7] Wright's political awareness was all the more striking because, unlike so many Australian writers of her time, she was not avowedly leftwing or influenced by socialism and communism. Her family was an established force in a traditionally conservative area of Australia, and she so clearly prized her domestic role as a wife and mother. Much like her contemporary Patrick White, Judith Wright's political turn was given more potency by the fact that she was not conventionally political.[8]

An Australian ecological awareness is laced with a consciousness of the incongruity of traditional European conventions of perceptions of the natural world applied to a landscape with different and inverted seasons, and flora and fauna such as kangaroos, gum trees, and kookaburras that did not conventionally fit in with Northern Hemisphere nature descriptions. Judith Wright was aware of the fecundity and diversity of the Australian landscape from having spent her formative years variously in Armidale, Sydney, Brisbane, and the Queensland mountains. In her work on behalf of the Wildlife Preservation Society of Queensland, she was surprised to work so much on coral reefs when she had been expecting to agitate largely on behalf of bushlands.[9] Yet the ecological in Australia always shades into the political, not only in the vulnerability of the Australian landscape to fire and drought and its scarcity of some resources (water) and abundance of other (minerals), but in the reality of the Indigenous presence on the land and the resistance of Aboriginal peoples to white domination and expansion. For Wright, these factors were further inflected by the danger of nuclear war and, even though she did not classify herself as a feminist, a concerted critique of human rhetoric of domination of nature which in many ways proceeded from her identity as a woman.

Wright's critique was multi-pronged: historical (as seen in her novel *All Our Generations* about her own family's frontier past, as well as in her poems); prophetic, warning of ecological catastrophe; and practical, making concrete critiques of existing conditions and urging specific reforms. During the Cold War, it would have been perfectly expected for the anti-nuclear issue to leap to the foreground in Wright's work. After McKinney's death in 1966, though, the Indigenous issue became more and more prominent in Wright's work. Central to this development was Wright's friendship with the Indigenous poet Kath Walker, later to take the name under which she is known today, Oodgeroo of the tribe Noonuccal. Wright had met Oodgeroo in 1963, when Wright had been the referee for Oodgeroo's first poetry book, published by Jacaranda Press. The two women had very different backgrounds, not only in racial terms but in their class background, as Oodgeroo's genuinely working class background—she had been in domestic service—contrasted with

Wright's status as the daughter of privileged pastoralists. But they shared a fruitful and collegial collaboration based on a common stake in the landscape, especially of Queensland, and an appreciation of each other's talents and willingness to employ them in the pursuit of social justice. Wright, though, understood the asymmetry in their relationship, in that Oodgeroo had a right to speak of Australia as hers, while Wright did not.

Partially because of the growing political torque of her work, by 1970 Wright had moved to an estate called "Edge" in Braidwood near Australia's capital, Canberra. Wright was always remarkable in her ability to write empathetically about very different landscapes: even the few poems she composed about New Zealand, based on a brief visit there, are singularly evocative and responsive to a very different terrain. Wright's Canberra-area poems chronicle both her new landscape and her adjustment to it:

> But far off southward
> a stony ridge lay waiting
> for me to know it. I move
> Closer towards the pole.[10]

In the latter third of the twentieth century, Wright became a household name in two very different capacities. Her poems, dense yet appraisable, became set on Australian secondary-school reading lists, and she became ever more prominent as an environmental and anti-nuclear activist, and above all for a definitive treaty between white Australians and the Indigenous people.[11] Although these two constituencies were very different, both exemplified Wright's growing cultural prominence as scold, goad, and sage. Nor was Wright shy of making explicitly political pronouncements in her later poems, saying in "The Eucalypt and the National Character" that we "are still of two minds about militarism and class-systems".[12] The "of two minds" here is characteristic, as Wright realizes that the change in consciousness of which her vision is a harbinger is still incipient and aborning. Wright's prominence necessarily made her the object of criticism. The Right criticized her for campaigning against mining and manufacturing, which might benefit those without her privilege and education, and for the subjugation of aesthetics to politics.[13] Later generations of Indigenous activists and white allies did not assign a central role to a once-and-for-all treaty as Wright did, emphasizing instead more complex and situated acts of reconciliation.[14] Some considered her, by the end of her life, both strident and quaint. But Wright's principled adamancy, the determination which led her to attend a reconciliation in March 2000 in Canberra weeks before her death at 85, was admired by nearly all.[15] Not only did her writing and life evoke the beauty of an Australian landscape in peril, but her activism—in deed, in art, and perhaps most importantly in philosophy—was prophetic of an altogether different Australia, divested of white privilege, that the future might hail her for having helped engender.

Notes

1 Judith Wright, *A Human Pattern: Selected Poems* (Manchester, UK: Carcanet 1992), p. 11.
2 Brigid Rooney, *Literary Activists: Writer-Intellectuals and Australian Public Life* (St. Lucia, Australia: University of Queensland Press, 2009), p. 185
3 Wright, *A Human Pattern*, p. 11.
4 Wright, *A Human Pattern*, p. 20.
5 Wright, *A Human Pattern*, p. 9.
6 Judith Wright, *The Generations of Men* (Melbourne, Australia: Oxford University Press, 1959), pp. 41–73.
7 Wright, *A Human Pattern*, p. 70.
8 Rooney, *Literary Activists*, p. 185.
9 Stephen Dovers, *Australian Environmental History: Essays and Cases* (Melbourne, Australia: Oxford University Press, 1994), pp. 246–9.
10 Wright, *A Human Pattern*, p. 211.
11 Robert Zeljer, "The Double Tree: Judith Wright's Poetry and Environmental Activism". Isle: *Interdisciplinary Studies in Literature and Environment*, 7(2) (2000): 55–65.
12 Wright, *A Human Pattern*, p. 197.
13 Vincent Buckley, *Essays in Poetry, Mainly Australian* (Melbourne: University of Melbourne Press, 1957), pp. 174–5.
14 Stuart Cooke, *Speaking the Earth's Languages: A Theory of Australian-Chilean Postcolonial Poetics* (Amsterdam, Netherlands: Rodopi, I 13), p. 36.
15 1 Jane Gleeson-White, *Australian Classics: 50 Great Writers and Their Celebrated Work.* (Crows Nest, Australia: Allen and Unwin, 2010), p. 210.

Wright's major writings

The Moving Image, Melbourne, Australia: Meanjin, 1946.

Woman to Man, Sydney, Australia: Angus and Robertson, 1949.

The Two Fires, Sydney, Australia: Angus and Robertson, 1956.

The Double Tree: Selected Poems 1942–76, Boston, MA: Houghton Mifflin, 1978.

The Cry for the Dead, Melbourne, Australia: Oxford University Press, 1981.

We Call for a Treaty, Sydney, Australia: Collins Fontana, 1985.

Born of the Conquerors: Selected Essays. Canberra: Aboriginal Studies Press, 1991.

A Human Pattern: Selected Poems, Manchester, UK: Carcanet 1992.

With Love and Fury: Selected Letters, Canberra: National Library of Australia, 2000.

Further reading

Arnott, G. *The Unknown Judith Wright*, Nedlands, Australia: University of Western Australia Press, 2016.

Brady, V. *South of My Days: A Biography of Judith Wright*, Pymble, Australia: Angus and Robertson, 1998.

Capp, F. *My Blood's Country: A Journey Through The Landscape That Inspired Judith Wright*, Crows Nest, Australia : Allen and Unwin, 2011.

Das. D. "Masterless Men in a Masterful Land: Judith Wright", *Rupkatha Journal on Interdisciplinary Studies in Humanities*, April 2010, 2(2), 145–53.

Kane, P. *Romanticism and Negativity in Australian Poetry*, Cambridge, UK: Cambridge University Press, 1996.

Sheridan, S. *Nine Lives: Postwar Women Writers Making Their Mark*, St. Lucia, Qld: University of Queensland Press, 2011.

Strauss, J. *Judith Wright*, Melbourne, Australia: Oxford University Press, 1995.

NICHOLAS BIRNS

BARRY COMMONER 1917–2012

Environmental pollution is an incurable disease. It can only be prevented. And prevention can only take place at the point of production.[1]

Described in 1970 by *Time* magazine as the "Paul Revere of ecology," Commoner followed Rachel Carson as, arguably, America's most prominent modern environmentalist.[2] He viewed the environmental crisis as a symptom of a fundamentally flawed economic and social system. A biologist and research scientist, he argued that corporate greed, misguided government priorities, and the misuse of technology undermined "the finely sculptured fit between life and its surroundings."[3]

In his first book, *Science and Survival* (1966), Commoner warned that "the age of innocent faith in science and technology may be over." He concluded: "Science can reveal the depth of this crisis, but only social action can resolve it."[4] Commoner insisted that scientists had an obligation to make scientific information accessible to the general public, so that citizens could participate in public debates that involved scientific questions. Citizens, he said, have a right to know the health hazards of the consumer products and technologies used in everyday life. Those were radical ideas in the 1950s and 1960s, when most Americans were still mesmerized by the cult of scientific expertise and such new technologies as cars, plastics, chemical sprays, and atomic energy.

Commoner first came to public attention in the late 1950s when he warned about the hazards of fallout caused by the atmospheric testing of nuclear weapons. He later used his scientific platform to raise awareness about the dangers posed by the petrochemical industry, nuclear power, and toxic substances such as dioxins. He was one of the first scientists to point out that although environmental hazards hurt everyone, they disproportionately hurt the poor and racial minorities because of the location of dangerous chemicals and because of the hazardous conditions in blue-collar workplaces. Commoner thus laid the groundwork for what later become known as the environmental justice movement. He linked environmental issues to a broader vision of social and economic justice, and called attention to the parallels among the environmental, civil rights, labor, and peace movements. He connected the environmental crisis to the problems of poverty, injustice, racism, public health, national security, and war.

Born in 1917, Commoner grew up in Brooklyn, New York, the child of Russian Jewish immigrants. He studied zoology at Columbia University and received a doctorate in biology from Harvard University in 1941. After serving in the Navy during World War II, Commoner was an associate editor for *Science Illustrated* and then became a professor at Washington University in St. Louis, Missouri. He worked at Washington University for thirty-four years, serving as associate professor of plant physiology, then chair of the Botany Department, and finally as university professor of environmental science as well as founder (in 1966) of the Center for the Biology of Natural Systems to promote research on ecological systems. In 1981, he moved the center to Queens College in New York City, where he became professor of earth and environmental sciences.

While serving in the Navy, Commoner discovered the disturbing unintended consequences of technology. He was put in charge of a project to devise an apparatus to allow bombers to spray DDT on beachheads to kill insects that caused disease among soldiers. The military wanted to remove the insects before troops landed. Commoner's crew discovered that the DDT sprayed from bombers effectively eliminated hordes of flies on the beach, but also that more flies soon came to feast on the tons of fish that the DDT had also killed. This lesson became a central theme for Commoner throughout his career: humans cannot take action on one part of the ecosystem without triggering a reaction elsewhere.

After the war, many scientists, including Albert Einstein, alarmed by America's use of the atomic bomb on Japan in 1945, began to rethink their role in society. They questioned whether dropping the bomb had been necessary for the United States to win the war. They were shocked by the scale of the damage in terms of both immediate deaths and long-term human suffering. And they worried about the potential for a prolonged arms race between the United States and the Soviet Union, which, they feared, could end in a nuclear war in which all humanity would be the losers. As Commoner told *Scientific American* in a 1997 interview:

The Atomic Energy Commission [AEC] had at its command an army of highly skilled scientists. Although they knew how to design and build nuclear bombs, it somehow escaped their notice that rainfall washes suspended material out of the air, or that children drink milk and concentrate iodine in their growing thyroids. I believe that the main reason for the AEC's failure is less complex than a cover-up but equally devastating. The AEC scientists were so narrowly focused on arming the United States for nuclear war that they failed to perceive facts—even widely known ones—that were outside their limited field of vision.[5]

Commoner and other scientists—including chemist Linus Pauling (a professor at the California Institute of Technology and a Nobel Prize winner)—believed they had a responsibility to sound the alarm about the potentially devastating effects of nuclear fallout. In 1956, when Adlai Stevenson ran for president as the Democratic Party nominee, he sought Commoner's advice and then called for the United States to take the lead in ending nuclear testing.

In 1958 Commoner and other scientists and activists formed the Committee for Nuclear Information with the goal of educating the public to understand how, in Commoner's words, "splitting a few pounds of atoms could turn something as mild as milk into a devastating global poison."[6] They started a new publication, *Nuclear Information* (later renamed *Scientist and Citizen*), to discuss the responsibility of scientists to the larger society. They drafted a petition, signed by 11,021 scientists worldwide, urging that "an international agreement to stop the testing of nuclear bombs be made now." These activities created a groundswell of public opinion that eventually helped persuade President John F. Kennedy to propose the 1963 Nuclear Test Ban Treaty.

Commoner's early experience with DDT led him to espouse what scientists call the "precautionary principle"—that new chemicals and technologies should not be introduced into society if there is reason to believe that they pose a significant public health risk. They should be approved only after it can be demonstrated that they are safe. Commoner warned about the risks to human health posed by detergents, pesticides, herbicides, radioisotopes, and smog. He argued that polluting products (such as detergents and synthetic textiles) should be replaced with natural products (such as soap, cotton, and wool). He alerted the public to the negative effects of nuclear power plants, toxic chemicals, and pollution on the economy, birth defects, and diseases like asthma.

In the 1970s Commoner spoke out against the view that overpopulation, particularly in the Third World, was responsible for the increasing depletion of the word's natural resources and the deepening ecological problems. The thesis was popularized by Paul Ehrlich (in his book *The Population Bomb*) and other scientists, but Commoner challenged those who echoed the ideas of the nineteenth-century British thinker Thomas Robert Malthus. As Commoner argued, it is rich nations that consume a disproportionate share of the world's resources. And it was their systems of colonialism and imperialism that led to the exploitation of the Third World's natural resources for consumption in the wealthy nations, making the poor even poorer. Without the financial resources to improve their living conditions, people in developing countries relied more heavily upon increased birthrates as a form of social security than did people in wealthier nations.

As Commoner wrote, "The poor countries have high birthrates because they are extremely poor, and they are extremely poor because other countries are extremely rich."[7] His solution to the population problem was to increase the standard of living of the world's poor, which would result in a voluntary reduction of fertility, as has occurred in the rich countries.

In *The Closing Circle* (1971), Commoner argued that our economy—including corporations, government, and consumers—needs to be in sync with what he called the "four laws of ecology":

- Everything is connected to everything else.
- Everything must go somewhere.
- Nature knows best.
- There is no such thing as a free lunch.

The Closing Circle helped introduce the idea of sustainability, a notion that is now widely accepted but was controversial at the time. As Commoner pointed out, there is only one ecosphere for all living things. What affects one, affects all. He also noted that in nature there is no waste. We can't throw things away. Therefore, we need to design and manufacture products that do not upset the delicate balance between humans and nature. We need to utilize alternative forms of energy, such as wind, solar, and geothermal power. And we need to change our consumption habits accordingly—to use fewer products with plastics (which are based on oil), aerosol cans (which harm the atmosphere), and industrial-grown food (which is produced with harmful chemicals).

In his best-selling book *The Poverty of Power* (1976), Commoner introduced what he called the "Three Es"—the threat to environmental survival, the shortage of energy, and the problems (such as inequality and unemployment) of the economy—and explained their interconnectedness. Industries that use the most energy have the most negative impact on the environment. Our dependence on nonrenewable sources of energy inevitably leads to those resources becoming scarcer, raising the cost of energy and hurting the economy.

Commoner was neither a back-to-the-land utopian nor a Luddite opposed to modern industrial civilization. He did not place the burden of blame on the consumers who buy these products or the workers who produce them. He believed that big business and their political allies dominate society's decision making, often leading to misguided priorities, a theme that paralleled the ideas of economist John Kenneth Galbraith and, later, consumer activist Ralph Nader. Commoner believed that the corporate imperative for wasteful growth is the root cause of the environmental crisis and must be corralled by responsible public policies demand by a well-educated public. As he told *Scientific American*: "The environmental crisis arises from a fundamental fault: our systems of production—in industry, agriculture, energy and transportation— essential as they are, make people sick and die."[8]

Commoner's proposals for addressing these problems reflect his lifetime of promoting a progressive agenda. He told *Scientific American*:

What is needed now is a transformation of the major systems of production more profound than even the sweeping post-World War II changes in production technology. Restoring environmental quality means substituting solar sources of energy for fossil and nuclear fuels; substituting electric motors for the internal-combustion engine; substituting organic farming for chemical agriculture; expanding the use of durable, renewable and recyclable materials— metals, glass, wood, paper—in place of the petrochemical products that have massively displaced them.[9]

In the 1970s, Congress passed laws for clean air, pure water, safer and healthier workplaces, and the protection of the environment, including bans on DDT and on lead in gasoline. Environmentalists made further gains during the next three decades, including Superfund and right-to-know laws, a federal law forbidding ocean dumping of plastic materials, the declining use of nuclear power plants, a presidential executive order regarding environmental justice affecting minority and low-income populations, and growing public awareness—in the United States and around the world—of the dangers of global climate change and the exploitation of fossil fuels. In 1990, a Gallup poll found that 76 percent of Americans identified themselves as "environmentalists." A 1996 report by the U.S. Environmental Protection Agency found that sulfur dioxide emissions had declined by 40 percent between 1970 and 1990 and that smog, carbon monoxide, and ozone levels had declined since the passage of the Clean Air Act in 1970.

Commoner acknowledged the environmental movement's victories, which he saw as evidence that society can prevent environmental hazards by changing the way we produce and consume. But in a 2007 interview with *The New York Times*, he warned that these measures did not go far enough. "Environmental pollution is an incurable disease. It can only be prevented. And prevention can only take place at the point of production. If you insist on using DDT, the only thing you can do is stop. The rest has really been sort of forgotten about."[10]

Many Americans embraced Commoner's ideas about workplace hazards, nuclear power plants, and recycling. But he grew frustrated by the influence of corporate America over both major political parties and by the failure of the mainstream environmental movement to join forces with other progressive movements to heed his warnings and challenge the basic tenets of the free-market system. In 1979 Commoner helped form the Citizens Party, hoping it would gain influence similar to that of the Green Party in Europe. The next year Commoner ran as the party's presidential candidate. He got on the ballot in twenty-nine states but received less than one-third of one percent of the national vote. Like most third parties in the American system, the Citizens Party wound up being a minor fringe force. Commoner did not run again for office, but he advised Jesse Jackson's two unsuccessful campaigns to win the Democratic Party nomination in 1984 and 1988.

In the 2007 interview with *The New York Times*, then-90-year-old Commoner remained the relentless radical:

> I think that most of the "greening" that we see so much of now has failed to look back on arguments such as my own—that action has to be taken on what's produced and how it's produced. That's unfortunate, but I'm an eternal optimist, and I think eventually people will come around.[11]

Notes
1 Thomas Vinciguerra, "At 90, an Environmentalist From the '70s Still Has Hope," *The New York Times*, June 19, 2007, www.nytimes.com/2007/06/19/science/earth/19conv.html
2 "Environment: Paul Revere of Ecology," *Time*, February 2, 1970.

3 Barry Commoner, *The Closing Circle: Nature, Man, and Technology*, New York: Knopf, 1971, p. 11.
4 Barry Commoner, *Science and Survival*, New York: Viking, 1966, p. 132.
5 Alan Hall, "Interview with Barry Commoner," *Scientific American*, June 23, 1997, www.yumpu.com/en/document/view/31851626/scientific-american-interview-with-barry-commoner
6 Hall.
7 Barry Commoner, "How Poverty Breeds Overpopulation and Not the Other Way Around," *Ramparts*, August–September 1975.
8 Hall.
9 Hall.
10 Vinciguerra.
11 Vinciguerra.

See also in this book
Bookchin, Carson, Ehrlich

Commoner's major writings
Science and Survival, New York: Viking, 1966.
The Closing Circle: Nature, Man, and Technology, New York: Knopf, 1971.
The Poverty of Power: Energy and the Economic Crisis, New York: Random House, 1976.
The Politics of Energy, New York: Knopf, 1979.
Making Peace With the Planet, New York: Pantheon, 1990.

Further reading
Egan, Michael. *Barry Commoner and the Science of Survival: The Remaking of American Environmentalism*. Cambridge, MA: MIT Press, 2007.
Gottlieb, Robert. *Forcing the Spring: The Transformation of the American Environmental Movement* (rev. edn), Washington, DC: Island Press, 2003.
Hall, Alan. "Interview with Barry Commoner," *Scientific American*, June 23, 1997. www.yumpu.com/en/document/view/31851626/scientific-american-interview-with-barry-commoner
Vinciguerra, Thomas. "At 90, an Environmentalist From the '70s Still Has Hope," *The New York Times*, June 19, 2007. www.nytimes.com/2007/06/19/science/earth/19conv.html

PETER DREIER

JAMES LOVELOCK 1919–

> The idea that the Earth is alive is at the outer bounds of scientific credibility. I started to think and then to write about it in my early fifties. I was just old enough to be radical without the taint of senile deliquency.[1]

It is an idea that has absorbed James Lovelock for more than thirty years, the idea that is encapsulated in the name 'Gaia'. The name itself was suggested by the novelist William Golding, a friend and at one time a neighbour, in the course of one of the long walks the two men used to take together in the Wiltshire countryside. In Greek mythology Gaia, or Ge, was the Earth. She sprang from Chaos and gave birth to Uranos, the Heavens and Pontus, the Sea. She was not a goddess. She preceded the gods and goddesses and provided the context, the environment if you will, in which the gods could exist. Her name lives on in those words in our language that begin with 'ge-' – ge-ography, ge-ology, ge-odesy, ge-ometry, and all the rest. The image is powerful and Lovelock's Gaia hypothesis is conceived on an appropriately grand scale.

James Ephraim Lovelock was born on 26 July 1919 in Letchworth Garden City, Hertfordshire. His father was a keen gardener with a highly developed moral awareness that owed little to formal religious belief, but appears to have been based on a mixture of folk Christianity and traditions and superstitions that used to be widespread in rural Britain. He communicated his love of the countryside to his son and, with it, an enthusiasm for walking. Later, James became a keen hill walker, and in 1999 he and his wife celebrated his eightieth birthday by walking along the Cornwall Coast Path. This runs along the tops of high sea cliffs, then plunges down narrow, steep-sided valleys only to climb again on the far side. It is strenuous walking for anyone.

When he left school, James Lovelock worked in a laboratory, studying chemistry in the evenings but by day learning the practical laboratory techniques that were to serve him well in later years. Eventually he left to become a full-time student at Manchester University, graduating in chemistry in 1941.

It was wartime and the young graduate was absorbed into the national effort, going to work for the Medical Research Council at the National Institute for Medical Research, in London. At the end of the war, in 1946, he went to work at the Common Cold Research Unit, in Wiltshire. There, he and his colleagues found that the search for a cure for colds was fruitless, but they were able to design ways to prevent their transmission. He remained at the Unit until 1951.

He received his PhD in 1948, from the London School of Hygiene and Tropical Medicine. This degree was in medicine. He received the degree of DSc in biophysics in 1959, from the University of London.

His path lay ahead of him, clearly defined. He could have remained a scientist employed by the Civil Service. Year by year his salary would have increased, his standard of living would have been fairly high, and eventually he would have retired with an index-linked pension that would have kept him in reasonable comfort. It was not enough. James wanted more. He once told me that his enthusiasm for science arose from the opportunities it provides for finding answers to questions. It can deliver intellectual freedom, however, only to those who are able to frame the questions for themselves. As a Government scientist, his researches would necessarily have been directed towards the resolution of matters of public interest. His imagination would have been constrained.

Still an employee of the National Institute for Medical Research, in 1954 he was awarded a Rockefeller Travelling Fellowship in Medicine. He spent it at Harvard University Medical School, in Boston, and in 1958 spent a year working at Yale as a visiting scientist. He resigned from the National Institute in 1961 in order to take up an appointment as Professor of Chemistry at Baylor University College of Medicine, in Houston, Texas.

His particular skill had always been intensely practical. He had a talent for constructing instruments that would measure with a fine sensitivity whatever anyone wished to measure. A time came when it seemed that this skill might provide him with a means of earning a living while leaving him sufficient free time to pursue his own interests – to find answers to his own questions. So, in 1964, he became a freelance research scientist. He holds more than fifty patents, most of them for instruments used in chemical analysis.

Many of these instruments have developed and refined the technique of gas chromatography. In this, the substance to be analysed is vaporized, and the vapour is mixed with a gas and then introduced into a stationary column filled with a finely powdered solid or liquid. Different components of the specimen react with the column at different rates. This separates them in a way that allows them to be identified – originally by their colour, hence the name of the technique.

In 1957 Lovelock invented the electron capture detector. This is still one of the most sensitive of all detectors. It revealed the presence of residues of organochlorine insecticides such as DDT throughout the natural environment, a discovery that contributed to the emergence of the popular environmental movement in the late 1960s. Later it registered the presence of minute concentrations of CFCs (chlorofluorocarbon compounds) in the atmosphere.

In the early 1960s, while living in Texas, James was a consultant at the Jet Propulsion Laboratory (JPL) of the California Institute of Technology, in Pasadena. At the invitation of NASA he had already helped with some of the instruments used to analyse lunar soil and he was then asked to advise on various aspects of instrument design for the team of scientists planning the two Viking expeditions to Mars. Instruments can help in finding answers to questions, but before a new instrument can be devised the question must be framed clearly. Asking the right question is often more difficult than finding the answer to it.

James was not directly involved with the question that has always been central to all Martian exploration: Is there, or has there ever been, life on Mars? Nevertheless, as he contemplated the ways in which the team proposed to seek answers, he found himself driven to ask more fundamental questions. The Viking experiments were based on the assumption that Martian biology would resemble that of the only living organisms of which we have any knowledge at all – the ones on Earth. But to James Lovelock this seemed to be a huge assumption with nothing to justify it. 'How can we be sure that the Martian way of life ... will reveal itself to tests based on Earth's life style?' ... 'What is life, and how should it be recognized?'[2]

When his colleagues at JPL asked how he would set about finding answers, the only thought he could offer was that living organisms must, in one way or another, increase the amount of order in the world around them. It was not much help, but the idea seeded itself in his brain.

No matter what its composition or biochemical pathways might be, any living organism must take certain chemical substances from its surroundings and use them to build and repair its own tissues. This will generate waste products that the organism will dispose of into its surroundings. Eventually this metabolic process will alter the composition of its surroundings, increasing the abundance of certain substances and depleting it of others. In this way, the organism will modify its own environment, giving it a chemical composition markedly different from the one it would have if it were allowed to reach a state of chemical equilibrium.

That difference, James maintained, should be detectable, and he and Dian Hitchcock, a philosopher employed to assess the logical consistency of NASA experiments, decided the place to seek it was in the atmosphere. The atmosphere has a much smaller mass than the solid or liquid components of a planet, and so perturbations to its composition would be more easily detectable. Also, the atmosphere is more easily accessible to investigators on another planet or in space.

When the atmosphere of Earth is compared with those of Mars and Venus the chemical disequilibrium of our atmosphere becomes immediately evident. It contains both methane and oxygen, for example. These react naturally to yield carbon dioxide and water, so some process must constantly replenish the methane, and at the pressure and temperature prevailing on Earth, it is only biological reactions that are capable of releasing methane. Were some process not releasing gaseous nitrogen, Earth's atmosphere would have lost any it had billions of years ago, as it was oxidized by lightning to stable oxides that are soluble in water and were washed to the surface by rain. In fact, nitrogen is released by denitrifying bacteria. Both other planets, in contrast, have atmospheres that are in chemical equilibrium.

The atmospheres of Mars and Venus consist mainly of carbon dioxide. Our atmosphere contains very little of this gas (about 0.04 per cent by volume). This has not always been the case. At one time our atmosphere contained much more carbon dioxide. It reacted to produce the carbonates that now form the chalks and limestones that are among the commonest of sedimentary rocks. James calculated that carbon has been removed from the atmosphere by this means at a rate far faster than could have been achieved by simple, inorganic chemical reactions. Living organisms played a major part, principally by building seashells from calcium carbonate, and carbonate rocks are predominantly of biological origin.

Since the time when the earliest organisms are believed to have appeared on Earth, the Sun has increased its output of energy by about 30 per cent. Carbon dioxide is a so-called 'greenhouse gas' and James concluded that a consequence of its progressive removal from the atmosphere was that the surface temperature remained fairly constant. The gas was removed by organisms and its removal maintained the climate most favourable to them. In other words, living organisms were regulating the global climate.

With this realization the basis was established of what was to grow into the Gaia hypothesis. It grew as James became increasingly persuaded of the extent to which biological regulation pervades the environment. In 1979, in his first book on the subject, he defined Gaia as 'a complex entity involving the Earth's biosphere, atmosphere, oceans, and soil; the totality constituting a feedback or cybernetic system which seeks an optimal physical and chemical environment for life on this planet'.[3] From this the concept developed of the Earth itself as a single, discrete, living organism, equipped with biological mechanisms for maintaining its overall homoeostasis.

The search then intensified for evidence to test the thesis. In 1987 it was established that cloud formation over the oceans is initiated by particles released by single-celled marine algae.[4] The suggestion of long-term climate regulation was confirmed in 1989.[5] Other predictions arising from Gaian theory have also been confirmed, and the status of the idea has advanced from hypothesis to theory.

Its reception has been mixed. Environmentalists, especially those of a more mystical bent, have embraced it enthusiastically, cheerfully overlooking some of its implications. Gaia, if it (she?) exists, has no great concern for the fate of organisms more complex than microorganisms. She would remain unmoved by the extinction of elephants, whales, tigers or humans. This is in keeping with her mythical origin, of course. The Greek Gaia was destroyer as well as creator: she buried people.

At this point the name became a handicap. It had never been meant as more than a metaphor and a preferable alternative to some ungainly acronym, but it was proving too evocative and what James intended as a rigorous scientific proposal began to look like sentimentalism. As one critic expressed it:

> The conflict between accepting what science teaches us and what the human heart would like to believe is well illustrated by James Lovelock's Gaia concept. It is a lovely thought, a tempting one too, because it is a form of religion and the human soul requires the comfort of a guided universe; it needs religion. Alas, it is also unnecessary, because the world as it was, has evolved, and now exists, is not explicable. It is merely very complex, and life plays a role in it, but not the main one.[6]

Some scientists warmed to the idea, however. In its less-extreme form it is hardly novel. The influence of living organisms on the cycling of minerals has been known for many years. Indeed, the cycles are described as biogeochemical cycles. What Gaia added was an over-arching, unifying concept leading to a new way to approach problems relating to the functioning of the planet. New questions could be asked about possible perturbations along the lines of 'How would the totality of living organisms respond?' Where environmental difficulties could be analysed they could be remedied with the help of living organisms. This is now a well-established technique, known as bio-remediation. It was used, for example, to clean up Alaskan beaches following the oil spill from the *Exxon Valdez*.

Such an approach can be described, not too fancifully, in something approximating to medical terminology. Environmental problems can be seen as 'ailments', the nature of which can be 'diagnosed' and to which 'therapies' can be applied. James's first doctorate was in medicine and one of his heroes is James Hutton (1726–97), one of the founders of modern geology, whose training was also in medicine. Hutton told a meeting of the Royal Society of Edinburgh in 1785 that the Earth was a super-organism and that its proper study should be physiology. The 'Gaian' study of the Earth is now called 'geophysiology'.

Extend the concept beyond this, however, and it remains controversial. Most evolutionary biologists reject it. Evolution occurs at the very local level of individuals. Genes spread and mutations are fixed in populations as individuals inheriting genes that confer a reproductive advantage produce more offspring than individuals who do not. It is difficult to see how this Darwinian process can link to planetary regulation. The even stronger idea, that the Earth itself is a single organism, finds little support from biologists.

Nevertheless, Gaia remains one of the most interesting and influential ideas of modern times, and its author has been rewarded for it, as well as for his other contributions to science. In 1974 he was elected a Fellow of the Royal Society and he has received many prizes for his contributions to chromatography, climatology and environmental sciences. In 1997 he was awarded the prestigious Blue Planet Prize for helping in the resolution of global environmental problems. He has received honorary doctorates in science from eight universities. In 1990 he was awarded a CBE, and in 2003 Her Majesty the Queen made him a Companion of Honour.

Despite his deep commitment to environmental protection, however, James has long been critical of environmentalist doctrine, which he has likened to a religion. He strongly opposes wind farms, for example, and equally strongly supports fracking shale for its natural gas.

James Lovelock and his wife Sandy now live in a cottage on the Dorset coast, where he continues to write.

Notes
1 *The Ages of Gaia*, p. 3.
2 *Gaia: A New Look at Life on Earth*, p. 2.
3 Ibid., p. 11.
4 R.J. Charlson, J.E. Lovelock, M.O. Andreae and S.G. Warren, 'Oceanic Phytoplankton, Atmospheric Sulphur, Cloud Albedo, and Climate', *Nature*, 274, 246–8, 1987.
5 David W. Schwartzman and Tyler Volk, 'Biotic Enhancement of Weathering and the Habitability of Earth', *Nature*, 340, 457–60, 1989.
6 Tjeerd H. Van Andel, *New Views on an Old Planet: A History of Global Change*, Cambridge, UK: Cambridge University Press, 1994, p. 402.

See also in this book
Darwin, Ehrlich, Midgley, Schumacher, Spinoza, Wilson

Lovelock's major writings
Gaia: A New Look at Life on Earth, Oxford, UK: Oxford University Press, 1979.
The Ages of Gaia, Oxford, UK: Oxford University Press, 1989.
Gaia: The Practical Science of Planetary Medicine, London: Gaia Books, 1991.
The Revenge of Gaia, London: Penguin, 2007.
The Vanishing Face of Gaia, London: Penguin, 2010.
A Rough Guide to the Future, London: Allen Lane, 2014.
Homage to Gaia, London: Souvenir Press, 2014.

Further reading
Allaby, Michael, *Guide to Gaia*, London: Optima, 1989.
Allaby, Michael and Lovelock, James, *The Great Extinction*, London: Secker & Warburg, 1983.
Allaby, Michael and Lovelock, James, *The Greening of Mars*, London: Andre Deutsch, 1984.
Joseph, Lawrence E., *Gaia: The Growth of an Idea*, New York: St Martin's Press, 1990.
Volk, Tyler, *Gaia's Body: Toward a Physiology of Earth*, New York: Copernicus, 1998.
Westbroek, Peter, *Life as a Geological Force: Dynamics of the Earth*, New York: W.W. Norton & Co., 1992.

MICHAEL ALLABY

MARY MIDGLEY 1919–

Naturalism ... [is] a word that I deeply distrust. It ought to mean a profound belief in *nature* as the green-and-brown living world that actually surrounds us, and in the forces behind that world. But this word is more often used to mean something almost opposite to this – a stark, reductive, materialistic nothing-buttery, a conviction that nothing is real except the ultimate particles of chemicals. ... My main interest has always been to bring forward evidence for the solid contribution of *nature* itself to the things we value in the world.[1]

A common theme in environmental philosophy is to emphasise the ways in which life on our planet is more complex, multifaceted, interconnected and valuable than we might previously have suspected. Beyond these superficial agreements, there is much that is debated. Should we regard non-human species as similar to us, or should we emphasise their differences? Is value a feature of the world or an anthropogenic feature of our species? Should we emphasise the global or the local? Often, our disagreements can spring from the fact that we are asking the wrong questions: we are relying, however unwittingly, on sets of simplistic oppositions or false dichotomies which hold us back in our attempts to think more clearly and our endeavours to bring about a better world. To get to grips with these problems, we must dig deep into the conceptual and historical roots of our thought. It is through this task that Mary Midgley finds her place in environmental thought.

Midgley was born Mary Scrutton in 1919. Although she was the daughter of a curate (later a college chaplain at Cambridge) she recalls being more wedded to the sprawling rectory garden – with its constant amusements and hiding places – than she was to the church. As she grew up, she developed parallel passions for the biological sciences and for the imaginative visions conveyed through art and literature. The idea of a supposed dichotomy between rational 'scientific' thought and the passion associated with artistic endeavour became a running theme in her later work.

As a student at Oxford, she befriended Elizabeth Anscombe, Philippa Foot, Iris Murdoch and Mary Warnock. This group of women have, between them, reshaped the landscape of contemporary moral philosophy. Each of these Oxford contemporaries has influenced Midgley's work. She shares Anscombe's suspicion of absolutist generalisations, Foot's conviction that human goodness should be understood in terms of our natural features as living animals, Iris Murdoch's emphasis on vision and attention, and Warnock's determination that the techniques of philosophy should be applied to the practical problems of social and political life.[2]

Midgley had a brief teaching career at Reading and Newcastle Universities, but it was only after taking early retirement that her writing took off. Her career has taken a very different shape to the typical route followed by many philosophers and, as such, she spent much of the first part of it reading and absorbing a broad range of work in philosophy and other disciplines before she published any of her major works. In addition to publishing eighteen books, Midgley has published hundreds of articles, both in academic journals and in the popular press, and has become one of Britain's leading public intellectuals. In 1995, Durham University awarded her an honorary DLitt, and she was made an honorary Doctor of Civil Law at Newcastle University in 2008. She is an honorary fellow of the Policy, Ethics and Life Sciences Research Centre at Newcastle

University, an honorary fellow at Somerville College Oxford, and was awarded the Edinburgh Medal by the City of Edinburgh in 2015.

Her influence has arguably been greater outside philosophy than within it, in her participation in public debate, as well as the dialogues that she has had with practitioners in the natural sciences. The areas in which she might be said to have most influence in relation to environmental themes are the philosophy of science, the study of animals, and the philosophy of politics and society. Midgley's philosophy embraces the lessons that we can learn from scientists, often making use of insights from psychology, anthropology, physics and evolutionary biology. However, she also emphasises the limits of scientific investigation. A major aspect of her approach is to determine the appropriate applications of scientific methodology, and to point out when a scientific approach is overstepping the mark. We can describe this as a warning against 'scientism'.

In a piece on Midgley, Ian Kidd characterises scientism as reflecting three central and interconnected 'urges' or 'impulses': the imperialistic urge, the salvific urge and the absolutist urge. The imperialist urge is 'a compulsion to extend the concepts, methods, and practices of scientific enquiry into areas in which their effectiveness is limited at best, nil at worst'.[3] The salvific urge is the temptation to treat science as providing the solution to all problems and predicaments. The absolutist urge is the impulse to provide complete 'totalising' visions of the world. Midgley's criticism of scientism is not a rejection of science, but a call for its practitioners (and, perhaps more often, its cheerleaders) to think more carefully about what science is for, and what it can and cannot achieve.

Midgley argues that it is tempting for scientists to build 'grand theories' that purportedly apply to all areas of life. Thus, the methodology that is applied in one area is taken to apply to all areas, and alternative visions are regarded as competitors. Midgley calls for perspective: the methodology that we apply in one field has developed in the way that it has because it is well suited to that particular purpose. Getting a better view of things will involve using different tools for different jobs.

Midgley's critique of scientism is closely connected with her thoughts on the role of imagination, creativity, vision and myth. Any area of life is, Midgley maintains, shot through with particular kinds of imagery and stories. For example, there is the mechanistic enlightenment picture of nature as a great lifeless machine, or the image of evolution as the outcome of a ruthless battle for individual survival. It is not a problem that we have imagery and myth-making in science – Midgley believes that science could not progress without imaginative visions – the problem comes when the pictures that we have chosen do not serve us (or our planet) well.

It is no use merely ordering the imagination to keep out of science. Science cannot do its work without it. What is wrong with the distortions is not that they connect science with some influential picture of the world, but that they have chosen a bad picture. They have to be dealt with, not by cutting science off from all such pictures, but by thinking harder about the rest of life.[4]

Midgley argues that 'bad pictures' have been responsible for enormous damage to the earth. There are several elements to this. For example, there is the idea of nature as inert matter, and the scientist as a separate and disembodied subject, unconnected to the material that she researches. There is the image of organisms as separate and competing self-interested individuals, with few significant connections. There is the picture of the earth's 'resources' as a random collection of matter, lacking a rational principle to connect it.

This raises the prospect of a second kind of imperialistic urge: not just the temptation for science to eclipse other ways of thinking, but the idea that the world itself is a collection of valueless stuff which can only be given meaning or significance by human activity. The implication is that there can be no great wrong or harm done by manipulating the fabric of the planet to serve human ends. It was her rejection of this view that drew Midgley to Gaia theory, James Lovelock's idea of the earth as a living and self-sustaining organism:

Such a lifeless jumble of would be no more capable of being injured than an avalanche would. Indeed, until quite lately our sages have repeatedly urged us to carry on a 'war against nature'. We did not expect the earth to be vulnerable, capable of health or sickness, wholeness or injury. But it turns out that we were wrong; the earth is now unmistakably sick. The living processes (or, as we say, 'mechanisms') that have so far kept the system working are disturbed, as is shown, for instance, by the surge of extinctions.[5]

Lovelock's idea has attracted a lot of interest and support, but also a great deal of controversy. Some are put off by the name. Invoking the image of a mother-Goddess strikes many as suspiciously unscientific and new age. Midgley sees this as a response to the seventeenth-century conviction that the earth is inert and lifeless. The problem is not that the concept of Gaia draws on myth and imagination, but that the myth and imagination that runs through other scientific pictures often goes unacknowledged. Gaia stands out because it is providing an alternative vision which emphasises the interconnection of all organic life and other physical processes on the planet, and challenges the idea that matter is inert.

Similar themes were present in Midgley's thought a long time before she became interested in Gaia theory, for example in her work on non-human animals. Animals were one of the subjects of her first book, *Beast and Man*, published in 1978. The other subject was, of course, ourselves. Many books on animal ethics consider animals as the objects of some kind of detached observation. The author first considers what a given moral theory takes to be the main criterion of moral regard: rationality, perhaps, or sentience. Following on from this, she will assess the evidence for the view that these concepts can be applied to non-human animals. If animals (or some animals) pass the moral status test, then they are allowed into the moral community.

What this method omits is any consideration of our relationships with animals, and how these have a role in our moral concepts. Midgley discusses how we use terms like 'beastliness' to condemn human bad behaviour. Our moral concepts developed, in part, by drawing a division between what is human and what is animalistic. It should not surprise us that this language does not help us very much when considering how we ought to relate to the non-human.

Midgley observes that over the course of human history, animals have always played a significant role in our lives and communities, but our attitudes to them have always been ambiguous, treating them at one moment as personal experiencing others, with whom we can have relationships, and at another as flesh or machines. To make ourselves feel better, we can assert the aspects of ourselves that separate us from other species of animals. Through doing this, we can deny our own physicality, appetites and vulnerabilities. In this way, an artificial 'species barrier' is erected, and it is maintained by the images that we have of ourselves as rational superior minds.

Midgley does not seek to deny that there are features of human beings that are unique, or that we differ in significant respects from other species of animal. Unfashionably, she does not wish to deny human nature. Rather, she paints a more nuanced picture, where our similarities to other animals, and our interactions with them, are brought further to the fore. As with her work on Gaia, Midgley provides us with a way of seeing the world in which living organisms are essentially connected, but she does so without denying or ignoring important distinctions. She emphasises the importance of disciplines like ethology and comparative psychology in paying close attention to the particularities of animals and our relationships with them.

At heart, Midgley is motivated by the conviction that philosophy is not a remote esoteric 'armchair' activity, but that it should be a public and practical concern, with its feet firmly planted in the soil. Philosophical problems arise, she maintains, because there is a social need for them when it turns out that our systems of thought are having negative implications for our lives on this planet. One of her most striking statements of this position appears in her article 'Philosophical Plumbing':

> Plumbing and Philosophy are both activities that arise because elaborate cultures like ours have, beneath their surface, a fairly complex system which is usually unnoticed, but which sometimes goes wrong. In both cases, this can have serious consequences. Each system supplies vital needs to those who live above it. Each is hard to repair when it does go wrong, because neither of them was ever consciously planned as a whole.[6]

One of Midgley's great insights is that our thought, like our planet and its ecosystems, is more messy, complex and interesting than the great systematisers generally acknowledge. To navigate both effectively requires not only the tools of rational and empirical investigation, but also compassion, attention and imagination.

Notes
1 M. Midgley, 'Afterword' in I.J. Kidd and L. McKinnell (eds) *Science and the Self: Animals, Evolution, and Ethics: Essays in Honour of Mary Midgley*, New York: Routledge, 2016, p. 230.
2 The work of this group of philosophers, and the themes that connect them, is a central focus of Durham University's 'In Parenthesis' project, which archives the work of women in philosophy and highlights their contributions. The project's website contains interviews, useful links and information, and news about the archive: https://womeninparenthesis.wordpress.com/.
3 I.J. Kidd, 'Doing Science an Injustice: Midgley on Scientism' in ibid. p. 152.
4 M. Midgley, *Wisdom, Information, and Wonder: What is Knowledge For?* London: Routledge, 1989, p. 85.
5 M. Midgley, *Gaia: The Next Big Idea*, London: Demos, 2001, p. 20.
6 M. Midgley, *Utopias, Dolphins, and Computers: Problems of Philosophical Plumbing*, London: Routledge, 1996, p. 1.

See also in this book
Darwin, Lovelock

Midgley's major writings

Beast and Man: The Roots of Human Nature, London: Routledge, 1978.

Animals and Why They Matter: A Journey Around the Species Barrier, Harmondsworth, UK: Penguin, 1983.

The Ethical Primate: Humans, Freedom, and Morality, London: Routledge, 1994.

Science and Poetry, London: Routledge, 2003.

The Owl of Minerva: A Memoir, London: Routledge, 2005.

The Essential Mary Midgley, David Midgley (ed.) London: Routledge, 2005.

Further reading

I.J. Kidd and L. McKinnell (eds) *Science and the Self: Animals, Evolution, and Ethics: Essays in Honour of Mary Midgley*, New York: Routledge, 2016.

ELIZABETH McKINNELL

MURRAY BOOKCHIN 1921–2006

Social ecology advances a message that calls not only for a society free of hierarchy and hierarchical sensibilities, but for an ethics that places humanity in the natural world as an agent for rendering evolution – social and natural – fully self-conscious and as free as possible ... We stand at a cross-roads of conflicting pathways: either we will surrender to a mindless irrationalism that mystifies social evolution ... or we will regain the activism, that is denigrated today, and turn the world into an ever-broader domain of freedom and rationality. This entails a new form of rationality, a new technology, a new science, a new sensibility and self – and above all, a truly libertarian society.[1]

Murray Bookchin is one of the most well-known and influential activist-theorists of radical green politics of the twentieth century. He was born in New York City on 14 January 1921, to Russian immigrant parents. In the 1930s he entered the communist youth movement, but by the late 1930s had become disillusioned with its Stalinist, authoritarian character. He was involved in organizing activities around the Spanish Civil War, and the fight against European fascism, and remained with the communists until the Stalin–Hitler pact of September 1939, when he was expelled for 'Trotskyist-anarchist deviations'. He was active in radical politics (both left-wing, anarchist, and ecological) from the 1930s, wrote extensively on revolutionary ecological politics, and was a driving force in the ecological movement in America for over thirty years.

After returning from service in the U.S. Army during the 1940s, he was an autoworker and became deeply involved in the United Auto Workers (UAW). In time, he became a left-libertarian anarchist and in the 1960s he was deeply involved in counter-cultural and New Left movements almost from their inception, and he pioneered the ideas of social ecology in the USA. His first American book, *Our Synthetic Environment* (written under the pseudonym Lewis Herber), was published in 1962, preceding Rachel Carson's *Silent Spring* by nearly half a year, while his first published piece of work was in 1952 on the socio-economic origins of environmental pollution and chemicals in food. In the 1960s he had already identified climate change as a major issue that the political left needed to address.

In the late 1960s Bookchin taught at the Alternative University in New York, one of the largest 'free universities' in the USA, then at City University of New York in Staten Island. Such was his intellect and body of published work, that he became a full professor without having an undergraduate degree. In 1974, he co-founded and directed the Institute for Social Ecology in Plainfield, Vermont, which went on to acquire an international reputation for its advanced courses in eco-philosophy, social theory and alternative technologies, all subjects which reflect his ideas. In 1974, he also began teaching at Ramapo College of New Jersey, becoming Full Professor of Social Theory, and retiring in 1983 in an emeritus status. He lived in semiretirement in Burlington, Vermont until his death in 2006.

Bookchin's main contribution to green politics has been the development of 'social ecology', a radical and revolutionary form of green political theory and action which he has developed and espoused since the 1960s. His earlier thinking laid the basis for this later development, particularly his focus on critical social theory (following Marcuse to some extent), the liberatory potential of eco-technology (in the tradition of Lewis Mumford), a focus on the material and political potential of urban living, and the creation of a 'post-scarcity society'. For Bookchin, echoing Ivan Illich, post-scarcity does not mean the Marxist 'abundance of material affluence' but rather 'a sufficiency of technical development that leaves individuals free to select their needs autonomously and to obtain the means to satisfy them'.[2] The extent to which he renounced his earlier commitment to Marxism can be seen in his well-known, acerbic, blunt and refreshingly irreverent essay 'Listen Marxist!'. A key feature of Bookchin (and which made him stand apart from strands of mainstream green politics was his urbanism and belief in the emancipatory power of technology and a disavowal of presenting sustainability as 'sacrifice'. As White puts it, for Bookchin:

> We need forms of ecological urbanism and eco-technological restructuring that aim for more than technocratic low-carbon

outcomes ... we should aspire to socio-technical forms which as far as possible restore a sense of 'selfhood and competency' to an 'active citizenry.'[3]

Social ecology can be described as a form of eco-anarchism, in which the cause of the ecological crisis lies in structures of hierarchy and power associated with the modern bureaucratic state and corporate capitalism. Bookchin has summarized social ecology as made up of 'an organic way of thinking ... dialectical naturalism ... a mutualistic social and ecological ethics ... the ethics of complementarity ... a new technics ... eco-technology; and ... new forms of human association ... eco-communities'.[4]

The main principles of social ecology are

1   that the domination of nature by humans has its roots in the historical emergence of patterns of hierarchy and domination within human society;
2   a dialectical approach to understanding the relationship between human society and the natural world. Underpinning many of his ideas is a reworking of dialectical thinking which combines Hegel's dialectical system of logic with ecological thinking in order to 'naturalize' the dialectical tradition. His 'dialectical naturalism' contrasts with Hegel's dialectical idealism and Marx's dialectical materialism;
3   a rejection of eco-centrism and the idea that humans are 'simply one species amongst others', and anthropocentric views which pit humans over, above or against nature. This is expressed by the notion of how 'first nature' (non-human world) 'grades into' 'second nature' (human culture), and how the latter is derived from the former;
4   a philosophy of nature in which values and practices such as freedom, subjectivity and mutualism are present in germinal form within nature and constitutive of its evolutionary *telos*;
5   a rejection of both the modern nation-state and corporate capitalism and a revolutionary-utopian vision of decentralized, ecologically sustainable, participatory democratic communities in which the economy is run on mutualist and co-operative lines.

Bookchin's work on social ecology developed into 'libertarian municipalism'. In the words of his partner and fellow theorist of social ecology, Janet Biehl, libertarian municipalism is 'the revolutionary forms of freedom that give organizational substance to the idea of freedom. In brief, libertarian municipalism seeks to revive the democratic possibilities latent in existing local governments and transform them into direct democracies.'[5]

For Bookchin, libertarian municipalism is defined as, 'a confederal society based on the co-ordination of municipalities in a bottom-up system of administration as distinguished from the top-down rule of the nation-state'.[6] It differs from bioregionalism in its concern with the issue of interaction between communities and the rejection of the bioregional model of small-scale, self-sufficient communities, promoted by other environmental thinkers such as Rudolf Bahro. The confederal nature of the arrangement means it is a voluntary political association of autonomous communities with sovereignty retained at the local level. Yet, the relativism that typifies some anarchist political arrangements is explicitly ruled out. As he puts it, 'Parochialism can ... be checked not only by the compelling realities of economic interdependence but by the commitment of municipal minorities to defer to the majority wishes of participating communities.'[7] Here economic-ecological interdependence goes hand in hand with political autonomy and self-determination. Autarky is not a central principle of social ecology, as it is for other radical green decentralist approaches such as bioregionalism.

Bookchin is well known as a polemical writer and has spent much time and energy criticizing those aspects of the ecological movement which he sees as based on flawed and dangerous political and moral principles, aims and analyses of the ecological crisis. His most vehement critiques have been levelled at deep ecology. In 1988 he stated that deep ecology was 'the same kind of ecobrutalism [that] led Hitler to fashion theories of blood and soil that led to the transport of millions of people to murder camps like Auschwitz', and at other times he has called deep ecology 'eco-lala'.[8]

Bookchin's polemical and uncompromising stance had led him to vehemently disagree and disown those within the broad 'social ecology' school with whom he disagrees. According to Clark:

> Although Bookchin develops and expands the tradition of social ecology in important ways, he has at the same time also narrowed it through dogmatic and non-dialectical attempts at philosophical systems-building, through an increasingly sectarian politics, and through intemperate and divisive articles on 'competing' ecophilosophies and on diverse expressions of his own tradition. *To the extent that social ecology has been identified with Bookchinist sectarianism, its potential as an ecophilosophy has not been widely accepted.*[9]

Bookchin's combative style often obscures the originality and 'anticipatory brilliance' of his thought, and while one could say that Bookchin's style is a classic example of how 'exaggeration is when the truth loses its temper', the dogmatism of his presentation for many also betrays a dogmatism in the content of his work, which of course stands at odds with his libertarian, anarchist thrust.

Bookchin leaves a mixed legacy: a combination of groundbreaking and impressive scholarly analysis, critique and prescriptions for explaining and combating the socio-ecological crisis, combined with an obsession with combating any perceived threat to his position from deep ecology and any other non-Bookchin forms of ecological thought and action. For many, even those broadly sympathetic to social ecology, this polemical and dogmatic propensity is damaging, to his legacy, to social ecology and to the wider cause of finding political and economic solutions to socio-ecological problems. Andrew Light suggests that 'the question is whether the approach to political ecology that Bookchin champions, including a tendency to make judgements about interlocutors based on a few extreme examples, is what we need today'.[10]

Bookchin has been enormously influential within the ecological movement in North America and Europe, and was one of the most invigorating and original thinkers about green politics. His political vision in centring the transition from unsustainability around an urban vision of a green future, the liberatory and democratic self-management potentials of technology, and the necessity of political leadership within a decentralized municipal green polity is one that will both endure and inspire.

According to Peter Marshall, Bookchin's main achievement is to have

> combined traditional anarchist insights with modern ecological thinking ... In this way he has helped develop the powerful libertarian tendencies within the Green movement. Just as Kropotkin renewed anarchism at the end of the last century by giving it an evolutionary dimension, so Bookchin has gone further to give it an ecological perspective. In his view, the creation of an anarchist society is now the only way to solve the threat of ecological disaster confronting humanity.[11]

Notes

1 *Remaking Society*, p. 204.
2 *Towards an Ecological Society*, p. 251.
3 Damian White, 'Murray Bookchin's New Life'.
4 Murray Bookchin, in Steve Chase (ed.), *Defending the Earth: A Dialogue between Murray Bookchin and Dave Foreman*, Boston, MA: South End Press, p. 131, 1991.
5 Janet Biehl, *The Politics of Social Ecology: Libertarian Municipalism*, p. viii.

6 Bookchin, 'Libertarian Municipalism', *Society and Nature*, 1:1, pp. 94–5, 1992.
7 Ibid., p. 97.
8 Bookchin, 'Social Ecology vs Deep Ecology', *Socialist Register*, 18 (3), p. 13, 1988.
9 John Clark, 'A Social Ecology', *Capitalism, Nature, Socialism*, 8 (3), p. 9, 1997, emphasis added.
10 Andrew Light, 'Introduction', in Andrew Light (ed.), *Social Ecology after Bookchin*, p. 4.
11 Peter Marshall, *Demanding the Impossible: A History of Anarchism*, London: Fontana Press, p. 602, 1993.

See also in this book
Bahro, Marx

Bookchin's major writings
Our Synthetic Environment, pseud. Lewis Herber, New York: Harper & Row, 1962.
Post-scarcity Anarchism, London: Wildwood House, 1971.
The Spanish Anarchists, New York: Harper & Row, 1977.
Towards an Ecological Society, Montreal/Buffalo, Canada: Black Rose Books, 1980.
The Modern Crisis, Philadelphia, PA: New Society Publishers, 1986.
Remaking Society, Montreal, Canada, and New York: Black Rose Books, 1990.
The Ecology of Freedom: The Emergence and Dissolution of Hierarchy, rev. edn, Montreal, Canada, and New York: Black Rose Books, 1991.
Urbanization without Cities: The Rise and Decline of Citizenship, Montreal, Canada: Black Rose Books, 1992.
The Philosophy of Social Ecology, rev. edn, Montreal/Buffalo: Black Rose Books, 1994.
Re-Enchanting Humanity: A Defence of the Human Spirit against Antihumanism, Misanthropy, Mysticism and Primitivism, London: Cassell, 1995.

Further reading
Barry, J. *Rethinking Green Politics: Nature, Virtue and Progress*, London: Sage, 1999.
Biehl, J. (ed.), *The Murray Bookchin Reader*, Montreal, Canada: Black Rose Books, 1997.
Biehl, J., *The Politics of Social Ecology: Libertarian Municipalism*, Montreal, Canada: Black Rose Books, 1998.
Biehl, J., *Ecology or Catastrophe: The Life of Murray Bookchin*, Oxford, UK: Oxford University Press, 2015.
Clark, J. (ed.), *Renewing the Earth: The Promise of Social Ecology*, Basingstoke, UK: Green Print, 1990.
Light, A. (ed.), *Social Ecology after Bookchin*, London: The Guilford Press, 1998.

Watson, D., *Beyond Bookchin: Preface for a Future Social Ecology*, New York and Detroit, MI: Autonomedia, 1996.

White, D, 'Murray Bookchin's New Life', Jacobin, available at: www. jacobinmag.com/2016/07/murray-bookchin-ecology-kurdistan-pkk-rojava-technology-environmentalism-anarchy/

White, D, *Bookchin: A Critical Appraisal*, London: Pluto Press, 2008.

JOHN BARRY

THICH NHAT HANH 1926–

> Everything outside us and everything inside us comes from the Earth. We often forget that the planet we are living on has given us all the elements that make up our bodies.[1]

Thich Nhat Hanh is a Zen Master, peace activist, international bestselling author and Nobel Peace Prize nominee. His teachings as an international thought leader on non-violence and environmentalism span over five decades. His perspectives on humankind's intimate relationship to the Earth are rooted in the tradition of Buddhist teachings. Through this lens he offers a unique perspective on how to approach environmental issues from a place of non-dualistic thinking and non-discrimination. He proposes that through deep looking we can see that we and the planet are one, and by reconnecting with the Earth through love, respect and understanding we will naturally protect the earth just as we would protect ourselves.[2]

Thich Nhat Hanh is considered the founder of 'Engaged Buddhism', which applies the teachings of Buddhism to everyday life, and social, political and environmental suffering and injustice. Understanding his personal history and work as a peace activist is key to understanding his approach to the environment. He was born in Hue Vietnam in 1926 and entered the monastery in 1942, becoming a novice monk at the age of sixteen at the Tu Hieu Temple. When the Vietnam War broke out in 1955 he was faced with the dilemma of continuing the contemplative life of a monastic inside the temple, or going out into the world to help with the suffering and trials of war. He chose to do both. This choice resulted in his founding of 'Engaged Buddhism' and became the roots of his lifelong work.

In 1961 he traveled to the United States, where he taught comparative religion at Princeton University. The following year he went to Columbia University, where he worked as a teacher and researcher of Buddhism. In 1964 he founded the School of Youth for Social Service (SYSS), a grass-roots relief organization made up of over 10,000 volunteers who worked throughout the Vietnam War to rebuild villages that were bombed, take care of orphans, continue education for children, and offer support to those in need.[3] The SYSS used the foundations of mindfulness and meditation in order to offer this service to their communities in a time of great suffering and fear.

In 1965 he issued a call for peace, urging both sides to reconcile, followed with a trip to the United States in 1966 where he continued his public appeals for reconciliation. Throughout this time his approach to peace was consistently one of non-discrimination and non-violence, refusing to condemn or choose either side in the conflict, resulting in his eventual denouncement and exile from Vietnam. Non-discrimination is a central teaching in Buddhism, one that Thich Nhat Hanh has applied in an engaged way throughout his life work, with an aim to break down barriers and mental formations that might inhibit the global community from working together for the betterment of ourselves and our planet. Many years after the war ended he reflected on his non-dualistic approach to conflict in his book *Together We Are One*:

> [T]hose who create discrimination and hatred are the victims of fear. It is fear that is the obstacle. In Vietnam, then and still now, I want to help people to be free from fear. Those who fought in Vietnam, those who fight now in Iraq and Afghanistan, are not my enemies. They are the ones I want to help. They are objects of my practice of compassion and understanding. I have no enemies.[4]

This profound statement shows how he applies the Buddhist principles of compassion and understanding to the real-world experiences of conflict and injustice. Even in the face of great atrocities, he proclaims that the individuals involved are not his enemies, but rather that the fear and anger at the root of their actions is the true enemy. This concept is one that he also applies to his views on environmental activism, suggesting not that we 'fight' climate change, but rather that we must focus on falling back in love with the Earth, and that only through this kind of compassionate action will we be able to offer the time and care needed to save our planet.[5] The notion of not labeling individuals, countries, groups or ideologies as either friend or enemy is a unique approach in the sector of activism and a major paradigm shift to the way that global politics and policy has traditionally functioned. It is a perspective that asks us to see all of humanity as one global community.

Another fundamental theme in his teachings is his emphasis on the importance of community building, called Sangha in Sanskrit. Building

Sanghas has been a major focus throughout his life as a global spiritual leader, and a central component of what inspired his friendship with Martin Luther King Jr., who nominated Thich Nhat Hanh for a Nobel Peace Prize in 1967. Thich Nhat Hanh met Dr. King in 1966, bonding over their mutual aspirations. Thich Nhat Hanh shares about their connection in his book *Good Citizen* stating:

> Like Martin Luther King Jr., the Buddha dreamed of relieving the world of suffering. It's a noble dream. Each of us can do something to help relieve the suffering of the world, but none of us can do it alone.[6]

In his letter nominating Thich Nhat Hanh for the Nobel Peace Prize Dr. King wrote about the impact this award could have, stating that 'His ideas for peace, if applied, would build a monument to ecumenism, to world brotherhood, to humanity.'[7] The Nobel Peace Prize was not awarded that year, but the impacts of Dr. King's nomination raised awareness to Thich Nhat Hanh's calls for peace in Vietnam and his profile as an international spiritual leader.

When Thich Nhat Hanh learned of Dr. King's assassination, he made a vow to continue building community and working for peace for the rest of his life. In 1982 he established the practice community of Plum Village in the South of France, which has grown to be the largest active monastery in the West, with over 200 resident monastics and over 8,000 visitors every year. He has also established centres in the United States in California, New York and Mississippi, in Paris (France), Hong Kong, Thailand, Australia and Germany. Visitors to his centres experience a wide range of practices, to help cultivate compassion and peace, that cover a diverse set of topics such as sitting, eating and walking meditation, reconciliation techniques, connecting to the earth and how to handle strong emotions – to name just a few.

Reflecting his Buddhist roots, all of Thich Nhat Hanh's practice centres have always been vegetarian, going vegan in recent years in response to research findings on the impact of the meat industry. In October 2007, he wrote his famous 'Blue Cliff Letter' where he invited practitioners to consider going vegetarian for 15 days a month and discussed the negative impact of the meat industry on the planet. In this letter he states, 'As a spiritual family and a human family, we can all help avert climate change with the practice of mindful eating. Going vegetarian may be the most effective way to stop climate change.'[8] In his 2014 'New Contemplations Before Eating', a set of reflections for the start of a meal, he updated the reflection: 'May we keep our compassion alive by eating in such a way that reduces the suffering of living beings' to also include the sentence 'stops contributing to climate change, and heals and preserves our precious planet'.[9]

In addition to building an international community to cultivate the practices and understanding of peace, Thich Nhat Hanh has continued to tirelessly work with global leaders to develop policies to support a healthier more peaceful world, for humanity and all living beings – including the earth. An example of this is the Manifesto 2000, a proposal he drafted with several Nobel Peace Prize winners. The Manifesto 2000 text looks at concrete steps to transform violence in the world to benefit children. The proposal was accepted by the UN who responded by issuing a decree, ARES 53129, declaring the year 2000 as the 'International Year for the Culture of Peace'. The Manifesto 2000 is a clear example of how Thich Nhat Hanh does not see a separation between the violence and peace we cultivate amongst humanity and the violence or peace that we offer to the environment. The third paragraph of this manifesto includes the environment as a direct recipient of the benefits that would come from reducing violence amongst humanity, 'because the culture of peace can underpin sustainable development, environmental protection, and the well-being of each person'. This statement captures succinctly his view that what benefits humanity as a species also directly benefits the environment.

Thich Nhat Hanh's teachings have had an immeasurable influence on the global community's perspective and approach to peace and the environment. In 2014, the United Nations Framework Convention on Climate Change (UNFCCC) approached him to request a brief statement. This statement was published on the UNFCCC's website, ahead of the Paris Climate Summit in September 2015. The statement, with the title 'Falling in Love with the Earth', was a call to action, addressing our modern culture of consumption, our alienation from the earth, the need to recognize our interconnection with all beings and above all the shared responsibility of each individual for our planet. In this statement Thich Nhat Hanh says,

> There's a revolution that needs to happen and it starts from inside each one of us ... Cherishing our precious Earth – falling in love with the Earth – is not an obligation. It is a matter of personal and collective happiness and survival.[10]

Through this statement, Thich Nhat Hanh offered world leaders a framework for mutual understanding and collective motivation. But perhaps even more important than this was the impact his teachings had on Christina Figueres, executive secretary of the UNFCCC, who led the climate talks. In an interview with the *Huffington Post* at the World

Economic Forum's annual meeting in 2016, Figueres shared that 'I don't think that I would have had the inner stamina, the depth of optimism, the depth of commitment, the depth of the inspiration if I had not been accompanied by the teachings of Thich Nhat Hanh.'[11] Figueres' experience is the perfect example of Thich Nhat Hanh's approach to activism, being that, through personal transformation and healing – the development of self-care and compassion, the understanding of interbeing and the practice of non-discrimination – individuals that make up our global community will have the qualities and perspectives needed to make the changes that are necessary to save our planet.

Throughout his life, Thich Nhat Hanh has consistently viewed inner and outer peace as inseparable from one another. His approach to creating peace in the world has always been grounded in his practice as a Zen Buddhist Monk. While he has published many books and articles that relate specifically to environmentalism, they have always been extensions of his lifelong teachings and translations of Buddhism to everyday life. There has always been one main thread running through every teaching: we must learn to take care of ourselves in order to take care of the world around us, and when we take care of ourselves we already take care of the world around us. From his teachings one can understand that this is done by cultivating understanding, compassion, non-discrimination and true love. Like many concepts in Buddhist philosophy this is both simple and the greatest of challenges.

In November of 2014, at the age of eighty-nine, Thich Nhat Hanh suffered a stroke, which left him paralyzed on the right side and unable to speak. After several months of treatment he returned to live and practice at Plum Village, his centre in the South of France. He continues to offer teachings through his presence and embodiment of mindfulness, and his many books, talks, poems and calligraphies continue to be an invitation for all of humanity to do the work needed to heal ourselves and heal the planet.

Notes

1 *Love Letter To The Earth*, p. 8.
2 Ibid., p. 14.
3 'About' (2016). Retrieved from plumvillage.org/about/thich-nhat-hanh/
4 *Together We Are One*. Berkeley, CA: Parallax Press, 2010, p. 88.
5 *Love Letter To The Earth*, p. 29.
6 *Good Citizens*. Berkeley, CA: Parallax Press, 2012, p. 121.
7 Martin Luther King Jr. 'Letter from MLK to the Nobel Institute' (1967). Retrieved from the King Centre archives thekingcenter.org
8 'Blue Cliff Letter' (2016). Retrieved from http://plumvillage.org/letters-from-thay/sitting-in-the-autumn-breeze/

9 'New Contemplations Before Eating' (2014). Retrieved from http://plumvillage.org/news/new-contemplations-before-eating/

10 'Falling in Love with the Earth' (2014). Retrieved from UNFCCC Newsroom. Retrieved from https://plumvillage.org/letters-from-thay/thich-nhat-hanhs-statement-on-climate-change-for-unfccc/

11 Joe Confino. 'This Buddhist Monk Is An Unsung Hero In The World's Climate Fight'. Retrieved from www.huffingtonpost.com/entry/thich-nhat-hanh-paris-climate-agreement_us_56a24b7ae4b076aadcc64321

See also in this book
Buddha, Cage

Thich Nhat Hanh's major writings
The Miracle of Mindfulness. Boston, MA: Beacon, 1975.
Peace Is Every Step. New York, NY: Bantam Books, 1991.
The World We Have. Berkeley, CA: Parallax Press, 2008.
Love Letter To The Earth. Berkeley, CA: Parallax Press, 2013.

Further reading
James, Simon P., *Zen Buddhism and Environmental Ethics.* Aldershot, UK: Ashgate, 2004.
McMahon, David, *The Making of Modern Buddhism.* Oxford, UK: Oxford University Press, 2008.
Willis, Jennifer S. (ed.). *A Lifetime of Peace: Essential Writings By and About Thich Nhat Hahn,* New York: Marlowe & Co., 2003.

ELLI WEISBAUM
(Advisors: John Bell and The Plum Village Community)

EDWARD OSBORNE WILSON 1929–

When the century began, people could still think of themselves as transcendent beings, dark angels confined to Earth awaiting redemption by either soul or intellect. Now most or all of the relevant evidence from science points in the opposite direction: that having been born into the natural world and evolved there step by step across millions of years, we are bound to the rest of life in our ecology, physiology, and even our spirit. In this sense, the way in which we view the natural world, Nature has changed fundamentally.[1]

Edward Osborne Wilson was born in Birmingham, Alabama, in 1929, the son of a travelling government accountant. His scientific career, which began with the study of ants and ultimately generated theories that were to influence profoundly concepts of biodiversity, sociobiology and, most recently, the unification of all knowledge, have earned him many of the highest academic honours. In 1996 he was described by *Time* magazine as one of America's twenty-five most influential people. By the time of his retirement from Harvard in 1997 he had become recognized as one of the greatest evolutionary biologists of the twentieth century.

Wilson describes his early life as 'blessed', although he was often beset by difficult emotional and physical circumstances. These included the divorce of his parents and an itinerant schooling where he attended fourteen different schools in eleven years. The loss of an eye in a fishing accident denied him access to a military career but left him with eyesight characteristics that he turned to his advantage in science. Gradual, partial loss of hearing during his adolescence influenced his choice of studies in scientific research, deflecting him away from ornithology towards the study of ants. He considers that three formative experiences during his youth influenced his later career and personal philosophy: an intimate knowledge of natural history that first developed in his childhood; an induction into military discipline and the virtues of hard work at the Gulf Coast Military Academy; and a Southern Baptist upbringing that left him with the conviction that religion and science might be reconciled by the understanding of the former by means of the latter.

After gaining Bachelor's and Master's degrees at the University of Alabama, Edward Wilson studied for his PhD at Harvard University, where he taught from 1953 until 1997. He was successively a Harvard Professor of Zoology, Curator of Entomology at the Museum of Comparative Zoology, Baird Professor of Science, Mellon Professor of the Sciences and Pellegrino University Professor. He is currently Pellegrino Professor Emeritus at Harvard and holds a position at Duke University, where he set up the E.O. Wilson Biodiversity Foundation. His scientific awards include the U.S. National Medal for Science, the Swedish Academy of Sciences Crafoord Prize, Germany's Terrestrial Ecology Prize, Japan's International Prize for Biology, the French Prix du Institut de la Vie and, more recently, the International Cosmos Prize.

Edward Wilson's influence is in no small part attributable to his skill as a writer whose elegant prose has confirmed his status as one of the finest communicators of science in our times. Several of his fluent, beautifully written books are at once important academic sources and accessible, engrossing works of popular scientific literature. They have also earned him many literary honours, including two Pulitzer Prizes, the Los Angeles

Times Book Prize, the Publishers' Marketing Association Benjamin Franklin Award, the Sir Peter Kent Conservation Book Prize and the John Hay Award from the Orion Society. He is the author, too, of a prize-winning novel, *The Anthill* (2008) – one of some dozen or more books he has written in the new millennium.

Even if his scientific career had been confined to the study of ants (myrmecology), Wilson's reputation as an outstanding biologist would be indisputable. His taxonomic and behavioural studies on ants have made him a leading international expert on these insects. *The Ants*, published in 1990 with Bert Hölldobler, was not only an authoritative study of their anatomy, taxonomy, ecology and social behaviour, but also a winner of the Pulitzer Prize, acclaimed as much for its detailed information and taxonomic keys for specialists as for its engaging accounts of ant social behaviour for the interested layman.

Inevitably, close field-based study of such a complex, diverse and widely distributed group of insects brought Wilson in close contact with the biodiversity of numerous temperate, sub-tropical and tropical ecosystems. In 1967 he and Robert MacArthur published *The Theory of Island Biogeography*, describing how the number of species in an isolated patch of habitat – whether a true oceanic island or an island of surviving natural vegetation in a once continuous tropical forest – could be determined with reference to a simple mathematical expression and distance to the nearest source of immigrant species. The theory showed that a balance between new species immigration and extinction of established species was eventually reached, and that the extent of biodiversity in such islands was determined by their size. The theory was successfully validated by denuding a small island in the Florida Keys of all animals and then following in detail the pattern of re-colonization.

Subsequently MacArthur and Wilson's theory of island biogeography has been criticized and modified, but remains immensely influential in the design of nature reserves, emphasizing the importance of conserving the largest possible patches of natural, undisturbed habitat. More controversially, the theory has been used to calculate probable rates of extinction, since it also provides a means of calculating species loss as habitats become fragmented, isolated and reduced in size. The development of this theory coincided with novel methods for measuring biodiversity, such as those of Terry Erwin,[2] who proposed vast increases in estimates of species diversity based on extrapolation from sub-samples of beetle biodiversity measured on a single tree species in the Panamanian rain forest. New estimates of total biodiversity were pitched at 10, 30 or even as many as 100 million species, when only about 1.5 million species have been scientifically classified. Wilson's work indicated that extinction rates due to habitat degradation and destruction were far higher than anyone had hitherto imagined. His writings have tirelessly warned of the disastrous consequences of the likely rapid loss of a large proportion of earth's biodiversity which, he warns, is 'the folly our descendants are least likely to forgive us'.

Edward Wilson's behavioural studies of ant societies were the foundation for a second great theme of his scientific career, the study of sociobiology. His proposal that there were genetically determined elements in human behaviour – that evolution has generated certain patterns of neural connections that predispose human behaviour towards certain courses of action – was instantly controversial. His book *Sociobiology* brought him into conflict with Richard Lewontin[3] and Steven Jay Gould, whose ideological predispositions abhorred any suggestion that nature rather than nurture could be a guiding force of human behaviour. In retrospect, Wilson's admission that 'at my core I am a social conservative, a loyalist. I cherish traditional institutions, the more venerable and ritual-laden the better' made it probable that there might be no easy accord with those of a more Marxist disposition in American society. The possibility that characteristics such as altruism or aggression in humans might be even partially governed by instinctive, genetically determined algorhythms had profound consequences for sociology, civil rights and justice. The reception for *Sociobiology* was at times abusive and even violent, as when Wilson was doused with water by protestors at a sociobiology symposium in Washington in 1978. Subsequently, accumulated circumstantial evidence and data from molecular biological studies has reinforced the notion that there are genetic components in human behaviour. Wilson's Pulitzer Prize-winning *On Human Nature* was to some degree a rebuttal of his detractors' politically motivated criticisms of sociobiology.

Wilson's rapid rise in academic status and public recognition coincided with developing tensions between Harvard's traditionalists in biology, whose work was based on the study of whole organisms, and the growing power of the reductionist molecular geneticists who sought to explain the complexity of nature through exploring its constituent molecules. Wilson, a whole-organism traditionalist through and through, with an upbringing that had instilled Old-World courtesies, civility and good manners in academic debate, magnanimously describes himself as 'being blessed with brilliant enemies' but admits to despising 'the arrogance and self-regard so frequently found amongst the very bright'. He made no secret of his personal dislike and professional admiration for Nobel Laureate James Dewey Watson, co-discoverer of the structure of DNA, who he describes in his autobiography *Naturalist* as 'the Caligula of biology'. Many years later, he was understandably offended by an intemperately hostile review by Richard Dawkins of his book *The Social Conquest of Earth* (2012).

Some people perceive a certain irony in Wilson's later work, described in *Consilience*, which seeks to unify all knowledge – including religion, economics and aesthetics – in terms of reductionist physical and biological principles. The term 'consilience' was originally coined by the nineteenth-century philosopher William Whewell, to describe the solving of problems by the combined use of inferences drawn from disparate sources, a process which is common practice in science. Harking back to the controversial concepts first outlined in *Sociobiology*, *Consilience* proposes that an understanding of the biological mechanisms underlying human behavioural characteristics, assembled during the evolution of the brain, will ultimately provide the framework for understanding the decisions that we make about our interactions with our environment and with each other. Predictably, this attempt to reduce the arts and social sciences to an understanding of genetic programming has not received a warm welcome amongst most practitioners in those disciplines, but perhaps it might prompt the re-examination of their intellectual legitimacy, in much the same way that whole-organism biologists were compelled to reconsider their future in the face of the molecular biological revolution. In the somewhat safer home territory of conservation of biodiversity, Wilson has proposed the concept of biophilia, which he defines as 'the innately emotional affiliation of human beings for other organisms' and believes may be resident in our genes. He has argued that biophilia governs our aesthetic response to the living world and acts as a powerful driving force in environmental ethics.[4]

E.O. Wilson's most recent books, such as *The Meaning of Human Existence* (2014), have often explored and defended the position he called 'scientific humanism', according to which humaneness, tolerance and other virtues can be promoted by a proper scientific understanding of the world and of our biological development and needs. While respectful of religions, whose origin he regards as explicable in terms of evolutionary advantage, he remarked in an interview with the *New Scientist* (21 January 2015), that religious belief is now an obstacle to progress, something that is 'dragging us down'.

Ernst Mayr, another of the twentieth century's outstanding evolutionary biologists, considers the most memorable lesson he learned from Darwin is that 'the most important thing in scientific research is not to add to the accumulation of facts, but to ask challenging questions and to try to answer them'.[5] Edward Wilson is one of a small cadre of contemplative evolutionary biologists, imbued with a deep knowledge of field natural history from an early age, who, in a career that has combined meticulous observational and experimental study with scholarship, has asked challenging questions, providing answers that have consistently generated controversy, and which by doing so have stimulated whole fields of scientific endeavour.

Notes
1   From the author's *Prelude*, in E.O. Wilson, *Naturalist*, p. xii.
2   Terry Erwin, 'Tropical Forests: Their Richness in Coleoptera and Other Arthropod Species', *Coleopterists' Bulletin*, 36 (1), pp. 74–5, 1982.
3   R.C. Lewontin, *The Doctrine of DNA*, pp. 87–104.
4   'Biophilia and the Environmental Ethic', in *In Search of Nature*.
5   Ernst Mayr, 'Understanding Evolution', *Trends in Ecology and Evolution*, 14 (9), pp. 372–3, 1999.

See also in this book
Darwin, Ehrlich, Humboldt, Uexküll

Wilson's major writings
MacArthur, Robert H. and Wilson, E.O., *The Theory of Island Biogeography*, Princeton, NJ: Princeton University Press, 1967.
The Insect Societies, Cambridge, MA: Belknap Press, 1971.
Sociobiology: The New Synthesis, Cambridge, MA: Harvard University Press, 1975.
On Human Nature, Cambridge, MA: Belknap Press, 1978.
Biophilia, Cambridge, MA: Harvard University Press, 1984.
Wilson, E.O. and Peter, Frances M., *Biodiversity*, Washington, DC: National Academy Press, 1988.
Hölldobler, Bert and Wilson, E.O., *The Ants*, Cambridge, MA: Belknap Press, 1990.
The Diversity of Life, Cambridge, MA: Belknap Press, 1992.
Naturalist, Washington, DC: Island Press, 1994.
In Search of Nature, Washington, DC: Island Press, 1996.
Consilience, London: Little, Brown & Co., 1998.
The Social Conquest of Earth, New York: Liveright, 2012.
The Meaning of Human Existence, New York: Liveright, 2014.

Further reading
Futuyma, Douglas J., *Evolutionary Biology*, 3rd edn, Sunderland, MA: Sinauer, 1997.
Kaufman, Whitley, 'The Evolutionary Ethics of E.O. Wilson', *The New Atlantis*, Spring/Summer 2013.

Leopold, Aldo, *A Sand County Almanac and Sketches Here and There*, New York: Oxford University Press, 1947.

Lewontin, R.C., *The Doctrine of DNA*, Harmondsworth, UK: Penguin, 1995.

Simberloff, Daniel S. and Wilson, Edward O., 'Experimental Zoogeography of Islands: Defaunation and Monitoring Techniques', *Ecology*, 50 (2), 267–78, 1969.

Williamson, Mark, *Island Populations*, New York: Oxford University Press, 1981.

PHILLIP J. GATES

YI-FU TUAN 1930–

I confess that I don't love or even much like nature if one means by that word biological life and little else ... I am ambivalent about organic striving, the clever maneuvers of the selfish gene. That so much of the universe is "mineral" consoles rather than dismays me.[1]

Yi-Fu Tuan is not a champion of "the environment" if that term is interpreted to imply an interest in conservation, preservation, and ecology. Nonetheless, he is an important contributor to the study of human–environment relations if one understands by this term a set of concerns focused on the experience of being in the world, particularly questions of environmental values, perceptions, and representations. Tuan's framing of these issues has greatly enriched geographical notions of space and place, offering inspiration to those who want to rethink human–environment relations from a humanistic vantage point.

Tuan labels himself a cosmopolite,[2] which is not surprising considering his background. Born in Tientsin in 1930, the son of a Chinese foreign minister, he grew up in genteel poverty during the Second Sino-Japanese War. Because of the war his family fled to Chongqing when he was three, then relocated to Australia when he was eleven. There he and his brothers were subject to racial harassment. It seems he adapted to his situation by stubbornly refusing to see himself as an outsider.[3] It was perhaps at this early age that he first began to look for commonalities in human experience and to appreciate the small pleasantries and kindnesses one finds almost anywhere. These preoccupations are central to the "cosmopolite's viewpoint" that became a notable aspect of his academic sensibility.

His university education began at University College London and Oxford, and was followed by graduate studies at the University of California, Berkeley. While his training was in physical geography, he segued into the study of environmental perception and values by examining early modern understandings of the hydrological cycle.[4] Two of his four books from the 1970s, *Topophilia*, and *Space and Place*, became classics and staked out a research terrain for humanist geographers, environmental psychologists, planners, and many others. The fact that his term topophilia (strong attachment to place) is still widely used in scholarly writing also attests to his importance. Following teaching positions at Indiana University, University of New Mexico, and University of Toronto, he took a position at the University of Minnesota in 1968, then relocated in 1983 to the University of Wisconsin, Madison, where he was John Kirtland Wright Professor and Vilas Research Professor before retiring in 1998.

In regard to our understanding of human–environment relations, Tuan's theoretical contribution has been to situate environmental values, perceptions and representations in relation to several loosely interrelated ideas: first, environmental perceptions are culturally and historically specific; second, the environmental impacts of a group do not necessarily reflect its professed environmental values; third, human environments are domesticated through a mixture of dominance and affection; fourth, human–environment relations always involve complex pulls between the contradictory ideals of cosmos and hearth; and fifth, some of life's most valuable lessons have been taught by human encounters with extreme and inhospitable environments. These emphases are interwoven in erudite and articulate ways throughout Tuan's extensive career from which I hope to indicate a few high points.

Tuan's early research is marked by suspicion of the ideal of harmonious human–environment relations. He demonstrated how such ideals coexist with practices that lead to environmental despoliation. He shared with many other scholars an interest in culturally engrained attitudes and philosophies about nature, but in addition he wanted to highlight the fact that attitudes do not necessarily predict or reflect the ways in which people will act. Thus in any given place there are discrepancies between human–environment relations as they are idealized and human–environment relations as they may be directly observed. For example, the classic Chinese garden was designed to immerse its observers in a naturalistic setting with weathered materials, asymmetrical plantings, and winding paths. The contrast with a seventeenth-century European garden and its geometrically arranged, spherical bushes ranged along its straight pathways could not be more striking. But naturalistic landscape aesthetics did not translate into more harmonious interactions with the natural environment; the Chinese civilization produced a serious amount of deforestation, habitat loss, and erosion.

Simultaneous with his interest in discrepancies between environmental attitudes and actions, Tuan explored teleological aspects of Western environmental philosophy—ways in which earth systems were interpreted as demonstrating the "wisdom of God."[5] He showed that, from the seventeenth through the nineteenth century, environmental processes beneficial to humankind were interpreted teleologically, as signs of divine providence. Rain fell for the good of humankind and oceans were salty "for the better support of navigable vessels" while these same oceans were large because if they were smaller the land would suffer from lack of rain.[6] *The Hydrologic Cycle and the Wisdom of God* remains valuable as a rigorous study of early modern European environmental thought.

During the 1980s the general shift toward academic interest in power relations resonated in Tuan's work. He linked the passion to control nature with the passion to control people. In either case control is best understood, he believed, not as a means but as an end. Controlling something is part of the process of forming an attachment to it. Tuan revealed affection as deeply contaminated by the wish to control and twist something to our purposes and whims. This obsession is demonstrated in *Dominance and Affection* by everything from the keeping and breeding of pet fish to the creation of bonsai trees, to the design of magnificent gardens, to the alternate pampering and abuse of women, children, and people with physical deformities.[7] This book is deeply disturbing but perhaps also oddly reassuring for reasons I will explain below.

By the 1990s Tuan broadened his focus to the distinctly geographical project of revealing how human experience is structured within a continuum of scales from the home or hearth to the universe or cosmos. This framework did not lead to familiar environmental arguments like pointing out the ecological value of forests and wetlands. Instead it illuminated natural environments as settings in which people encounter the grandeur and distance epitomized by the cosmos as well as the comfort and reassurance of the hearth.[8] Cultural attitudes toward nature mix these scales. He takes the United States and China as two contrasting cultures, each with its own peculiar blend of localism and cosmopolitanism, and its peculiar ways of altering and adapting to natural environments.

In the past decade Tuan has engaged in yet another way with environmental thought by expressing his strong objections to "environmentalist" efforts to understand and mitigate human impacts on ecological systems. Tuan's judgment is quite clear:

> [C]onsider the extraordinary popularity, not only in academia but in society at large, of such conservative, housekeeping notions as environmentalism, ecology, sustainability, and survival. The issues

they raise and the vocabulary they use may differ, but since they all attempt to make the earth a stable and livable home, they all come down to being "home economics."[9]

At the end of *Romantic Geography* he clarifies: "too much concern for the nitty-gritty details of housekeeping—what I have called 'home economics'—can lead to a bureaucratic frame of mind."[10] He worries that if we disenchant "cosmos" by taking care of a wetland or forest we may mistakenly start to think of the cosmos itself as something that must be subjected to our housekeeping efforts—and we thereby will disenchant nature. He dreads the idea of treating nature like a set of accounts we have to manage. It appears that what he values most in nature is not nature itself but rather the role nature has played in human life as a symbol or manifestation of the boundless, the infinite, and the sacred.

Rather than stop with this critique of environmentalism he offers an alternative grounded in humanism. In *Romantic Geography* he investigates the ways in which environments such as mountains, deserts, and icefields have forced us to encounter human strength and weakness, survival versus annihilation—dichotomies that make more poignant the diverse relationships between place and self, as evident even in small acts like lighting a camp stove. For Tuan, the appeal of harsh and inhospitable environments lies in the human qualities that are brought out by valiant efforts to gain a foothold in these places. He therefore celebrates the heroism of explorers who ventured to the ends of the earth "to experience, even at the risk of death, something vast and intoxicating such as might be found at the North or South Pole, the highest mountain, the deepest trench, the densest forest, or the bleakest desert."[11] To climb a mountain or brave the Arctic ice is to make oneself a living testament to the human power to overcome adversity. A self-avowed romantic, he prefers conquest of nature to its preservation because in the name of conquest we encounter the extremes (e.g. cosmos and hearth, dominance and affection, space and place) that give richness and meaning to the human condition.

In summary, these are the main elements of Tuan's environmental thinking: a precautionary awareness that a society's professed environmental ideals quite often deceive; a commitment to the systematic analysis of environmental perceptions; an unsettling insight that what we find worthy of affection is nearly always what we have dominated and domesticated; a fascination with the scale continuum from the very small to the very large; a rejection of environmentalism on the argument that it disenchants the world; and a belief that struggling against nature brings out some of the finest human qualities.

We have here a key thinker on the environment who is definitely not an environmentalist, at least not in the modern sense of the word. His sensibility is occasionally more in line with nineteenth-century geographers than with other geographers of the late twentieth and early twenty-first century. This is not to say that Tuan is unmoved by natural environments, but the "nature" he loves is the part most difficult to destroy. "Over time, I was forced to conclude that, for me, beauty has to be inhuman—even inanimate—to be a balm to the soul. Thus my love for the desert."[12] He even sees the hard-edged sterility of the desert in himself, as "the objective correlative of the person I am, absent the social façade."[13] Of course if "environment" is taken to mean the whole of the cosmos, then the inanimate most certainly predominates over the animate, and a Tuanian affection for the mineral environment, however odd, would inspire one to feel topophilia for most of the universe.

Reflecting on Tuan's body of work, I believe *Dominance and Affection*'s subtitle "*The Making of Pets*" offers a bit more hope for conservationists and preservationists than he realizes. Recalling that people have a tendency to feel affection for something only after they have dominated it, domesticated it, and in essence turned it into a pet, then this implies that if we have entered a human-driven geological era, the Anthropocene, there is something hopeful implied by this insight. Hope lies in the fact that as people become more aware of their collective power to destroy or preserve natural systems the earth may be treated with affection and indulgence. It may become less like something to conquer *and* less like an account to manage, ultimately revealing itself as something more like a large and unruly pet.

Notes

1 *Who am I?* p. 52.
2 *Cosmos and Hearth*, pp. 133–188.
3 *Who am I?* p. 19.
4 Tuan, *The Hydrologic Cycle and the Wisdom of God: A Theme in Geoteleology*. Toronto: University of Toronto Department of Geography and University of Toronto Press, 1968.
5 *The Hydrologic Cycle and the Wisdom of God*.
6 Ibid. p. 75.
7 *Dominance and Affection*.
8 *Cosmos and Hearth*.
9 *Romantic Geography*, p. 6.
10 *Romantic Geography*, p. 177.
11 *Romantic Geography*, p. 159.
12 *Who am I*, p. 55.
13 *Place, Art, and Self*, Santa Fe, NM: Center for American Places, p. 19, 2004.

See also in this book
Lopez, Merleau-Ponty

Tuan's major writings
Topophilia: A Study of Environmental Perception, Attitudes, and Values, New York: Columbia University Press, 1974.
Space and Place: The Perspective of Experience, Minneapolis, MN: University of Minnesota Press, 1977.
Dominance and Affection: The Making of Pets, New Haven, CT: Yale University Press, 1984.
Cosmos and Hearth, Minneapolis, MN: University of Minnesota Press, 1996.
Who am I? An Autobiography of Emotion, Mind, and Spirit, Madison, WI: University of Wisconsin Press, 1999.
Romantic Geography: In Search of the Sublime Landscape, Madison, WI: University of Wisconsin Press, 2013.

Further reading
Adams, P.C., Hoelscher, S., and Till, K.E., eds. *Textures of Place: Exploring Humanist Geographies*, Minneapolis, MN: University of Minnesota Press, 2001.
Daniels, S., DeLyser, D., Entrikin, J.N., and Richardson, D. eds. *Envisioning Landscapes, Making Worlds: Geography and the Humanities*. London and New York: Routledge, 2012.
Harrison S., Pile S., and Thrift N. *Patterned Ground: Entanglements of Nature and Culture*. London: Reaktion Books, 2004.

PAUL C. ADAMS

PAUL EHRLICH 1932–

'Nothing less is at stake than the fate of human civilization'[1] is Paul Ehrlich's motto both now and for much of his academic career. Of all the fields of the natural sciences, it might be expected that biology might produce the most thinkers on environmental matters, and the entry on Aldo Leopold is another example of this. But of all the recent (post-1960) contributors to the provision of information and to participation in public debate, Ehrlich is one of the most prominent. Born in 1932, he took his first degree at the University of Pennsylvania and his PhD at the University of Kansas (1957); an appointment as Professor of Biology at Stanford University in 1966 was the first of a series of posts in that institution. From this secure base he has published a series of books and papers, travelled widely and engaged in numerous debates and acts of public service. His contributions to environmental thought and action have brought him numerous honours including election in 2012 to Fellowship of the Royal Society in London and being honoured with a BBVA Foundation of Frontiers of Knowledge Award in Conservation Biology.

Although Ehrlich has had a high public profile in the USA and in certain world forums, most people are influenced by his published work. There are perhaps four strands to this: (1) basic research in the natural sciences, and in particular on the population ecology of birds and butterflies; (2) advocacy on the subject of human population growth, with a strong neo-Malthusian outlook which suggests that many, if not most, problems of the human species are the result, immediately or indirectly, of rapid population growth; (3) human ecology: the connection of human activities to the biophysical systems of the planet in areas such as biodiversity and agriculture; and (4) widely read popular works and student texts on population–resource–environment linkages.

Category (1) includes work on birds, butterflies and coral reefs in the classic team mode of the natural sciences,[2] but it is worth noting that a 1965 paper on the co-evolution of butterflies and plants has become a Citation Classic in the ISI *Current Contents* series.[3] The ways in which the central concern with basic biology (which has acted as a grounding for all the other work throughout) include a concern for the extinction of species, the conservation of both tropical and temperate forests, and even the effect of scientific study upon butterfly populations.[4] The key point is that although Ehrlich became mostly known for his advocacy – and indeed polemic – on environmental concerns, his attention to basic science has remained constant to the present day.

As a result of rapid immigration and industrial anabasis, coupled with an affluent and well-educated population, California in the 1960s became a centre of 'alternative' thinking about population–resource–environment relations. The 'hippy' movement with its attention to communal lifestyles and illegal substances was one strand, but another was a more intellectual and factually well-informed questioning of the gospels of growth and development as they appeared in that state, in the USA, in the industrial nations, and finally in the world as a whole. One pointer was the volume of essays edited by S. von Ciriacy-Wantrup and J.J. Parsons,[5] which brought many of the issues into focus, another the radical questioning of 'growth' by the geographer D.B. Luten,[6] yet another the expansion of the influence of the Sierra Club (which is based in the San Francisco Bay area) as an environmental campaigning body rather than a mountaineers' organization. In this zeitgeist the strongly expressed views of Ehrlich on population, for example, were not seen as extreme, and indeed the outlooks developed in category (2) fitted well into the relatively radical sets of ideas being developed at the time.

Thus it was that the publication by the Sierra Club of *The Population Bomb* in 1968 propelled Ehrlich from a base in which notions of population control in the affluent countries were not seen as necessarily controversial, to a wider public discussion in which they certainly were. The USA was described as the world's largest consumer and so strong were its effects, example and influence that, '[W]e must have population control at home ... by compulsion if voluntary methods fail. We must use our political power to push other countries into programs which combine agricultural development and population control.'[7] The book created considerable interest worldwide and has been reprinted and translated into several languages. The uncompromising neo-Malthusian message, combined with some startling prophecies (the Prologue's second sentence starts, '[I]n the 1970's the world will undergo famines – hundreds of millions of people are going to starve to death ...'[8]), not only presented a series of challenges to development-minded agencies in the USA and internationally, but was sufficiently well expressed to propel Ehrlich to the status of a media-figure and global guru. In particular, it confronted the orthodox position of the Roman Catholic Church on chemical and physical methods of contraception (mathematics was, however, allowed), although these were not particularly strongly obeyed in most developed countries: growth rates in, for example, Latin America were then very high. The term 'Vatican roulette' inspired the inclusion in the book of the text of letters to the then Pope and the local Archbishop suggesting that the Church modify its position: the letter to Paul VI seems not to have been passed to his successor. Famines did occur in the 1970s, though mostly in zones of civil strife rather than in areas with especially rapid population growth (of course competition for resources of any kind may have a demographic component), and there have been some notable downturns in population growth rates though the highest in Africa, for example, are not associated with an especially Catholic culture, and AIDS has rather transformed the demographics of several African nations.

The bulk of *The Population Bomb* was however devoted to extending the ideas of Malthus in the sense that it was not the absolute size of the population that mattered, but its relation to its resource base. So the foundations were laid in that book for explorations of the linkages of population growth to the new world of intensive agriculture, of high rates of per capita mineral and energy use, of the production of environmental contaminants and even of the crowding of recreation space. Small wonder therefore that such ideas were contested by those whose 'boosterist' heritage came under attack, and by those whose stance was fundamentally in favour of population growth as producing a responsive innovation in technological development and who in the end saw each extra human as the producer of a resource rather than a consumer. The refinement of the neo-Malthusian argument has, however, been a continuing theme of subsequent years, with more and more attention being paid to the social context of population growth and the contexts in which policy decisions are made about, for example, U.S. aid to family planning programmes overseas. These more developed ideas were brought together in *The Stork and the Plow: The Equity Answer to the Human Dilemma*,[9] though the use of the definite article in the subtitle perhaps suggests that there is still held to be a central relationship which determines most if not all of the others. The forcing function of population in all those linkages was underlined by a paper that used energy consumption as a surrogate for human impact on the environment to calculate the optimum population size.[10] This came out at 1.5×10^9 people (1.5 billion) using 4.5 TW of energy. The population in 1999 was 6 billion and the energy consumption in the order of 15 TW, so the difference is large.

The more detailed exploration of the relationships between human populations, resource use and environmental impact has been explored by Ehrlich (usually with co-authors and most frequently with Anne Ehrlich) in a number of papers in relatively specialized journals, as well as in sources with a wider circulation. These comprise category (3) of his output. The topics include, but are not confined to, food security and production[11] and the nuclear winter debate.[12] Inevitably, during the 1990s the term 'sustainability' enters the discussion and an integrated attempt to bring together several aspects of the relations of population, technology and environment can be found in the 1992 paper where the social dimensions of the perceived problems are linked to those provided by more mainstream ecological science: '[S]ound science ... can give minimal guidance at best regarding the issues surrounding the question of the kinds of lives people would choose to live.'[13] Their bottom line, not one popular with either democratic governments, large corporations or dictatorships, is that technology cannot make biophysical carrying capacity infinite, though there is presumably a stage somewhere when a vastly increased world population is one half of a food–humans monoculture. The working-out of detail in the topics of food, energy, wildlife, toxicology, water and minerals is at the heart of a number of books which are aimed at college students in the USA and which convey the Ehrlich world-view as well as a great deal of factual material,[14] as well as popular books designed to raise awareness among lay people.[15]

As a result of the study of these connections, Ehrlich was often ready to make significant predictions. The putative famines of the 1970s were accompanied by suggestions that smog in Los Angeles and New York might kill 200,000 people (predicted in 1969), that England would not exist in the year 2000 (said in 1969) and that accessible minerals would be facing depletion before 1985 (dated 1976). These and other prophetic scenarios have been held up as errors and overconfident predictions, for example by Simon in Myers and Simon (1995) and by Gardner (2010). Ehrlich has consistently defended his fundamental ideas, maintaining that facts and science prove them to be correct. While Ehrlich conceded that 'the biggest tactical error in *The Bomb* was the use of scenarios; stories designed to help me think about the future' (Ehrlich and Ehrlich 2009), he insisted that these were never intended as predictions. He admits that the scenarios were way off, especially in their timing, but insists that they deal with such issues as water shortages and famines that people should have been thinking about in the 1960s; events that occurred or now still threaten.

Category (4), widely read student texts and semi-popular works, does not need extended discussion here except to note that throughout this period Ehrlich has been concerned to disseminate his work to as many people as possible. In part, this seems like the action of any advocate who is convinced of his or her case, but it also seems to stem from the fundamental and laudable trait of scientists to expose their work to sceptical audiences. There is no lack of audience for the latter in Ehrlich's case, of course, and the anti-Ehrlich viewpoints have had no shortage of outlets, both in academia and especially in business publications. Books with titles like *The End of Affluence*[16] (published in the U.S. bi-centenary year of 1976) strike at the vitals of the American way of enthusiasm. So there has been, and continues to be, a 'brownlash' of anti-environmentalist rhetoric designed to show that everything is getting better and better: one commentator's summary in 1997 was 'technology has thwarted Ehrlich's projections, and you needn't be Nostradamus to know it always will'.[17] The response to much of this polemic is in the book of that year which takes up the theme of some previous publications, namely the reception of the findings of the human ecology-environmentalist strand in US thought during the post-1960 period.[18]

A few considerations might strike the non-American and guardedly sympathetic commentator. One is the persistently North American tone of the debate, both pro and contra. In early works, a global set of scenarios was often discussed, but the emerging tone from a period of intensive reading is one of a rootedness in the discourses of the world's richest nation. The particular diversities of the many poorer countries seem to be elided. In part at the beginning, this appears to be the consequence of a biologist's view of humans as behaviourally homogenous diversivores, and although there are some strenuous efforts to encompass the social context of change, the cultural context is often given a rather minimal position as if the whole debate over the social construction of 'environment' in the post-structuralist sense had not happened (some movement in the direction of the management of cultural change, largely in North America, is given in a book which stresses that the human mind and its features are mismatched with the world as it now is[19]). It is perhaps then surprising that there seems to be a consistent underestimation of the role of technology in the human–environment relationship. While the consumption of commercial energy may be a good broad-scale indicator of the penetration of technology, it undervalues agents of change such as the microelectronics that make possible vast and immediate transfers of capital. If it is accepted that technology and its associated cultural metaphysics of acceptability have been at the heart of the great changes in ecology and economy such as the spread of agriculture from its hearths and the dissemination of industrialism based on fossil fuels, then to deny it a central and high-profile role in current and near-future metamorphoses tends to the eccentric. True, it may not in the very long run allow humanity to escape certain biophysical constraints, but it may buy time (as did the Green Revolution), and in a world that is often held to be best described by versions of chaos theory rather than linear equations, there is no telling what synergisms may emerge. Not many commentators, after all, forecast the 'soft' revolutions of the late 1980s in Eastern Europe and the place of 'green' thinking that was one of the factors in those mass convulsions.

Ehrlich's contribution to environmental thought since the later 1960s has been characterized by his energetic lack of fear. To engage for half a century in continued controversy with a powerful opposition, while still producing basic science, is an example of stamina which deserves every plaudit. At the age of 81 he continued to give dire predictions for global civilization: 'I believe and all of my colleagues believe that we are on a straightforward course to a collapse of our civilization.'[20] Even those not convinced by all the arguments and those for whom a high-profile role is not part of their personality have to engage with the central question of what numbers of the human species the earth could support at what quality of life for them and for other species as well. Ehrlich's role can perhaps be measured by the fact that this question is now always part of the schedule in any serious environmental debate or research programme.

Notes

1 P.R. Ehrlich, 'Recent Developments in Environmental Sciences', address at presentation of the H.P. Heineken Prize for Sciences, 25 September 1998, accessed at http://dieoff.com/page157.htm

2 For example, P.R. Ehrlich, A.E. Launer and D.D. Murphy, 'Can Sex Ratio Be Defined or Determined? The Case of a Population of Checkerspot Butterflies', *American Naturalist*, 124, 527–39, 1984.

3 P.R. Ehrlich and P.H. Raven, 'Butterflies and Plants: A Study in Coevolution'; see ISI, 'Citation Classics', *Current Contents*, 37, 16, 1984.

4 S. Harrison, J.F. Quinn, J.F. Bauman, D.D. Murphy and P.R. Ehrlich, 'Estimating the Effects of Scientific Study on Two Butterfly Populations', *American Naturalist*, 137, 227–34, 1991.

5 S. von Ciriacy-Wantrup and J.J. Parsons (eds), *Natural Resources: Quantity and Quality*, Berkeley and Los Angeles, CA: University of California Press, 1967.

6 T.R. Vale, *Progress Against Growth: Daniel B. Luten on the American Landscape*, New York and London: Guilford Press, 1986.

7 *The Population Bomb*, prologue; revised and updated as P.R. Ehrlich and A.H. Ehrlich, *The Population Explosion*.

8 *The Population Bomb*, prologue.

9 P.R. Ehrlich, A. Ehrlich and G.C. Daily, *The Stork and the Plow: The Equity Answer to the Human Dilemma*, New York: Putnam, 1995.

10 G.C. Daily, P.R. Ehrlich and A. Ehrlich, 'Optimum Human Population Size', *Population and Environment*, 15, 469–75, 1994.

11 For example, G.C. Daily and P.R. Ehrlich, 'Population, Sustainability, and Earth's Carrying Capacity'.

12 For example, P.R. Ehrlich, A.H. Ehrlich and H.C. Mooney (eds), *The Cold and the Dark: The World After Nuclear War*, New York: Norton, 1984.

13 G.C. Daily and P.R. Ehrlich, 'Population, Sustainability and Earth's Carrying Capacity', p. 770.

14 For example, P.R. Ehrlich and A.H. Ehrlich, *Population Resources Environment: Issues In Human Ecology*, San Francisco, CA: Freeman, 1970.

15 For example, P.R. Ehrlich and R.L. Harriman, *How To Be a Survivor*, New York: Ballantine Books, 1971.

16 P.R. Ehrlich and A.H. Ehrlich, *The End of Affluence: A Blueprint for Your Future*, Rivercity, MA: Rivercity Press, 1976.

17 S. Milloy, 'Doomsayer Paul Ehrlich Strikes Out Again', accessed at http://fumento.com/environment/bomb.html

18 P.R. Ehrlich and A.H. Ehrlich, *Betrayal of Science and Reason*.

19 R. Ornstein and P.R. Ehrlich, *New World, New Mind. Changing the Way We Think to Save Our Future*, London: Methuen, 1989.

20 Clarke A. https://vtdigger.org/2013/05/01/biologist-paul-ehrlich-gives-dire-prediction-for-global-civilization/

See also in this book
Darwin, Goethe, Leopold, Wilson

Ehrlich's major writings
The Population Bomb, New York: Sierra Club/Ballantine Books, 1968.

Ehrlich, P.R. and Ehrlich, A.H., *The Population Explosion*, New York: Simon & Schuster, 1990.

Ehrlich P.R. and Ehrlich A.H., *Healing the Planet*, New York: Addison-Wesley, 1991.

Ehrlich, P.R. and Ehrlich, A.H., *Betrayal of Science and Reason: How Anti-Environmental Rhetoric Threatens Our Future*, New York: Putnam, 1996.

Ehrlich P.R. and Ehrlich, A.H., *The Dominant Animal: Human Evolution and the Environment*, Washington, Island Press, 2008.

Ehrlich, P.R. and Ehrlich, A.H., 'The Population Bomb Revisited', *Electronic Journal of Sustainable Development*, 1 (3): 63–71.

Ehrlich, P.R. and Ornstein, R.E., *Humanity on a Tightrope: Thoughts on Empathy, Family, and Big Changes for a Viable Future*, Rowman and Littlefield, Lanham, Maryland, USA, 2010.

Further reading

Gardner, G., *Future Bubble: Why Expert Predictions Fail – and Why We Believe Them Anyway*, Toronto, Canada: McClelland and Stewart, 2010.

Johnson, S., *The Politics of Population: The International Conference on Population and Development, Cairo 1994*, London: Earthscan, 1995.

Lutz, W. (ed.), *The Future Population of the World: What Can We Assume Today?*, London: Earthscan for IIASA, 1995.

Myers, N. and Simon, J.L., *Scarcity or Abundance? A Debate on the Environment*, New York and London: Norton, 1995.

Simon, J.L. (ed.), *The State of Humanity*, Oxford, UK and Cambridge, MA: Blackwell, 1995.

Simon, J.L., *The Ultimate Resource 2*, Princeton, NJ: Princeton University Press, 1996.

United Nations Environmental Program, *Global Outlook 2000*, London: Earthscan, 1999.

Vitousek, P., Mooney, H.A., Lubchenco, J. and Melillo, J.M., 'Human Domination of Earth's Ecosystems', *Science*, 277, 494–9, 1997.

IAN G. SIMMONS

HOLMES ROLSTON III 1932–

> Duties arise to the individual animals and plants that are produced as loci of intrinsic value within the system.
>
> *(Environmental Ethics, p.188)*

Holmes Rolston III is widely recognized as the 'father' of environmental ethics as an academic discipline. More so than any other, he has shaped the essential nature, scope and issues of the discipline.

Throughout Rolston's nine books and many articles, he holds that intrinsic value entails duties. Especially influential were Rolston's early, ground-breaking article in the journal *Ethics* (1975), his comprehensive formulation of his ethical theory in the book *Environmental Ethics* (1988), and his impressive compendium and assessment of environmental ethics for the new millennium in *A New Environmental Ethics* (2012). In 1997, in recognition of Rolston's contributions to natural theology, he gave the prestigious Gifford Lectures at the University of Edinburgh in Scotland, published under the title *Genes, Genesis and God* (1999). In 2003, he received the Templeton Prize for 'progress or discoveries about spiritual realities', awarded by Prince Philip in Buckingham Palace. Teleological theism is also a theme in *Three Big Bangs: Matter-Energy, Life, Mind* (2010).

Holmes Rolston III was born 19 November 1932, the son and grandson of Presbyterian ministers, whose names he shares. Except for summers spent in Alabama on his mother's parents' farm, Rolston spent his childhood in the Shenandoah Valley in Virginia, where his father was a Presbyterian minister and respected theologian. In these rural places, Rolston grew to love nature and to value simplicity. The Maury River flowed in front of the family home, which was nestled in the woods, and the Blue Ridge Mountains shaped the horizon. The house lacked electricity, and water came from cisterns.

As an undergraduate at Davidson College, Rolston wanted to study nature and so completed his degree in physics (BS, 1953). Planning to be a Presbyterian minister like his father and grandfather, Rolston next obtained a divinity degree from Union Theological Seminary in Richmond, Virginia (BD, 1956), and then a PhD in philosophical theology at the University of Edinburgh in Scotland (1958). For the next decade, he was a minister in the Appalachian Mountains in Virginia near the Tennessee and North Carolina borders. He and his wife, Jane, have two children, a daughter and a son.

In his spare moments while serving as minister, Rolston attended classes at East Tennessee State University, and explored the biology, mineralogy and geology of the southern Appalachian Mountains, becoming a recognized naturalist and bryologist. He also worked as an activist to conserve wildlife, to preserve Mount Rogers and Roan Mountain, and to maintain and relocate the Appalachian Trail.

Rolston felt a need to study philosophy in an attempt to explain the values he found in nature and to resolve the intellectual conflicts between his religious faith and the non-theistic naturalism of the biological sciences. Leaving his beloved Virginia, he studied philosophy of science at the University of Pittsburgh. There he began to formulate his theory of the intrinsic value of nature and his objections to the naturalistic fallacy. After finishing a degree in philosophy of science in 1968, Rolston was appointed Professor of Philosophy at Colorado State University, Fort Collins, where during the ensuing decades he achieved international academic recognition. He has given invited lectures on all seven continents. In addition to his many academic achievements, he has continued his ordained status in the local Presbytery.

Five concepts frequently recur throughout Rolston's environmental writings: (1) the intrinsic value of nature, which value is non-anthropocentric and even anti-anthropocentric since it is independent of and apart from humankind; (2) ecological-systemic holism; (3) the derivation of duties to nature from the intrinsic value of nature, which logically entails, Rolston argues, the denial of the naturalistic/is-ought fallacy; (4) the intrinsic value of species as forms of life; and (5) biocentrism, that is, the intrinsic value of and derivative duty to respect every individual living organism. Prominent in Rolston's later religious works is the controversial claim that evolution, at least on some tracks, results in progress and is best explained by some form of teleological theism.

Central to Rolston's theory of environmental ethics are the concepts 'intrinsic value' and 'holism'. Aldo Leopold proposed holism under the rubrics of 'community' and 'land ethic'. Holism is a familiar concept in ecology, and has become a key component in many contemporary theories of environmental ethics. In Rolston's theory, ecological wholes are intrinsically valuable. His ethic is explicitly an ethic of duties derived from intrinsic value.

Rolston clearly identifies two 'rules' or 'principles': the Homologous Principle and the Principle of Value Capture.[1] He also uses at least four other principles, for a total of six. Others may need to be added. These six principles are:

1　The Homologous Principle: Follow Nature
2　The Value-Capture Principle
3　The Organic Principle: Respect for Life
4　The Species Principle: Preserve 'Forms' of Life
5　The Ecosystemic Principle
6　The Three 'Environments' Principle: Urban, Rural and Wilderness.

By 'nature', Rolston generally means *non-human* nature. He carefully distinguishes 'nature' and 'culture'. Culture is an artefact made possible by human self-awareness and linguistic rationality, found to a degree much exceeding any other species, and which make possible the cumulative acquisition and transfer of knowledge, information, science, technology, ethics, religion, and a host of other achievements. In contrast to 'deliberative' culture, nature is 'spontaneous' and 'non-reflective'.[2] Natural processes are law-like, orderly though also probabilistic, even chaotic, and open to historical novelty, as evidenced in the creativity in evolving ecosystems. Natural selection, combining with genetics, results in the genesis of value.

Rolston acknowledges that humans are in and part of nature, enfleshed or incarnate in vital respects. The biology of human bodies, for instance, is fully natural. He often says that humans (and human culture) 'emerged out of' nature. For Rolston, 'wilderness' is a synonym for the environment of nature wherever it is free of human interventions. Wilderness, rural culture and urban culture make up the present world's three kinds of 'environments', each having its own particular intrinsic goods.[3]

Understanding Rolston's metaphysical commitments is essential to understanding his ethic. His explicit commitments are deeply biological and evolutionary. Yet, he parts company with contemporary theoretical evolution when he denies that nature operates by 'nothing but chance'.[4] Rolston's philosophy, in addition to being deeply biological, is also deeply theistic. The ultimate explanation for the origin, order and historical novelty in nature, for this genesis, is God.[5]

Rolston's denial of total chance is consistent with his Organic Principle, which is the assertion that every individual organism, from the simplest cell to the most complex multi-cellular organism, is intrinsically valuable and, therefore, worthy of appropriate respect. Unlike inorganic things, living organisms have 'vitality'. Every living organism has four features: (1) each individual has an identity; (2) it defends itself; (3) it functions with an end (telos); and (4) it has within its DNA information that is passed on, or communicated, to others via reproduction. By virtue of these traits, organisms are centres of valuing; even when unconscious, what happens to them 'matters'. In addition, natural organic evolution is often projective in value in the sense that the values are captured and carried forward in time, producing increases in both (a) numbers (quantity) of individuals and species, and (b) complexity (quality) of the forms of life.[6]

Denying the is-ought fallacy, Rolston argues for a naturalistic ethic in which morality – including both values and duties – can be derived from the holistic character of the ecosystem. 'Substantive values', Rolston contends, 'emerge only as something empirical is specified as the locus of value.'[7] All values are objectively grounded and supported by the possibilities and limitations within the earth's ecosystems.

Rolston concedes that some *concepts* of value important in holism, namely, the Leopoldian concepts of beauty, stability and integrity, are human and perhaps non-natural. Nevertheless, such *values* are a product of the inter-relationship of human persons with an objective environment. What counts as beauty, stability and integrity emerges from the interaction of world and human appreciation. Rather than being located solely in human persons, values are collectively relocated in human persons in the environment. The value of the ecosystem is not imposed on it but is discovered already to be there: 'we find that the character, the empirical content, of order, harmony, stability is drawn from, no less than brought to, nature'. Because the substantive, empirical content is in nature, and in nature independent of human and other valuing beings, the value is appropriately and most clearly called 'intrinsic value'. Rolston asserts that 'here an "ought" is not so much *derived* from an "is" as discovered simultaneously with it'.[8]

As a theory of value, ecological holism claims that multiple levels of value, whether a gene, an individual, a species or a collective ecosystem, are morally relevant and valuable. Rolston argues that value is both in the thing and in the system directly and intrinsically, not just indirectly – or instrumentally – as the thing or system is related to humans or other beings who are rational, sentient, conative or alive.

To use a term favoured by Rolston, the value that emerges at the evolutionary ecosystem level is 'systemic'.[9] Rolston asserts that systemic value is intrinsic. In addition, he seems to hold that systemic intrinsic value is qualitatively richer than – greater than – the intrinsic value of the component parts and sub-systems, whether these components are considered as discrete things or sub-systems, even if their discrete intrinsic values are totalled. The value of the whole is greater than the sum of the parts; the systemic intrinsic value of the whole exceeds the net sum of the intrinsic values of the individuals, species, and sub-systems making up the whole system. Moreover, when the system is compared to any component part or sub-system, the qualitatively richer intrinsic value of the whole system seems to entail that, whenever the health or integrity of the system is threatened, the parts are expendable. The system as a whole captures lower intrinsic values and qualitatively enhances them, thereby exceeding the net sum of their individual intrinsic values.

In support of his notion of natural systemic intrinsic value, Rolston cites research in evolutionary history. The explanation for the accumulated diversity of species in nature is systemic: natural processes include trends to produce greater diversity and complexity of life forms. This generalization seems to be true, despite the catastrophic extinctions in the fossil record, after which nature rebounds. The natural tendency of earth's ecosystems is to increase species diversity – and to do so without any evident limit. Rolston calls such natural value 'systemic'. Natural systemic values are also intrinsic values, and as such they entail duties and obligations, Rolston argues.[10]

Systemic value does not prohibit instrumental use of the component parts, provided the health and integrity of the system are not threatened. According to Rolston's Principle of Value Capture, any human action should not destroy anything of intrinsic value unless the action produces something else of equal or greater intrinsic value.

Evolutionary adapted fit tends to integrate intrinsic values in individuals and species within the habitats of the ecosystems they inhabit, Rolston contends. The pressure is toward good adapted fit. Conflicts between individuals using resources and ecosystems are more a problem for culture, not nature. In other words, Rolston claims that a feature of evolution is the generation of increasingly greater kinds and amounts of intrinsic value. When predators kill prey, for instance, they contribute to greater emergent value – more skilled prey and predators. Even with parasites, evolution is producing greater diversity of life forms. Except for human intrusions that shut down such evolutionary progress, values are enhanced and increased in nature.

Rolston argues that because humans are only members – one of many members – of the biotic community, holism is non-anthropocentric, if not anti-anthropocentric. Moral value is attributed to the natural environment considered as an ecological-systemic whole, independently of humans and human interests. In contrast, anthropocentric-humanistic approaches treat ecosystems only as resource values to be exploited for human ends. A scientifically enlightened humanist would have no reason not to use the planet as a mere resource according to long-term ecological science and the highest humanistic values.

Rolston rejects the anthropocentric view that ecology is merely enlightened and expanded human self-interest. We preserve the environment, not merely because it is in our best long-term economic, aesthetic and spiritual self-interest, but because there is no firm boundary between what is essentially human and what is essentially ecosystem. Human and environmental interests merge; egoism becomes 'ecoism'. Since the boundary between the individual and the ecosystem is diffuse, 'we cannot say whether value in the system or in the individual is logically prior'. The individual is not suppressed but enriched.[11]

A scientific ecological fact is that complex life forms evolve and survive only in complex and diversified ecosystems. If 'human' as we know it is to survive, we must maintain the oceans, forests and grasslands. To convert the planet entirely into cultivated fields and cities would impoverish human life. Humans too need 'ecosystem services.' We also ought to preserve ecosystems to enable the further evolution of the planet, including that of human mental and cultural life.[12]

Echoing Leopold, Rolston maintains that normatively right actions – our duties – are those actions that preserve ecosystemic beauty, stability and integrity. Preserving the ecosystemic status quo, however, is not always entailed because humans can improve and transform the environment for their interests, as with agriculture. Borrowing a metaphor from contemporary physics, Rolston holds that integrity is a function of a 'field' interlocking species and individuals, predation and symbiosis, construction and destruction, aggradation and degradation.

Since human life-support is part of the ecosystem, domestication is enjoined in order maximally to utilize the ecosystem. Biosystemic welfare allows alteration, management and use. 'What ought to be does not invariably coincide with what is.'[13]

Regarding species, Rolston contends that our duties are to the species as forms of life rather than to the individual members of the species. The species is the form; whereas, the individual member re-presents the form. 'The dignity resides in the dynamic form; the individual inherits this, instantiates it, and passes it on.' Biologically and ecologically, the individual is subordinate to the species.[14]

Although extinctions do occur in nature, natural extinctions are open-ended, usually producing diversification, new ecological niches and opportunities, new species and ecological trade-offs. In contrast, extinctions caused by humans are dead ends destroying diversity, producing monocultures and shutting down evolution.

Concerned about recent enthusiasm for humans managing earth in an Anthropocene Epoch, Rolston cautions: 'We must learn to manage ourselves as much as the planet. ... Be a resident on your landscape. ... We do not want to live a de-natured life on a de-natured planet.'[15]

Notes

1 *Environmental Ethics*, pp. 61, 79, *passim*.
2 *Conserving Natural Value*, p. 4, *passim*.
3 *Philosophy Gone Wild*, pp. 40–6; *A New Environmental Ethics*, pp. 48–52.
4 *Environmental Ethics*, p. 207.
5 See *Genes, Genesis and God* and *Three Big Bangs*.
6 *Environmental Ethics*, chapter 6.
7 *Philosophy Gone Wild*, p. 19.
8 Ibid., pp. 19–20.
9 *Environmental Ethics*, pp. 186–9; *Conserving Natural Value*, pp. 68–100.
10 Ibid., pp. 155–7. Rolston cites D.W. Raup and J.J. Sepkoski, *Science*, 215, 1501–3, 1982. See *Genes, Genesis and God* and *Three Big Bangs*.
11 *Philosophy Gone Wild*, p. 25.
12 Ibid., pp. 22–4.
13 Ibid., p. 25.
14 Ibid., p. 212.
15 'After Preservation?', p. 30.

See also in this book
Callicott, Leopold

Rolston's major writings

John Calvin Versus the Westminster Confession, Richmond, VA: John Knox Press, 1972.

'Is There an Ecological Ethic?', *Ethics*, 85, 93–109, 1975.

Religious Inquiry: Participation and Detachment, New York: Philosophical Library, 1985.

Philosophy Gone Wild: Essays in Environmental Ethics, Buffalo, NY: Prometheus, 1986.

Science and Religion: A Critical Survey, Philadelphia, PA: Temple University Press/ Random House, 1987. Reprinted with a new introduction by Rolston entitled, 'Human Uniqueness and Human Responsibility: Science and Religion in the New Millennium', Philadelphia, PA: Templeton Foundation Press, 2006.

Environmental Ethics: Duties to and Values in the Natural World, Philadelphia, PA: Temple University Press, 1988.

Conserving Natural Value, New York: Columbia University Press, 1994.

'Feeding People Versus Saving Nature', in William Aiken and Hugh LaFollette (eds), *World Hunger and Morality*, 2nd edn, Englewood-Cliffs, NJ: Prentice-Hall, pp. 248–67, 1996.

Genes, Genesis and God: Values and Their Origins in Natural and Human History, The Gifford Lectures, University of Edinburgh, 1997–8; New York: Cambridge University Press, 1999.

'What is a Gene? From Molecules to Metaphysics', *Theoretical Medicine and Bioethics*, 27, 471–497, 2006.

Three Big Bangs: Matter-Energy, Life, Mind, New York: Columbia University Press, 2010.

A New Environmental Ethics: The Next Millennium for Life on Earth, New York: Routledge, 2012.

Archives and digital archives, including a full bibliography, are in the library at Colorado State University. Many items are available on-line at:

https://dspace.library.colostate.edu/handle/10217/100484
http://hdl.handle.net/10217/38997
http://hdl.handle.net/10217/38998

Further reading

Preston, Christopher J., *Saving Creation: Nature and Faith in the Life of Holmes Rolston III*. San Antonio, TX: Trinity University Press, 2009. An intellectual biography.

Preston, Christopher J., 'Epistemology and Intrinsic Values: Norton and Callicott's Critiques of Rolston', *Environmental Ethics*, 20, 409–28, 1998.

Preston, Christopher J. and Ouderkirk, Wayne (eds), *Nature, Value, Duty: Life on Earth with Holmes Rolston, III*. Dordrecht: Springer, 2007. A *festshrift*.

Kheel, Marti, 'From Heroic to Holistic Ethics: The Ecofeminist Challenge', in Greta Gaard (ed.), *Ecofeminism: Women, Animals, and Nature*, Philadelphia, PA: Temple University Press, pp. 243–71, 1993.

Rolston, Holmes, 'After Preservation? Dynamic Nature in the Anthropocene' in Ben A. Minteer and Stephen J. Pyne (eds), *After Preservation in the Age of Humans*, Chicago, IL: University of Chicago Press, pp. 32–40, 2015.

The Monist, 75, 2 (April), 1992; topical issue on 'The Intrinsic Value of Nature'.

JACK WEIR

SEYYED HOSSEIN NASR 1933–

The crisis of the natural environment is an external reminder of the crisis within the souls of men and women who, having forsaken Heaven in the name of Earth, are now in danger of destroying Earth as well.[1]

Since the mid-twentieth century, Seyyed Hossein Nasr has contributed a philosophical and theological worldview in contrast with secular materialism, which he blames for environmental crises. Like many other environmental writers in the late twentieth century, he locates the source of environmental problems in problematic worldview. Asserting a return to the religious from the secular, to the spiritual from the material, to the internal from the external, Nasr considers nesting human self-understanding in a sacralized natural order to be foundational for ecological ways of relating to nature. Nasr's work reflects his commitments to inter-religious conversation, grounded in a perennialist or traditionalist view, identifying key commonalities between various religious traditions. Often cited as a leading voice of Islamic environmental thinking, his critique of "Western" development also suggests his concern for globalization's economic and cultural transformations, disrupting a variety of communities that perceive nature religiously. Writing on the subject of religion and ecology in 1966 before Lynn White, Jr.'s infamous critique of Christianity's human-centered worldview, Nasr has received many accolades in the West for his inter-cultural, environmental, Islamic scholarship. Thirty years later, he produced another significant publication in religion and ecology, *Religion and the Order of Nature* (1996), which appeared the same year as the Harvard University conference series began on World Religions and Ecology. Spanning more than fifty years, nearly fifty books, and multifarious articles, Nasr's work is highly acclaimed among Western philosophers and Islamic intellectuals, encompassing Islamic science, art, philosophy, and Sufism, as well as perennial philosophy and the environmental crisis.

Born in Tehran, Iran, Nasr grew up in an intellectual household, where the great minds of the time came to dine. Nourished in his youth on Islamic philosophical tradition and Persian culture, Nasr moved to America in 1945 at the age of thirteen. Attending school in a new country and a new language, a few years later he graduated valedictorian and accepted entrance to Massachusetts Institute of Technology. Gravitating toward the study of physics in order to understand the intricacies of the universe, Nasr found a starkly non-religious atmosphere, which failed to encompass the full scope of his intellectual commitments. In 1954, he completed his physics degree with honors at MIT, continue his studies in a more philosophical area, the history of science, for the Ph.D. at Harvard University. During his early career, he filled academic and administrative roles in Iran, Lebanon, and the US. Since 1984, he has served as a professor of religion at George Washington University in Washington, DC.

Two of Nasr's major contributions in religion and ecology coincide with significant historical nodes in this area of study. In 1966, Nasr gave a lecture series at the University of Chicago, later published as *Man and Nature: The Spiritual Crisis of Modern* Man (1967). According to a colleague of Nasr, in the audience during these talks was none other than Lynn White, Jr., who listened to Nasr before publishing his pivotal article, "The Historical Roots of Our Ecological Crisis" (1967). Nasr's lectures locate the crisis of ecological degradation with the scientistic disentanglement of human beings from a cosmological framework of meaning. He traces Western historical changes in the late patristic, medieval, and Renaissance eras, which serve to demystify nature, slowly but steadily reducing the natural world from the prime, enlivened location of spiritual self-understanding to inert matter, instrumentally studied with mechanistic relationality. For example, Nasr writes, "Renaissance man ceased to be the ambivalent man of the Middle Ages, half angel, half man, but now a totally earth-bound creature. He gained his liberty at the expense of losing his freedom to transcend his terrestrial limitations."[2]

Although he cites Copernicus, Sir Francis Bacon, and other familiar people on the historical chopping block of scientific, religious, and ecological meaning, Nasr begins earlier than most by citing a challenge to metaphysical doctrines in a geo-centric, relational orientation to cosmology. Nasr highlights Persian Islamic philosopher Avicenna (Ibn Sina, c. 980–1037), whose cosmology includes angelic presences met in the journey of the soul through the seven heavens. Nasr cites criticism by William of Auvergne and others: "By neglecting the Avicennian souls of the spheres, these scholars had to a certain extent already secularized the Universe and prepared it for the Copernican revolution."[3] Nasr's 1966 lectures go beyond a narration of Western intellectual history, also offering a study of other ecologically significant world religious traditions, and challenging dogmatic aspects of scientific meaning production. Nasr concludes that "modern man" requires rebirth spiritually in order to enhance a simultaneous sense of connection with the divine and the divinely created world.

Nasr published a major work in 1996, the same year as another significant historical moment in the study of religion and ecology. From 1996 to 1998, Harvard University hosted a conference series on World Religions and Ecology. The publications from these conferences represent a major corpus for the field. Nasr's book, *Religion and the Order of Nature* (1996), does comparative work to frame his study of religious cosmologies toward resacralization of nature. Themes arise in parallel with his 1966 work, but he adds further dimension from the thirty years of unfolding conversation religiously, inter-religiously, and intellectually.

Nasr highlights the importance of curbing greed via ascetic perspective and practices shown in a variety of religious traditions. Nasr highlights St. Francis' celebration of animals, nature, and the cosmos while limiting consumption:

> How can one forget that the recently anointed patron saint of ecology, St. Francis, was an ascetic if there ever was one. That did not prevent him from addressing the birds and considering the Sun and the Moon as his kin. The modern world needs nothing more than that so-called world-denying mysticism that is nothing other than its ascetic aspect that seeks to control the passions and to slay the dragon within, without which the greed that drives the current destruction of nature cannot be controlled.[4]

Appreciating a Franciscan sensibility for kinship with other-than-human animals and celestial bodies, curbing consumerist greed with ascetic practices, and cultivating a mystical framework for everyday actions combine to support Nasr's stance on how to solve the ecological crisis through religious means.

Nasr strongly critiques secular thinking, relating it to colonial, religious, political, and economic impositions:

> It is the secularized worldview that reduces nature to a purely material domain cut off from the world of the Spirit to be plundered at will for what is usually called human welfare, but which really means the illusory satisfaction of a never-ending greed without which consumer society would not exist.[5]

He asserts that the divine precedes nature, can be contemplated through nature, and that spiritual self-understanding can be mediated through an understanding of sacrality through the divinely ordained order in the living world.

Nasr's work demystifies science and reanimates nature, in the sense that the fundamental problem has been with perception of a world, which in its essence never lost its sacral quality. Nasr criticizes the scalpel cut dividing "pure" science from realms of religious and spiritual meaning. Beginning during his undergraduate years in physics, he continues for over fifty years with a consistent critique of scientific reductionism, which undergirds an extractive materialism related to environmental crises. For Nasr, the resacralizing of nature has less to do with changing nature, but rather restoring the lens of contemporary inheritors of modern ways of seeing and knowing to view anew life's divine origin when observing nature. For this purpose, Nasr targets the "scientism" of scientific thinking, a problematic ideology reducing life to its component parts, completely measurable, knowable, and materially expressed. Without respect for the creator of that which is created, he posits, the ultimate meaning of reality is forgotten. Lacking sufficient orientation to what nature is, exploitative perspectives are not only acceptable, but lauded as efficient and effective. Further, Nasr argues that peace among people can be achieved through attending to the same spiritual problems encompassed by an exploitative worldview toward nature.

For Nasr, God is the primordial, primary, unifying aspect of life, providing an essential, prime, and perfect non-material ideal in a pre- and post-existent Platonic-like realm, which then informs the whole of nature as created and mirroring the first principle. This perspective reflects Islam's first pillar, which all Muslims assert in the *shahadah*, a verbal expression to bear witness to God's unity. For Nasr and others, all of nature acts as a mirror to see beyond the material to the eternal. Although some might critique such dualistic thinking as anti-earthly or denigrating the bodily, Nasr calls not only for a re-animating of nature, but also of the human body:

> To rediscover the body as the theater of Divine Presence and manifestation of Divine Wisdom as well as an aspect of reality that is at once an intimate part of our being and a part of the natural order is to reestablish a bridge between ourselves and the world of nature beyond the merely physical and utilitarian.[6]

Nasr highlights cosmological doctrines connecting the human body with the earth, both reinvigorated by a reframing of reality from mechanistic to sacralized locations of meaning. In a similar vein, he translated an increasingly popular medieval Islamic treatise from a mystical group called the Ikhwan al-Safa, or the Brethren of Purity, which includes poetic associations between the human body and nature's elements: "The body itself is like the earth, its bones like mountains, its marrow like mines, the abdomen like the sea, the intestine like rivers, the veins like brooks, the flesh like dust and mud."[7] Nasr consistently invokes the importance of analogy and symbolism to elucidate human relationality in a sacralized world. Further, Nasr names the importance of traditional halal slaughter for participating spiritually in nature, and thus, food practice can be a means of reconnecting in a religiously significant manner.[8]

To conclude, Nasr is one of the most important figures in Islamic philosophy working at the intersection of Western and Islamic intellectual traditions, highlighting the vitality of religious meaning in response to ecological devastations. Rather than giving up in the face of such devastation, Nasr invites people of the "West," as well as people touched by the mechanistic mode of thinking, yet closer to traditional perspectives, to return to their roots, to historical doctrines that nest humanity in both divine and natural relationship. He reflects on a perennialist understanding of God's presence in multiple religious traditions, like streams that source from the same origin in mountainous depths. He further envisions a response to environmental ills that is based in humanity's ability to see anew its relation with the whole of nature, a kind of contemporary gnosis:

> Nature has been already sacralized by the Sacred Itself, and its resacralization means more than anything else a transformation within man, who has himself lost his Sacred Center, so as to be able to rediscover the Sacred and consequently to behold again nature's sacred quality.[9]

Notes

1   Seyyed Hossein Nasr, *Religion and the Order of Nature*, New York: Oxford University Press, 1996, p. 6.

2   Seyyed Hossein Nasr, *Man and Nature: The Spiritual Crisis in Modern Man*, rev. edn, Chicago, IL: Kazi Publications, 2007, p. 64.

3   Nasr, *Man and Nature*, p. 62.

4   Nasr, *Religion and the Order of Nature*, p. 217.

5   Nasr, *Religion and the Order of Nature*, p. 271.

6   Nasr, *Religion and the Order of Nature*, pp. 261–2.

7   From Vol. III of the *Rasa'il Ikhwan al-Safa*, trans. Seyyed Hossein Nasr, *Introduction to Islamic Cosmological Doctrines*, New York: Random House/ Shambala, 1978, pp. 101–2, also in Nasr, *Religion and the Order of Nature*, p. 254.

8   Seyyed Hossein Nasr, "Ecology," in *A Companion to Muslim Ethics*, ed. Amyn
     B. Sajoo, pp. 79–90. New York: I.B. Tauris, 2010, p. 83.
9   Nasr, *Religion and the Order of Nature*, p. 271.

See also in this book
Patriarch Bartholomew, Saint Francis of Assisi, White

Nasr's major writings (related to ecology)
"Ecology," in *A Companion to Muslim Ethics*, ed. Amyn B. Sajoo, 79–90. New
     York: I. B. Tauris, 2010.
"The Ecological Problem in Light of Sufism: The Conquest of Nature and the
     Teachings of Eastern Science," in *Sufi Essays*, 2nd edn, Seyyed Hossein Nasr,
     ed., pp. 152–63, Albany, NY: State University of New York Press, 1991.
"Islam and the Environmental Crisis," *The Islamic Quarterly* 34, (4) (1991):
     217–34.
"Islam and the Environmental Crisis," *Journal of Islamic Science* 6, (2) (1990):
     217–34.
"Islam and the Environmental Crisis," in *Spirit in Nature*, Steven C. Rockefeller
     and John C. Elder, eds, pp. 83–108, Boston, MA: Beacon Press, 1992.
"Islam, the Contemporary Islamic World, and the Environmental Crisis," in
     Islam and Ecology: A Bestowed Trust, ed. Richard C. Foltz, Frederick M.
     Denny, and Azizan Baharuddin, pp. 85–105, Cambridge, MA: Harvard
     University Press, 2003.
Man and Nature: The Spiritual Crisis in Modern Man, rev. edn, Chicago, IL: Kazi
     Publications, 2007.
Religion and the Order of Nature, New York: Oxford University Press, 1996.

Further reading
William Chittick, *Science of the Cosmos, Science of the Soul: The Pertinence of Islamic
     Cosmology in the Modern World*, Oxford, UK: Oneworld Publications, 2007.
Sachiko Murata and William C. Chittick, *The Vision of Islam*, St. Paul, MN:
     Paragon House, 1994.
Tarik M. Quadir, *Traditional Islamic Environmentalism: The Vision of Seyyed Hossein
     Nasr*, Lanham, MD: University Press of America, 2013.

SARAH E. ROBINSON-BERTONI

WENDELL BERRY 1934–

> We must achieve the character and acquire the skills to live much poorer than we do.[1]

Born in 1934 to John and Virginia Berry, Wendell Berry attended Millersburg Military Institute as a boy, then the University of Kentucky, where he earned A.B. and M.A. degrees in English (1956, 1957). While there he met Tanya Amyx; they married in 1957. Berry went to Stanford University as a Wallace Stegner Fellow, then to Italy on a Guggenheim Fellowship (1961–2), before returning to the US to teach English at New York University. In 1964 Berry returned to Kentucky. He taught at the University of Kentucky from 1964–77, and again from 1987–93, all the while farming and improving the same land he purchased when he left New York.

Berry's environmentalism is distinctive in a few ways that bear mentioning. One is that he dislikes the word "environment." Almost from the beginning of his life as a writer, and then increasingly in his unsolicited but inevitable position as a public figure, he has objected to "environment" because it implies that we are separate from nature when in fact we are in a state of uninterrupted exchange with it. "We tend to think," Berry told an interviewer in 1991, "that there can be a distinction between people and the air they breathe, for instance, or people and the food they eat, or people and the water they drink." He called this distinction "absurd." "There is no line that you can draw between people and the elements they depend on. That is why the term 'environment' is so bothersome to me. 'Environment' is based on that dualism, the idea that you can separate the human interests from the interests of everything else. You **cannot** do it. We eat the environment. It passes through our bodies every day."[2]

This matters to Berry because a unity and an order are at stake. We cannot diminish nature or ourselves without diminishing the other. And yet we have too willingly carved the world up into gardens here and garbage dumps there, as if nature herself were not a unity but, instead, a collection of sealed-off shipping containers. Nature—our mother *and* our judge—does not pardon this transgression.

But the unity and order are not external to us only: "our surroundings," Berry wrote in the 1970s, "from our clothes to our countryside, are the products of our inward life—our spirit, our vision—as much as they are products of nature and work."[3]

> [And] things that appear to be distinct are nevertheless caught in a network of mutual dependence and influence that is the substantiation

of their unity. Body, soul (or mind or spirit), community, and world are all susceptible to each other's influence, and they are all conductors of each other's influence. The body is damaged by the bewilderment of the spirit, and it conducts the influence of that bewilderment into the earth, the earth conducts it into the community, and so on.[4]

"Environment," then, is the perfect word for people who live at what he once called the far end of a broken connection. In the 1970s, in a long essay titled "Discipline and Hope," he said there is really only one concern: "the life and health of the world. If there is only one value, it follows that conflicts of value are illusory, based upon perceptual error."[5] No amount of fragmentation, whether articulated or experienced, will change that.

A more important feature of Berry's thought is his dismay that environmentalists have shown insufficient interest in good farming. Proper consideration must be given to what he calls the "third landscape," the landscape that is neither abused nor left alone but put to "kindly use." By this he means, principally, land that is farmed well by men and women who consult the land to see what it can be expected to provide while still retaining its agricultural potential.

The Unsettling of America (1977), which might still be Berry's most important book, was written out of this concern precisely. In it Berry argued vehemently against the "Get Big or Get Out" doctrine that characterized official U.S. agricultural policy after WWII, and he defended, instead, land put to "kindly use."[6] But, even more, he attempted to show how the ecological crisis is a crisis not only of agriculture but also of character and culture, due in large measure to a tendency toward specialization, which in a money economy results in people who are unequal to their needs—neither competent in the range of things that are necessary nor knowledgeable of them. As Berry would later say, industrialized agriculture allowed many people to rise above the need to feed themselves, even though none of them had ever risen above the need to eat. The post-war migration from farm to city sponsored by machinery and cheap credit resulted in an abused countryside and a drastic change in the eyes-to-acre ratio. There were too few eyes on the land to notice the abuse.

This change in ratio intimates another feature of Berry's particular kind of environmentalism: that, just as our farms must be small enough for the available eyes on them to keep watch over, so our solutions to the abuse of nature must be small enough for us to manage: "[O]ur understandable wish to preserve the planet must somehow be reduced to the scale of our own competence."[7] The keyword for Berry here, as elsewhere, is "scale." He is seldom impressed by the Planet-Savers with their Big Ideas and New Technologies. A preference for what is big leads almost inevitably to destruction, as it clearly has in industrial farming. And "new" too often the means the process of dividing a solution into two or more new problems—as when, for example, you remove animals from the farm, and therefore remove the free fertilizer they provide, and then find that you must import expensive chemical fertilizer to offset the loss. Our aims should be more modest—limited, for example, to caring for all the human and natural neighborhoods nearby, which is something we can actually do. "That will-o'-the-wisp, the large-scale solution to the large-scale problem, serves mostly to distract people from the small, private problems that they may, in fact, have the power to solve."[8]

If all of this has the sound of someone who prefers bottom-up as opposed to top-down solutions, that is because Berry almost invariably does. He does not for this reason believe that organizations and governments should do nothing. Clearly they too should enact the offices of stewardship as well. But Berry doubts that the affection necessary for stewardship is likely to come out of institutions and organizations. Affection issues from people, not from boardrooms. Big plans and new technologies, by contrast, issue from whatever distant saviors stand to profit by them.

They also tend to persuade us that they will solve our problems without making any demands on our behavior. Thus the car-of-the-future promises to solve the fossil-fuel problem without asking us to break our addiction to driving. Large-scale solutions, in addition to being invariably expensive, are ultimately inadequate because they do not solve the moral and intellectual problems in each of us that have caused the crises in the first place. Among those problems Berry identifies the following:

1   A failure to distinguish between needs and desires. When we attempt to satisfy our unlimited desires—which differ from real needs—without regard for the earth's limited capacity to fulfill them, we meet not only with frustration but with manufactured filth. "To have even the illusion of infinite quantity," Berry says, "we would have to debase both the finite and the infinite."[9]

2   A failure to reconcile ourselves to our condition. The curse of Adam, which is that he shall work by the sweat of his brow, is not ours to overturn. What are people *for*?, he asks.

   Is their greatest dignity in unemployment? Is the obsolescence of human beings now our social good? One would conclude so from our attitude toward work, especially the manual work necessary to the long-term preservation of the land, and from our rush toward mechanization, automation, and computerization.[10]

3   A failure to understand what kind of creatures we are. We tend to violate our nature in two ways: we aspire above our condition and presume to behave as gods—that is, as spirits unencumbered by the limits of the flesh—and then inevitably sink below our condition and behave as beasts. Culture and religion acknowledge our animal nature but remind us that we are "capable of living, not only within natural limits, but also within cultural limits, self-imposed."[11] Nor can we live artfully without respecting, as art does, formal limits.

4   A failure to understand energy. We squander energy in the form of ancient sunlight and refuse to work by the power of *contemporary* sunlight, which is clean, democratic, and free. We hand over our work to machines, and machines not only usurp us; they operate on the assumption that the energy they need is infinitely available. Meanwhile, we accrue costs in obesity and poor health, costs that do not get entered into the ledger.

5   A failure to keep honest books. We are quick to add up the benefits of economic activity but slow to subtract the costs. What is the net gain of computers, he asks in *Life is a Miracle*, after we have subtracted the costs for having made, used, and thrown them away? Our indicators of wealth, such as the Gross Domestic Product (GDP), are incapable of accounting for *health*. Divorce, disease, and sickness are good for the GDP. But these are costs that must be accounted for. And, like all costs, they will have to be paid at some point by someone.

Instead of the industrial economy, which "takes, makes, uses, and discards," Berry would place an agrarian economy, which "takes, makes, uses, and *returns*."[12] An agrarian economy works, as nature herself does, by wasting nothing. It extracts nothing that it does not give back in the form of decaying matter, which nourishes new life. The cycle is one of death and resurrection, on and on and on, as long as nature lasts. "Fertility is always building up out of death into promise."[13]

There is much more to Berry's thought, but this brief summary must limit itself to two more features, first among them, his insistence on the primacy of *place*—of knowing where you are and of being willing to stay there in the face of a powerful social prejudice for upward mobility. We care for places, Berry insists, only by long association with them; those who are always on the move, by contrast, cannot be counted on to care for any place in particular, for they are never anywhere long enough to develop the necessary affection for it. Berry himself has attempted to be worthy of where he is, to know his place well and belong to it fully. "Whereas most American writers—and even most Americans—of my time are displaced persons," he says, "I am a placed person."[14] His repatriation was an enterprise in arriving

> at the candor necessary to stand on this part of the earth that is so full of my own history and so much damaged by it, and ask: What *is* this place? What is in it? What is its nature? How should men live in it? What must I do?[15]

This emphasis on place accounts in part for Berry's dissatisfaction with education presently conceived, which can never teach a proper ecology so long as it also teaches students that where they are is never quite as good as where they might be. That doctrine cannot teach affection, and it will not lead to proper care.

This essay neglects certain aspects of Berry's thought, for example his critique of science: by its own authority it has "crowned and mitred itself."[16] It also neglects his critique of technology: "People who are willing to follow technology wherever it leads are necessarily willing to follow it away from home, off the earth, and outside the sphere of human definition, meaning, and responsibility."[17] But it has attempted something like a portrait of Berry for those who would strike out and follow him into the varied landscape of his work.

Once there, they will see what kind of moral guide they have:

> We must learn to discipline ourselves, to restrain ourselves, to need less, to care more for the needs of others. We must understand what the health of the earth requires, and we must put that before all other needs. If a catastrophic famine is possible, then let us undertake the labors of wisdom and make the necessary sacrifices of luxury and comfort.[18]

Notes

1 *What Are People For?* (San Francisco, CA: North Point Press, 1990), p. 201.
2 *Conversations with Wendell Berry*, ed. Morris Allen Grubbs (Jackson, MS: University Press of Mississippi, 2007), p. 41.
3 *The Unsettling of America* (San Francisco, CA: Sierra, 1977), pp. 11–12.
4 Ibid, p. 110.
5 *A Continuous Harmony* (1970; Washington, DC: Shoemaker & Hoard, 2004), p. 157.
6 *The Unsettling of America*, p. 30.
7 *What Are People For?*, p. 200.

8 Ibid., p. 198.
9 *The Unsettling of America*, p. 78.
10 *What Are People For?*, p. 125.
11 *What Matters? Economics for a Renewed Commonwealth* (Berkeley, CA: Counterpoint, 2010), p. 45.
12 *Home Economics* (San Francisco, CA: North Point, 1987), p. 124.
13 *The Long-Legged House*, p. 204.
14 Ibid, p. 140.
15 Ibid, p. 199.
16 *Life is a Miracle*, p. 18.
17 *Standing By Words* (1983; Washington, DC: Shoemaker & Hoard, 2005), pp. 59–60.
18 *The Unsettling of America*, pp. 65–6.

See also in this book
Balfour, Carson, Thoreau

Berry's major writings
A Continuous Harmony. 1970; Washington, DC: Shoemaker & Hoard, 2004.
Citizenship Papers. Washington, DC: Shoemaker & Hoard, 2003.
Home Economics. San Francisco, CA: North Point, 1987.
Jayber Crow. New York: Counterpoint, 2000.
Life is a Miracle. Washington, DC: Shoemaker & Hoard, 2000.
The Unsettling of America: Culture and Agriculture. San Francisco, CA: Sierra Club, 1977.
What Are People For? San Francisco, CA: North Point, 1990.
What Matters? Economics for a Renewed Commonwealth. Berkeley, CA: Counterpoint, 2010.

Further reading
Angyal, Andrew. *Wendell Berry*. New York: Twayne, 1995.
Baker, Jack R. and Jeffrey Bilbro. *Wendell Berry and Higher Education: Cultivating Virtues of Place*. Lexington, KY: University of Kentucky Press, 2016.
Merchant, Paul. *Wendell Berry*. Lewiston, ID: Confluence, 1991.
Mitchell, Mark and Nathan Schleuter (eds). *The Human Vision of Wendell Berry*. Wilmington, DE: ISI, 2011.
Oeschlaeger, Fred. *The Achievement of Wendell Berry*. Lexington, KY: University of Kentucky Press, 2011.
Peters, Jason (ed.). *Wendell Berry: Life and Work*. Lexington, KY: University of Kentucky Press, 2007.
Shuman, Joel James and L. Roger Owens. *Wendell Berry and Religion: Heaven's Earthly Life*. Lexington, KY: University Press of Kentucky, 2009.

Smith, Kimberly K. *Wendell Berry and the Agrarian Tradition: A Common Grace.* Lawrence, KS: University Press of Kansas, 2003.

Wirzba, Norman (ed.). *The Essential Agrarian Reader.* Lexington, KY: University Press of Kentucky, 2003.

JASON PETERS

JANE GOODALL 1934–

> Only if we understand, will we care. Only if we care, will we help. Only if we help shall all be saved. The least I can do is speak out for those who cannot speak for themselves. The greatest danger to our future is apathy.[1]

Dame Jane Goodall, DBE and United Nations Messenger of Peace, is an iconic, holistic, biocentric, eclectic, and courageous environmentalist. I constantly hear from people around the world "I want to be just like Jane Goodall." If even a small fraction of them achieved a small part of what Goodall has accomplished the world surely would be a much better place for (nonhuman) animals, people, and the environment. Goodall has had major global effects on numerous people, animals, and habitats through her original research and her tireless efforts to make the world a better place for all beings. She not only is concerned with environments, but also with the nonhumans who live in these landscapes. Goodall views environments as dynamic and constantly adapting entities. She is a vegetarian for ethical and health reasons.

Even today, many who are concerned with environmental and conservation issues take an anthropocentric point of view and place human interests first and foremost (for example, the "new conservationists") and forget that other animals are key members of diverse landscapes and their lives have to be taken into account when environmental problems and issues are discussed and attempts are made to solve the issues at hand. Humans are not the only show in town. Goodall has been a, some would say *the*, central figure in bringing animals to the table, always stressing an inclusive biocentric point of view. It is holistic as well, embracing the well-being of humans, animals, and flora with the same passion, and never-say-never attitude.

Goodall also has been extremely concerned with ethical issues surrounding the keeping of animals in captivity and also how they are treated in their wild homes, and to this end she and I also cofounded

Ethologists for the Ethical Treatment of Animals: Citizens for Responsible Animal Behavior Studies[2] in July 2000.

One might say Goodall was obsessed with going to Africa from her early years living in Bournemouth (UK). Her mother, Vanne, who accompanied Goodall on her maiden voyage to Gombe, supported Goodall's deep interests in animals. It's said that a stuffed chimpanzee named Jubilee stimulated her love for other animals. Goodall worked hard to earn the money to make her dream come true, and she first went to Kenya in 1957. There, she worked as a secretary and eventually called and met the renowned anthropologist, Louis Leakey. Because of his interests in learning about the behavior of early humans, he wanted to learn more about the behavior of nonhuman great apes. Goodall went back to London to study primate behavior and after Leakey raised the necessary money he sent her off to the Gombe Stream Reserve (now called the Gombe Stream National Park) in Tanganyika (now Tanzania) in 1962 for what they both thought would be a short study. Goodall went into the field without a degree and eventually received her Ph.D. from the University of Cambridge (UK). The research at Gombe continues today after 57 years.

Back in the early 1970s when I was a graduate student, I'd heard about Goodall going off to live with the chimpanzees of Gombe. And, I also knew of her first husband, the renowned National Geographic photographer, Hugo van Lawick-Goodall. I'd already read Goodall's groundbreaking *Animal Behaviour* monograph called "The Behaviour of Free-living Chimpanzees in the Gombe Stream Reserve." It was clear she was well on the way to making a difference in how animals were studied and the ways in which people referred to and would come to view these beings as emotional sentient individuals.

In fall of 1971, an unexpected visitor came to my home in St. Louis, Missouri, where I was a graduate student at Washington University. It was Hugo. While Hugo was there we had long chats about animal behavior, the importance of observing identified individuals over long periods of time, and what Jane was accomplishing, despite a large number of skeptics, mostly males. Jane's seminal observations of David Greybeard making and using a tool were met with skepticism until she showed a video of this amazing behavior. Based on these observations, Leakey said, "Now we must redefine tool, redefine Man, or accept chimpanzees as humans."[3] These observations began a field that is flourishing today, namely, the study of the manufacture and use of tools by a wide variety of nonhuman animals.

In addition to this groundbreaking seminal observation, Goodall's major influences on the field of animal behavior include her naming the chimpanzees she studied, talking freely about their wide-ranging emotions, and stressing their individual personalities. She also always felt that every single individual counts, not only among the animals she was studying but also when working with people who were concerned with saving other species and their homes. At the time, naming animals and talking about emotions and personalities were not standard operating procedure in studies of animal behavior, most of which were conducted in artificial situations in various sorts of captive settings. I was told, "Naming animals is too subjective and it'll influence how data are explained," individual differences and emotions are "noise in the system," and talking about animal personalities is fraught with error and taboo.

I well remember mentioning Goodall's research and views, and my doctoral committee accepted them. At the time, most researchers had very mechanistic views of other animals, engaged in normative thinking about them, and frowned on naming or assigning personalities. They preferred to talk about "the dog," "the coyote," "the chimpanzee," or "the elephant," and ignored individual variation and personalities. Goodall's views shattered these narrow views of other animals as reflex machines or automatons, and subsequent research has shown how right she was and how wrong they were.

Goodall refused to change the ways in which she referred to the chimpanzees. In the end, her courage and conviction worked and she received he Ph.D. And, over the past fifty seven years, Goodall has been proven to be right on the mark—animals are subjects, not objects, and their unique individual personalities are extremely important to study and to embrace. Their cognitive and emotional lives must be taken into account when we interact with them and when we make decisions that affect their lives. Of course, viewing other animals as sentient individuals needed to be factored into the ways in which environmentalists viewed and solved environmental conflicts, but it was slow going. Goodall's inclusive environmentalism significantly changed "business as usual" in which the interests of humans routinely and unquestionably trumped those of other animals.

A list of Goodall's accomplishments in the environmental sciences is staggering. In 1977 she founded the Jane Goodall Institute,

> a global nonprofit focused on inspiring individual action to improve the understanding, welfare and conservation of great apes and to safeguard the planet we all share ... [its] mission is based in Dr. Jane Goodall's belief that the well-being of our world relies on people taking an active interest in all living things.[4]

In 1986, an international conference called "Understanding Chimpanzees" was held in Chicago. It was at this gathering Goodall decided that she simply had to do more for chimpanzees than study them. Of course, she realized that research is essential, but she had and still has a solid team of researchers at Gombe. Goodall began her globe-trotting campaign dedicated to conservation of chimpanzees and other animals. She still travels more than eleven months a year.

Goodall also is deeply dedicated to working for and with humans, and her community-based conservation and development effort called the Tanganyika Catchment Reforestation and Education (TACARE) project launched in 1994 is an excellent example of this global work. TACARE "partners with local inhabitants to create sustainable livelihoods while promoting environmental protection. TACARE achieves conservation results by first consulting communities about their needs and priorities, working together to collaboratively design the future rather than imposing outside solutions."[5]

Goodall's concern with protecting sentient and other animals is legendary. She was an early compassionate conservationist. Compassionate conservation is a rapidly growing interdisciplinary field. The basic guiding principles for compassionate conservation are first do no harm when dealing with human–animal conflicts; all individuals matter; we must strive for peaceful coexistence between humans and nonhumans; and humans and nonhumans are all stakeholders whose interests have to be factored into solutions when conflicts arise. Given her interests in viewing animals as unique individuals with unique personalities, Goodall is the prototypical compassionate conservationist, embracing the holistic ideas that all individuals matter, all individuals are stakeholders, and that we must strive for peaceful coexistence among nonhumans and humans.

Being pragmatic and positive are two very good ways to work for a better future for all. Goodall embodies both. She realizes that we live in an increasingly human-dominated world in which the nonhumans need all the help they can get. She also realizes that for the vast majority of people, human interests outweigh and trump those of nonhumans. Nonetheless, Goodall has always worked hard to be sure that she is doing all she can do so that all beings are able to live in peace and safety. Goodall says,

> I like to envision the whole world as a jigsaw puzzle … If you look at the whole picture, it is overwhelming and terrifying, but if you work on your little part of the jigsaw and know that people all over the world are working on their little bits, that's what will give you hope.[6]

Goodall also has a strong spiritual side. She notes,

From my perspective, I absolutely believe in a greater spiritual power, far greater than I am, from which I have derived strength in moments of sadness or fear. That's what I believe, and it was very, very strong in the forest.[7]

Goodall remains positive and hopeful while facing situations that would make others shudder and from which they would turn away. For Goodall, hope rather than despair makes action possible. Her global Roots & Shoots program, founded in Dar as Salaam in 1991 with sixteen teenagers, now has more than 10,000 groups in more than 120 countries.[8] In Roots & Shoots groups of people of all ages and cultures are educated to face innumerable and challenging environmental problems head-on, often with a focus on local issues. By doing this they can rewild themselves, reconnect with nature and other animals, and encourage others to do the same. Humane education is key for a better future for all, and learning about the problems at hand and how they can be solved while taking into account the interests of all stakeholders is the leading face of Goodall's activism. She truly believes that *everyone* can make a difference.

All in all, Jane Goodall is an iconic environmentalist who looks at other animals, people, and the environment from many different perspectives. She clearly is one of the most influential scientists and spokespersons for other animals for all time. Goodall works and travels selflessly and tirelessly, remains positive and pragmatic, and embraces, biocentrism and holism. This is no easy task and somehow Goodall has been able to do this for many decades, and still remains optimistic and indefatigable in her early 80s.

Notes

1   www.quotationspage.com/quotes/Jane_Goodall/
2   www.ethologicalethics.org
3   www.janegoodall.ca/about-chimp-behaviour-tool-use.php
4   www.janegoodall.org
5   https://blogs.esri.com/esri/esri-insider/2011/08/15/jane-goodalls-tacare-geodesign-in-action/
6   www.passagetoafrica.com/articles/273-the-interview-jane-goodall
7   http://quotecourt.com/jane-goodall-quote-265043
8   www.rootsandshoots.org

See also in this book
Darwin, Singer, Uexküll

Goodall's major writings

"The Behaviour of Free-living Chimpanzees in the Gombe Stream Reserve," *Animal Behaviour Monographs*, 1968, 1 (3), 161–311. Available at: www.sciencedirect.com/science/article/pii/S0066185668800032

Innocent Killers (with Hugo van Lawick-Goodall), Boston, MA: Houghton Mifflin, 1971

The Chimpanzees of Gombe: Patterns of Behavior, Cambridge, MA: Belknap Press, 1986.

Reason for Hope: A Spiritual Journey, New York: Grand Central Publishing, 2000.

The Ten Trusts What We Must Do for the Animals We Love (with Marc Bekoff), San Francisco, CA: HarperOne, 2002.

Harvest for Hope (with Gary McAvoy and Gail Hudson), New York: Grand Central Publishing, 2006.

Hope for Animals and Their World (with Thane Maynard and Gail Hudson), New York: Grand Central Publishing, 2011.

Seeds of Hope: Wisdom and Wonder from the World of Plants (with Gail Hudson), New York: Grand Central Publishing, 2015.

Further reading

Marc Bekoff (ed.), *Ignoring Nature No More: The Case for Compassionate Conservation*, Chicago, IL: University of Chicago Press, 2013.

Marc Bekoff, *Rewilding Our Hearts: Building Pathways of Compassion and Coexistence*, Novato, CA: New World Publishing, 2014.

Dale Peterson, *Jane Goodall: The Woman Who Redefined Man*, Boston, MA: Mariner Books, 2008.

Dale Peterson and Marc Bekoff (eds), *The Jane Effect: Celebrating Jane Goodall*, San Antonio, TX: Trinity University Press, 2015.

MARC BEKOFF

RUDOLF BAHRO 1935–1997

To bring it down to the basic concept, we must build up areas liberated from the industrial system. That means, liberated from nuclear weapons and from supermarkets. What we are talking about is a new social formation and a new civilisation.[1]

Rudolf Bahro was a communist dissident, an early member of the (West) German Greens and a leading proponent of spiritual Green political thought and action. Bahro originally became well known as the author of *The Alternative in Eastern Europe*, which he wrote during the 1970s while he was a dissident Marxist and party member in the former East

Germany. This work was described by Herbert Marcuse as 'The most important contribution to Marxist theory and practice to appear in several decades'. In it Bahro argued that Eastern Europe's non-capitalist, communist development path has been shaped and corrupted by the same growth and materialist aims as Western capitalism. In 1977, the ruling communist government sentenced him to prison for his dissident activities and writings, and in 1979 he was deported to what was then West Germany, in part due to international protests at his imprisonment.

Soon after he arrived in West Germany, Bahro became involved with the nascent German Greens (Die Gruen), affirming that 'red and green go well together',[2] and urged communist groups to dissolve themselves and work within the Die Gruen. As such he was strongly identified with the 'eco-socialist' wing of the Green movement, arguing for a synthesis of Green and socialist ideals and aims, though the latter was a resolutely non-Marxist form of socialism. He was clear that such a rapprochement required the critical reconstruction of socialist politics, a central aspect of which was a rejection of the productivist and 'materialist abundance' dimensions of Marxist socialism, and the emergence of what Bahro called a 'historical compromise' between the labour movement and new social movements (environmental, feminist, peace), and a rejection of Marxist 'class politics' and proletarian revolution. While a vehement critic of capitalism and consumerism, Bahro's view (which had much in common with Antonio Gramsci's 'anti-hegemony' political strategy) was that what was required to defeat capitalism and create a more sustainable, just, democratic and peaceful social order, was a 'rainbow coalition' of all anti-capitalist social forces, and not just the labour movement and the industrial working class. Thus at this stage, Bahro's politics shared much with that of Andre Gorz's 'red-green' position, and indeed that of Murray Bookchin.

An example of where Bahro differed from Marxists was in relation to the creation of a more egalitarian and just world order in terms of the present inequalities between the developed 'North' and un/underdeveloped 'South'. The classic Marxist view would be that what is needed is 'communism on a world scale', a constitutive aspect of which would be the raising of the living standards and lifestyles of the 'Third world' to 'First world' levels. Against this 'cornucopian' view, Bahro, expressing a common Green view that this Marxist myth is both ecologically unsustainable (i.e. physically impossible) and spiritually undesirable/unworthy, proposed that, 'with a pinch of salt one might say ... the path of reconciliation with the Third World might consist in our becoming Third World ourselves'.[3]

Throughout the early 1980s, Bahro became an increasingly vocal and public critic of the 'realo' wing of the German Greens (those who became generally committed to competing for, winning and exercising parliamentary power) and became a leading spokesperson for the 'fundi' or fundamentalist wing of the party. The 'fundi–realo' split within the German Greens, a division which also emerged in other Green parties and Green political thinking, owes much to the passion and conviction with which Bahro railed against what he saw as the corruption and co-option of the radical and emancipatory potential of Die Gruen by 'the system'. He ultimately left the party in 1985. He and his companion Christine Schröter called for an end to all animal experimentation. The party agreed, but decided to make exceptions in the case of medical research, which was unacceptable to Bahro's uncompromising position.

In the mid-1980s, in keeping with his disillusionment with Die Gruen and 'normal' democratic politics, he began to speak less in political terms and more in religious-spiritual terms, asking that 'the emphasis [be] shifted from politics and the question of power towards the cultural level … to the prophetic level … Our aim has to be the "reconstruction of God"'.[4] Bahro had come to the view that if the Greens were to address the ecological crisis by radically changing society, they had to focus their efforts on psychological, cultural and spiritual levels. As he put it: 'If we take a look in history at the foundation on which new cultures were based or existing ones essentially changed, we always come up against the fact that in such times people returned to those strata of consciousness which are traditionally described as religious.'[5] For Bahro, personal inner transformation was a necessary and desirable part of, and precondition for, the wider social and cultural transformation of Western civilization away from its ecologically destructive path.

Green politics must be based on spiritualistic values, in Bahro's view, because, as Eckersley points out, for Bahro 'the challenge of ecological degradation is primarily a cultural and spiritual one and only secondarily an economic one'.[6] Echoing the redemptive character of deep ecology, Bahro's vision of Green politics is of a life- and earth-affirming spirituality and the primary aim of Green politics is to be uncompromising in bringing about radical cultural and psychological change from which political and economic change will follow. A central part of Bahro's analysis focuses on the failures of Western civilization and the Enlightenment as a whole, and his argument is that nothing less than change at the level of civilization will prevent what he calls the 'logic of exterminism' within the 'mega-machine' (a term he borrows from Lewis Mumford) from destroying the planet and humanity along with it.

Bahro's 'social ecology' is a combination of 'völkisch spirituality', deep ecology, 'post-industrial utopianism',[7] 'anti-consumerism'[8] and what Eckersley calls 'ecomonasticism',[9]: a form of Green politics in which the strategy is of 'withdrawal and renewal' or 'opting-out' from the life-denying logic of the industrial 'mega-machine', and the creation of 'Liberated Zones'. These Liberated Zones provide protection for alternative ecological practices and values, places within which experiments in sustainable living can take place, and finally bases from which ecological, cultural and spiritual renewal and change will come. This ecomonastic perspective he shares with Theodore Roszak. As Ferris notes, Bahro's view is that 'Greens should opt out of industrial society by adopting a new lifestyle and living in small self-sufficient communes. Eventually, the communes would demonstrate a qualitatively better way of life that others would wish to adopt.'[10] His ultimate vision of the 'sustainable society' is of an 'ecoanarchist' federation of communes, comprised of small-scale, face-to-face communities which produce and consume the vast majority of what they require, a way of 'living lightly' on the planet.

In 1989, Bahro co-founded an educational centre and commune near Trier, the Lernwerkstatt (an 'ecological academy for one world'), whose purpose is to synthesize spirituality and politics, to put the ecomonastic ideal into practice. It put on cultural events and weekend workshops on various New Age themes, including deep ecology, ecofeminism, Zen Buddhism, holistic nutrition, Sufism, and other alternative theories, therapies and practices. Bahro also held a professorship at Humboldt University in Berlin in 'social ecology', but Bahro's work is not to be confused with the resolutely non-spiritual social ecology conceived and developed by Murray Bookchin.

Critics, both within and outside the Green movement, were concerned at the nationalistic tone of his thinking which was in contradiction with the Green internationalist slogan of 'act locally, think globally'. For example, in the early 1980s peace movement, he alarmed many by enunciating nationalistic arguments against the deployment of Pershing missiles, while in he campaigned against the reunification of Germany after the fall of the Berlin Wall in 1989. Some accused him of flirting with fascism, authoritarian spirituality and linking ecological politics to rightwing/conservative/nationalistic values and principles. He has been portrayed as believing that the ecological crisis is resolvable only through authoritarian, non-democratic means. He calls for a spiritually based and hierarchically elitist 'salvation government' or a 'god-state' (Gottesstaat) 'that will be run by a "new political authority at the highest level": a "prince of the ecological turn"'.[11] Bahro's apocalyptic analysis leads him to suggest that what is required is a 'rescue government', which would be an emergency or crisis government which while possessing absolute power, and thus a non-democratic political order, would be a transitional rather than permanent political arrangement. Standing above Bahro's later analysis of and political prescriptions for the ecological crisis seem to be modern, Green descendants of Plato's Guardians – dedicated, knowledgeable and wise elite stewards who will guide society in the right direction away from ecological, spiritual and cultural disaster, who govern without any democratic input from the people. His thought also echoes aspects of the early ecoauthoritarian diagnosis of the ecological crisis put forward by William Ophuls,[12] and his call for an 'ecological Leviathan', as well as some deep ecological arguments that 'What is required is a new type of warrior – a person who is intense, centred, persistent, gentle, sincere, attentive and alert.'[13]

A rather startling example of the distance he had travelled since his early 'red-green' position is his statement that

> The most important thing is that ... [people] take the path 'back' and align themselves with the Great Equilibrium, in the harmony between the human order and the Tao of life. I think the 'esoteric'-political theme of 'king and queen of the world' is basically the question of how men and women are to comprehend and interact with each other in a spiritually comprehensive way. *Whoever does not bring themselves to cooperate with the world government [Weltregierung] will get their due.*[14]

While the mystical/spiritual-cum-cultural focus of Bahro's analysis here would find some support in the wider Green movement, the anti-democratic and indeed anti-political sentiment would not. His call for a return to the something that was lost is close to other 'green mavericks', such as the British ecological writer and activist (and founder of *The Ecologist* magazine) Ted Goldsmith's prescription of a return to 'the way',[15] and Paul Shepard's 'back to the neolithic' suggestion.[16] However, it is the increasingly authoritarian dimension of Bahro's thought and strategy for dealing with the ecological crisis that many Greens find most worrying, allied to his 'apocalyptic' framing of the socio-ecological crisis, which finds contemporary resonance in ecological movements such as the 'Dark Mountain Project'[17].

Ultimately, Bahro's legacy is a mixed one: a formative influence within the largest and most successful Green party in the world, the German Greens; a central figure in the fundi/realo split within the latter, and at the same time 'exporting' this 'radical/reformist' dichotomy to the theory and practice of the global Green movement as a whole; a defender of the position that Green politics and action is resolutely 'beyond left and right', and committed to a utopian, total transformation of the current socio-political order. However, for some, Bahro's thought and action could be said to be a study in someone who starts on the left and moves progressively to the right, moves from political activism to spiritual contemplation. At the same time, and consistent with the evolution of his thought, Bahro put the cultural, psychological and spiritual aims of the Green movement on the map as central substantive and strategic issues that had to be dealt with as essential to the resolution of our socio-ecological crisis.

Notes

1 *Building the Green Movement*, p. 29.
2 Bahro, quoted in Werner Hülsberg, *The German Greens: A Social and Political Profile*, London: Verso, p. 93, 1988.
3 *Building the Green Movement*, p. 88.
4 *From Red to Green*, pp. 220–2.
5 *Building the Green Movement*, p. 90.
6 Robyn Eckersley, *Environmentalism and Political Theory*, p. 164.
7 Boris Frankel, *The Post-industrial Utopians*.
8 Ralf Fücks, *Green Growth, Smart Growth*, London: Anthem Press, p. 52.
9 Eckersley, op. cit., p. 164.
10 John Ferris, 'Introduction', in Helmut Wiesenthal, *Realism in Green Politics: Social Movements and Ecological Reform in Germany*, Manchester, UK: Manchester University Press, p. 13, 1993.
11 *The Logic of Salvation*, p. 325.
12 William Ophuls, *Ecology and the Politics of Scarcity*, San Francisco, CA: Freeman, 1977.
13 Bill Devall, *Simple in Means, Rich in Ends: Practising Deep Ecology*, London: Green Print, p. 197, 1990.
14 Bahro, quoted in Jutta Ditfurth, *Feuer in die Herzen: Plädoyer für eine Ökologische Linke Opposition*, Hamburg, Germany: Carlsen Verlag, pp. 207–8, 1992, emphasis added.
15 Edward Goldsmith, *The Way: 87 Principles for an Ecological World*, London: Rider, 1991.
16 Paul Shepard, 'A Post-Historic Primitivism', in Max Oelschlaeger (ed.), *The Wilderness Condition*, Washington, DC, and Covelo, CA: Island Press, 1994.
17 Dougald Hine and Paul Kingsnorth (eds), *Dark Mountain: Volume 1*, London: The Dark Mountain Project, 2010.

See also in this book
Bookchin, Marx

Bahro's major writings

The Alternative in Eastern Europe: Towards a Critique of Real, Existing Socialism, London: New Left Books, 1979.

Socialism and Survival, London: Heretic Books, 1982.

From Red to Green, London: Verso, 1984.

Building the Green Movement, London: Heretic Books, 1986.

The Logic of Salvation. Logik der Rettung: Wer kann dieApokalypse aufhalten? – Ein Versuch über die Grundlagen ökologischer Politik, Stuttgart, Germany, and Vienna: Transaction Press, 1987.

Avoiding Social and Ecological Disaster· The Politics of World Transformation, Bath, UK: Gateway Books, 1994.

Further reading

Dobson, A., *Green Political Thought*, 2nd edn, London: Routledge, 1995.

Eckersley, R., *Environmentalism and Political Theory*, London: UCL Press, 1992.

Frankel, B., *The Post-industrial Utopians*, Oxford: Basil Blackwell, 1987.

Fücks, R. *Green Growth, Smart Growth: A New Approach to Economics, Innovation and the Environment*, London: Anthem Press, 2015.

JOHN BARRY

RODERICK NASH 1939–

WE THEREFORE, resolve to act. We propose a revolution in conduct toward an environment which is rising in revolt against us. Granted that ideas and institutions long established are not easily changed; yet today is the first day of the rest of our life on this planet. We will begin anew.[1]

Born and raised in Midtown Manhattan in New York City, Roderick Frazier Nash has become one of the world's foremost authorities on wilderness and the environmental history of the American West. Nash's father, Jay B., made his mark on academe as a professor of public health, physical education, and recreation at New York University, and published several books on the subjects. After attending Harvard University and earning an undergraduate degree in history, Roderick Nash enrolled in the doctoral program at the University of Wisconsin, Madison. Steeped in the strong tradition of Western American history forged by pioneers such as Frederick Jackson Turner, the Wisconsin history department included many noted scholars when Nash entered in the early 1960s. Although thoroughly grounded in traditional American history, Nash wanted to pursue a non-traditional, organic path in his graduate research. His principal advisor, Merle Curti, encouraged such a broad research design, and the result was a dissertation on American attitudes toward wilderness throughout the nation's history.

After completing his doctoral degree in 1964, Nash became an assistant professor at Dartmouth College in New Hampshire. While quickly distinguishing himself as an excellent teacher, he continued to refine and expand his dissertation. In 1967, Yale University Press published Nash's *Wilderness and the American Mind*. Destined to become one of the most influential twentieth century books on environmental history, *Wilderness and the American Mind* explores the most adverse attitudes toward the frontier and wilderness that came across the Atlantic with the European colonists.[2] Conversely, the book explores the American reverence for nature that resulted in the world's first national parks and wilderness preservation legislation. American ambivalence toward nature sparked pitched battles over environmental issues that Nash skillfully analyzes through biographical sketches of John Muir, Robert Marshall, Aldo Leopold, and David Brower. Nash devoted an entire chapter to the controversial damming of Hetch Hetchy Valley in California's Yosemite National Park for a water supply for the city of San Francisco, and used that event to begin the history of the modern preservation movement. *Wilderness and the American Mind*, now in its fifth edition, remains the definitive work on the philosophy and practice of wilderness activism in the United States, and its expansion to the world stage. *The Los Angeles Times* in the 1990s listed it in the top one hundred most influential books of the last quarter-century, while environmental activist Dave Foreman called the book "a must-read for anyone who wants to understand wilderness and the conservation movement."[3]

Coinciding with publication of *Wilderness and the American Mind* in 1967 was Nash's move west to become a professor of history at the University of California, Santa Barbara (UCSB). Only in its second decade of existence, UC Santa Barbara was a rapidly expanding campus in rapidly growing California. Nash was one of many young, talented scholars attracted to the university's majestic setting on the Pacific Coast just west of Santa Barbara. As Nash developed courses in American environmental history, fate stepped in when oil from a drilling pipe rupture on an offshore platform bubbled 70,000 barrels of crude into the ocean and onto the beaches of Santa Barbara.[4] The Santa Barbara Oil Spill of January 1969 not only blackened the beaches of the city and UCSB causing millions of dollars in damages, it also helped to jump-start the environmental movement.

Following the oil spill, several professors at UCSB resolved to change the direction of the nation's commercial and industrial development. In 1970, shortly after passage of the National Environmental Policy Act and one year to the day after the 1969 spill, they gathered at the site where the oil made landfall. Before a national television audience, Nash read the Santa Barbara Declaration of Environmental Rights, written by him and other organizers of the event. Recalling Thomas Jefferson's prose in the Declaration of Independence and conservationist philosopher Aldo Leopold's Land Ethic, the Santa Barbara declaration decried the fouling of the waters, air, and land of the planet, and proclaimed that human beings and all other forms of life had the *right* to a clean and healthful environment. The declaration's words echoed in subsequent federal and state legislation, including the Clean Water Act, the Endangered Species Act, and the California Environmental Quality Act.[5] At the same time, the UCSB professors persuaded their chancellor to create a new program to study the issues of ecology and the environment. In 1970, the first classes were developed in the new environmental studies program, now the oldest college curriculum of its kind in the United States. Nash chaired the program for its first five years, held a joint appointment in history and environmental studies until his retirement from the university in 1994, and taught courses in wilderness, environmental history, environmental ethics, popular culture, and environmental philosophy.

Roderick Nash's interest in environmental history and intellectual history led to an array of books, articles, and other publications. His interest in the society and popular culture of the United States during the 1920s resulted in *The Nervous Generation: American Thought, 1917–1930*.[6] This popular overview of American society during the Roaring Twenties was widely used in college courses, and republished in a second edition in 1990. Nash also proved to be an able generalist of American history. In 1975, he published the first edition of *From These Beginnings: A Biographical Approach to American History*. Through short biographies of individuals ranging from Puritan leader John Winthrop to first lady Eleanor Roosevelt, *From These Beginnings* tells the story of American history through the eyes of individuals. Currently in its eighth edition (Nash added former student Gregory Graves as his co-author starting with the fourth edition), *From These Beginnings* continues to be a popular textbook for college curricula around the nation and the world.[7]

While his intellectual interests are eclectic, Nash's primary focus has been environmental. To the present day, he has published numerous articles on various subjects of environmentalism, always providing the historical context for issues of the moment. His book, *The Rights of Nature: A History of Environmental Ethics* (1989), blends the views of pivotal individuals dating back to Greco-Roman times with early modern European philosophies, and culminates with analyses of twentieth-century radical movements such as Earth First!, the Animal Liberation Front, People for the Ethical Treatment of Animals, and Greenpeace. *The Rights of Nature* traces the evolution and expansion of ethics from early societal organizations like the tribe, to entire nations, and notes a movement toward ethical extension to all living things. Nash's skillful interweaving of history and philosophy drew strong acclaim for *The Rights of Nature*.[8] In the words of former secretary of the interior and environmental activist Steward Udall, "Roderick Nash has written another classic. This exploration of a new dimension in environmental ethics is both illuminating and overdue."[9]

Secretary Udall's reference to "another classic" points to the phenomenal success of Nash's *Wilderness and the American Mind*. Continually in print from its first publication in 1967 to the present day, Nash's classic transcended academics and also had a strong trade sales record. The fifth edition, published by Yale University Press in 2014, includes additional material on the importance of wilderness from the international perspective, as well as a new epilogue on Island Civilization: the concept of cordoning off civilization into islands to protect large areas of wild land. In the prologue to the fifth edition, Nash writes: "American attitude toward wilderness is old and complex. It was always the biggest fact on every frontier. The short summary of what we are going to explore in this book is that the wilderness idea was born in a context of fear and loathing, and that the old biases have died hard. Appreciation of wilderness is recent and incomplete."[10]

Throughout his career, Nash has been an environmental activist in both word and deed. Never an "armchair enthusiast" (Nash's own phrase), he has been an avid hiker, backpacker, rafter, skier, and sailor. Nash was a pioneer whitewater boater in the American West. He has navigated the Grand Canyon of the Colorado River more than sixty times since the 1960s. Nash was the first person to descend the Tuolumne River below Hetch Hetchy Reservoir in Yosemite National Park in a raft. As a professor in environmental studies, he organized classes that took students into the field, including Canyonlands National Park and down the Colorado River through the Grand Canyon. He regularly organized backpacking trips into California wilderness areas for students in his classes and in the environmental studies program, while also leading rafting trips on the Colorado and San Juan rivers. Based on his travels and journeys, Nash has written extensively on the geography and environmental threats to Arizona, Utah, and Colorado. With his colleague, Robert Collins, he wrote *The Big Drops* (1978), an account in word and picture on their shared experiences rafting ten rivers of the American West. He reissued the book in 1989 under his own name to reflect on river adventures in the dory, *Canyon Dancer*.[11] Presently, Nash spends extensive time cruising his tugboat, *Forever Green*, in the Gulf of California, the central coast of California, and the inland passages to British Columbia and Alaska, while also skiing one hundred days a year near his winter home in Crested Butte, Colorado.

Through meticulous research, skillful analysis, and clear writing in dozens of works, Roderick Nash has blazed new trails in American intellectual and environmental history. He has also contributed to numerous documentaries, and has been an expert witness on a variety of natural resources issues. Nash has told the story of attitudes toward wilderness, and has also expanded that concept to capture the dynamics of modern environmentalism. His scholarship, teaching, and mentoring are legendary, as are his significant contributions to the evolving field of environmental history.

Notes

1 From *The Santa Barbara Declaration of Environmental Rights* of 1970 – read to a national television audience by Roderick Nash following the Santa Barbara Oil Spill of January 1969, quoted in Roderick Nash, ed., *The American Environment: Readings in the History of Conservation*, 2nd edn (Reading, MA: Addison-Wesley Publishing Co., 1976), p. 300.

2 Roderick Nash, *Wilderness and the American Mind*, 1st edn (New Haven, CT: Yale University Press, 1967).

3 Quoted in Roderick Nash, *Wilderness and the American Mind*, 5th edn (New Haven, CT: Yale University Press, 1967), back cover.

4 See Nash, ed., *The American Environment*, pp. 298–306, for more on the Santa Barbara Oil Spill and aftermath.

5 See Nash, ed., *The American Environment*, pp. 298–306 for more on the Santa Barbara Declaration of Environmental Rights.

6 Roderick Nash, *The Nervous Generation: American Thought, 1917–1930* (New York: Rand McNally and Co., 1970).

7 Roderick Nash and Gregory Graves, *From These Beginnings: A Biographical Approach to American History*, 8th edn (New York: Pearson Longman Publishers, 2008), Volumes One and Two.

8 Roderick Nash, *The Rights of Nature: A History of Environmental Ethics* (Madison, WI: University of Wisconsin Press, 1989).

9 www.amazon.com/dp/0299118444?_encoding=UTF8&isInIframe=1&n=283155&ref_=dp_proddesc_0&s=books&showDetailProductDesc=1#ifr ame-wrapper

10 Quoted in Roderick Nash, Wilderness and the American Mind, 5th edn (New Haven, CT: Yale University Press, 2014), p. xxi.

11 Roderick Nash and Robert O. Collins, *The Big Drops: Ten Legendary Rapids of the American West*, Revised Edition, 1989, first publication 1978 (Boulder, CO: Johnson Books, 1989).

See also in this book
Leopold, Muir, Olmsted

Nash's major writings
[With Robert O. Collins] *The Big Drops: Ten Legendary Rapids of the American West*, (Boulder, CO: Johnson Books, 1st edn 1978, Revised edn, 1989).
The Rights of Nature: A History of Environmental Ethics (Madison, WI: University of Wisconsin Press, 1989).
[With Gregory Graves] *From These Beginnings: A Biographical Approach to American History*, 8th edn (New York: Pearson Longman Publishers, 2008), Volumes One and Two.
The Nervous Generation: American Thought, 1917–1930 (New York: Rand McNally and Co., 1970).
Wilderness and the American Mind (New Haven, CT: Yale University Press, 1st edn, 1967, 5th edn, 2014).
Ed., *The American Environment: Readings in the History of Conservation*, 2nd edn (Reading, MA: Addison-Wesley Publishing Co., 1976).

Further reading
Edward Abbey, *Desert Solitaire* (Tucson, AZ: University of Arizona Press, 1988).
Aldo Leopold, *A Sand County Almanac, and Sketches Here and There* (New York: Oxford University Press, 1989).
John McPhee, *Encounters with the Archdruid* (New York: Farrar, Strauss, and Giroux, 1971).
Joseph Sax, *Mountains Without Handrails: Reflections on the National Parks* (Ann Arbor, MI: University of Michigan Press, 1980).

GREGORY GRAVES

VAL PLUMWOOD 1939–2008

The logic of domination and the deep structures of dualism create 'blind spots' in the dominant culture's understanding of its relationship to the biosphere, understandings which deny dependency and community to an even greater degree than in the case of human society. The distorted perceptions and mechanisms of denial which arise from the master rationality are an important reason why the dominant culture which embodies this identity in relation to nature cannot respond adequately to the crisis of the biosphere and the growing degradation of the earth's natural systems.[1]

Val Plumwood began her work on environmental philosophy in collaboration with her then husband Richard Routley (later Sylvan, also an important environmental philosopher) in the early 1970s when the ecological crisis of the modern West was becoming more obvious. Within the framework of analytical philosophy in which they were both trained, the most obvious tools with which to explain the crisis were those of the ethics of value and respect. Routley and Plumwood argued that one reason why the dominant global culture of the West had been able to expand and conquer indigenous cultures as well as nature itself was that it lacked their respect-based constraints on the use of nature – a thought that cast Western triumphalism in a rather different and more dangerous light. This lack of respect had, as one of its main philosophical expressions, the deep-seated Western conviction that only humans could be of any direct ethical significance or value.[2]

The view that value and moral consideration were confined to humans, for which Plumwood and Routley coined the term 'human chauvinism', was supported by the assumption (too deeply embedded to be thought of as needing explicit statement or defence) that the natural world could have only indirect or instrumental value, as a means to human ends, or as mattering to human beings.

They exposed the arrogance of this assumption, and tried to clarify concepts like respect and instrumental value so as to make available alternative modes of valuing and respecting nature as an independent other. They met objections that there was no rational alternative to purely instrumental values by showing that this involved an infinite regress. They argued that, since its supposed 'naturalness' and inevitability were ultimately based on fallacies that closely paralleled those of philosophical egoism, human chauvinism had no more legitimacy than human-group chauvinism.[3]

For Plumwood, however, value was too narrow a focus for explanation and activism. While she agreed that we must value the natural world more highly, she held that value and the failure to discern and accord it provides only a very incomplete explanation of environmental failure. This last, she held, stems equally from what she called 'the standpoint of mastery': overconfidence, failure to recognize the other's agency and limits, and other kinds of insensitivity that come from the dominant rationalist and colonizing frameworks by which historically we have understood and created human/nature relations.

After Plumwood's collaboration with Routley ended in 1981, she refined their initial, relatively unanalysed concept of 'human chauvinism' by exploring the implications of seeing human-centredness as a parallel concept to androcentrism and eurocentrism.[4] From this she came to the conclusion that human-centred stances are subject to similar blind spots and distortions of conception and perception; for example, seeing the other as radically separate and inferior, the background to the self as foreground, as one whose existence is secondary, derivative or peripheral to that of the self or centre, and whose agency is denied or minimized.

A major feature of human-centred frameworks is the denial of human dependency on nature. In 'Anthrocentrism and Androcentrism' Plumwood drew on many aspects of women's oppression to theorize 'hegemonic otherness', the condition, in androcentric frameworks, in which women appear as appendages to men and in which their agency is treated as lesser or denied altogether. She saw a strong parallel here with the treatment of nature's agency in human-centred frameworks. When nature as agent and collaborative partner is similarly denied, she claimed, we get blind spots about human dependence and vulnerability, which help to make such frameworks dangerous and misleading.[5]

In her book *Feminism and the Mastery of Nature* (1993), in which many of the main themes of her work are articulated, Plumwood argued the ecofeminist thesis that the West's problems with the non-human world (now global problems) must be understood in the context of its larger dualist problematic. Traditions excluding non-humans from the spheres of ethics, mind, culture and reason are matched on the other side by traditions excluding humans from the realm of nature and animality, to form what she called a 'hyperseparation' between humans and nature which is entrenched in the dominant traditions of the West. She saw the drive to hyperseparation as part of a colonizing conceptual dynamic which places the human colonizer radically apart from, and above, those he conceives as part of the subordinated realm of nature.

Plumwood identified human/nature dualism as a key part of the system of gendered dualisms that have helped to shape Western world-views. These include reason/nature, human/animal and mind/body dualisms, which have been historically central to environmental thinking, as well as male/female, reason/emotion and civilized/primitive, which are associated with other forms of colonization. Thus she linked the treatment of non-humans to the treatment of women and other groups, such as indigenous people, who have also been considered part of the inferior realm of nature. The influence of dualistic approaches that separate the truly human from, and inferiorize, the ecologically situated body and the perishable order of biological life is traced in the Greeks, and is especially clear in the work of Plato (a major influence in the development of 'flesh-inferiorising dualisms' in Christianity).[6]

For Plumwood, dualism has deeply marked both concepts of nature and concepts of reason. The environmental crisis, she argued, should be seen as a crisis of dualistic reason, a form of rationality expressed especially in the contemporary global market, which conceives rationality as self-interest in opposition both to the emotions (including care for others) and to the ecologically situated body. Challenging this form of rationality challenges current social and political systems and means, bringing economic and ecological rationality together. Plumwood argued from an eco-socialist perspective against the treatment of animals and nature as property under capitalism and for the ecological virtues of more egalitarian and democratic social systems.[7]

Employing this analysis of human-centredness and of human/nature dualism, Plumwood argued that the dominant problematic of modern environmental ethics is set up in an anthropocentric way, focusing on establishing the qualifications of non-humans for moral consideration (usually through establishing some basis of similarity to the human), rather than on the problems of a human-centred ethical system and epistemology which excludes them. In many current forms of academic environmental philosophy, she claimed, the question of value seems to be taken to be the only issue. Moreover, valuing itself is too often treated as the stance of someone looking on at nature as a separate and passive entity; evaluating, ranking and assessing its existence, somewhat in the manner of a property appraiser. Such a stance, she claims, assumes that it is our human prerogative to order the world in terms of some generalized species-ranking, to assign or withhold 'value' according to whim or to degrees of rationality or consciousness.[8]

Equally problematic for Plumwood is the sort of environmental ethic that extends moral consideration only to some 'higher' animals on the basis of their similarity to the human (especially their similarity in consciousness) in the fashion of some animal rights positions.[9] This type of environmental ethics is rejected as neo-Cartesian and implicitly human-centred, a minimum-change position that relocates, rather than cancels, the radical break between the human and non-human. Plumwood argued that it is based implicitly on sameness, extending a human model of reason or consciousness, and is therefore both human-centred in a damaging sense and unable to acknowledge that difference does not mean inferiority.

In *Feminism and the Mastery of Nature* Plumwood delineated a position in environmental ethics that, on the one hand, is distinct from the conventional neo-Cartesianism that would extend moral consideration just to conscious beings ('like us'), and, on the other hand, from deep ecology. She agreed with deep ecology that it is a major problem that the modern West positions itself as 'outside nature'. But Plumwood's analysis sought to explain the West's misunderstanding of its ecological embedment not as the outcome of a separation from nature that departs from the deep ecological ideal of unity between humanity and nature, but as an aspect of dualistic hyperseparation, in which normative human identity excludes features traditionally associated with nature and the animal sphere.[10] Plumwood's underlying metaphysics, which she called 'weak panpsychism', refuses the demand to base ethical consideration on sameness that underlies both moral extensionism and deep ecology, the two conventional choices. Rejecting also the pure appeal to difference as the source of value which appears in some post-modernist positions, Plumwood insists on both continuity with, and difference from, the human as sources of value and consideration.

According to Plumwood, both sameness and difference are required to counter human/nature dualism. In the Western tradition especially, she admitted the need to stress continuity between self and other, human and nature, in response to the gulf created by the hyperseparated and alienated models of nature and of human identity that remain dominant. These models define the truly human as (normatively) outside of nature and in opposition to the body and the material world, and conceive nature itself in alienated and mechanistic terms as lacking elements of mind. In her last writings she described her position as 'philosophical animism', recognizing goal-directed, intentional, activity, not just in humans, but in animate and inanimate nature, though moral consideration was not to be confined to the intentional even in this broad sense.[11]

But she contended that we also need to stress the difference and divergent agency of the other in order to defeat that further part of the colonizing dynamic that seeks to assimilate and instrumentalize the other, recognizing and valuing it only as a part of or similar to the self, or as means to the self's ends. Vague concepts of unity and identity of the sort stressed by deep ecology, she argued, provide very imprecise and inadequate correctives to our historical denial of continuity with and dependency on nature. She maintained that ethical theories based on unity cannot provide a good model of mutual adjustment, communication and negotiation between different parties and interests, and are unhelpful in the key areas where we need to construct dialogical, ethical relationships.

Plumwood was not opposed to spirituality and the sacred per se but thought that dominant forms of Western spirituality have located the sacred in the wrong place – above and beyond a fallen earth. An ecological spirituality would need to relocate it in the immediate world around us, as in much indigenous spirituality. In this location the sacred can be experienced as in and of the earth, but need not, and should not, be overly singularized and centralized as it is in some forms of Gaia theory. In *Environmental Culture*, she saw such a 'materialist spirituality' as one component of a strong environmental culture which needs to be developed over a wide range of areas to counter the excesses of the dominant form of rationality.

To defeat human/nature dualism, Plumwood argued that we need to revise conceptions of human virtue which are based on excluding, from the ideal human character, the supposedly oppositional elements to reason, especially emotionality, embodiment and animality. She also emphasized 'counter-hegemonic virtues', ethical stances which can help to minimize the influence of the oppressive ideologies of domination and self-imposition that have formed our conceptions of both the other and ourselves. She advocated the adoption of philosophical strategies and methodologies that maximize our sensitivity to other members of our ecological communities and openness to them as ethically considerable beings in their own right, rather than strategies that minimize ethical recognition or that adopt a dualistic stance of ethical closure that insists on sharp moral boundaries and denies the continuity of planetary life.

Among such strategies, she stressed the need for communicative virtues of listening and attentiveness to the other to help counter the backgrounding which obscures and denies what the non-human other contributes to our lives and collaborative ventures. Openness and attentiveness involve giving the other's needs and agency more weight, being open to unanticipated possibilities and aspects of the other, and re-conceiving and re-encountering the other as a potentially communicative and agentic being, as well as an independent centre of value and an originator of projects that demand respect. These counter-hegemonic virtues, she claimed, help us to resist the reductionism of the dominant mechanistic conceptions of the non-human world, and to revise both our narrow epistemic objectives of prediction and control and our denial of non-human others as active presences and ecological collaborators in our lives.

Notes
1 *Feminism and the Mastery of Nature*, p. 194.
2 See Routley, 'Is There a Need for a New, an Environmental Ethic?', pp. 205–10; Plumwood's 'Critical Notice of Passmore's *Man's Responsibility For Nature*', pp. 171–85.
3 Routley and Plumwood, 'Human Chauvinism and Environmental Ethics', in D. Mannison, M. McRobbie and R. Routley (eds), *Environmental Philosophy*, Canberra: Philosophy Department, Australian National University, pp. 96–190, 1979; 'Against the Inevitability of Human Chauvinism', in Goodpaster and Sayre (eds), *Ethics and the Problems of the 21st Century*, pp. 36–59.
4 Plumwood, 'Anthrocentrism and Androcentrism: Parallels and Politics', *Ethics and the Environment*, pp. 119–52 and 'Paths Beyond Human

Centredness: Lessons from Liberation Struggles', in *An Invitation to Environmental Philosophy*, pp. 69–106.

5 Human vulnerability in the face of non-human agency was brought home to her dramatically in 1985 when she was attacked and nearly killed by a crocodile in Northern Australia. She has written memorably about the experience, now rare for humans, of being hunted for food in 'Being Prey', *Terra Nova*, pp. 33–44 and *The Eye of the Crocodile*.

6 'Prospecting for Ecological Gold among the Platonic Forms', *Ethics and the Environment*, pp. 149–68.

7 See 'The Crisis of Reason, the Rationalist Market, and Global Ecology', *Millennium. Journal of International Studies*, pp. 903–26; 'Has Democracy Failed Ecology? An Ecofeminist Perspective', *Environmental Politics*, 4, 134–68, 1995; 'Ecojustice, Inequality and Ecological Rationality', *Debating the Earth: The Environmental Politics Reader*, ed. J. Dryzek and D. Schlosberg, Oxford, UK: Oxford University Press, 1998.

8 See 'Self-Realization or Man Apart? The Naess–Reed Debate', in N. Witoszek and A. Brennan (eds), *Festschrift for Arne Naess*, pp. 206–12.

9 See 'Intentional Recognition and Reductive Rationality: A Response to John Andrews', *Environmental Values*, 17, 397–421, 1998; *Feminism and the Mastery of Nature*.

10 In addition to *Feminism and the Mastery of Nature*, see two later essays, 'Self-Realization or Man Apart?' and 'Deep Ecology, Deep Pockets, and Deep Problems: A Feminist Eco-Socialist Analysis', in A. Light, E. Katz and D. Rothenburg (eds), *Beneath the Surface: Critical Essays on Deep Ecology* pp. 59–84.

11 'Nature in the Active Voice', *Australian Humanities Review*, 46, 113–29, 2009. The last time I spoke with her, shortly before her death, she referred to her position as 'materialist animism' but I don't think this useful term ever appeared in print.

See also in this book
Lovelock, Passmore, Wright J.

Plumwood's major writings

'Critical Notice of Passmore's *Man's Responsibility For Nature*', *Australasian Journal of Philosophy*, 53, 171–85, 1975.

'Against the Inevitability of Human Chauvinism', with R. Routley, in *Ethics and the Problems of the 21st Century*, ed. K.E. Goodpaster and K.M. Sayre, South Bend: Notre Dame University Press, 1978.

Feminism and the Mastery of Nature, London: Routledge, 1993.

'Anthrocentrism and Androcentrism: Parallels and Politics', *Ethics and the Environment*, 1, 119–52, 1996.

'Being Prey', *Terra Nova*, 1, 33–44, 1996; reprinted in *Terra Nova: New Environmental Writing*, Cambridge, MA: MIT Press, 1999.

'Prospecting for Ecological Gold among the Platonic Forms', *Ethics and the Environment*, 2, 149–68, 1997.

'The Crisis of Reason, the Rationalist Market, and Global Ecology', *Millennium. Journal of International Studies*, 27, 903–26, 1998.

'Paths Beyond Human Centredness: Lessons from Liberation Struggles', *An Invitation to Environmental Philosophy*, ed. A. Weston, Oxford, UK: Oxford University Press, pp. 69–106, 1998.

'Self-Realization or Man Apart? The Naess–Reed Debate', *Festschrift for Arne Naess*, ed. N. Witoszek and A. Brennan, Savage, MD: Rowman & Littlefield, pp. 206–12, 1998.

'Deep Ecology, Deep Pockets, and Deep Problems: A Feminist Eco-Socialist Analysis', *Beneath the Surface: Critical Essays on Deep Ecology*, ed. A. Light, E. Katz and D. Rothenburg, Cambridge, MA: MIT Press, 1999.

Environmental Culture: The Ecological Crisis of Reason, London: Routledge, 2002.

'Nature in the Active Voice', *Australian Humanities Review*, 46, 113–29, 2009. Reprinted in R. Irwin (ed.) *Climate Change and Philosophy: Transformational Possibilities*, London: Continuum, 2010.

The Eye of the Crocodile, ed. L. Shannon, Canberra: ANU EPress, 2012.

Further reading

Reviews of *Feminism and the Mastery of Nature* by P. Hallen, *Australasian Journal of Philosophy*, 73, 645–6, 1995; by L.M.B Harrington, *Journal of Rural Studies*, 12, 319–20, 1996; by J. Cook, *Environmental Values*, 6, 245–6, 1997.

Reviews of *Environmental Culture*: by C. Cuomo, *Notre Dame Philosophical Reviews*, 2002.11.3, 2002; by P. Hallen, *Ethics and the Environment*, 7, 181–4, 2002; by J.G. Hintz, *Human Ecology Review*, 10, 77–8, 2003; R. Twine, *Environmental Values*, 12, 535–7, 2003.

Hawkins, Ronnie, 'Ecofeminism and Nonhumans: Continuity, Difference, Dualism, and Domination', *Hypatia*, 13, 158–97, 1998.

Hyde, Dominic, *Eco-Logical Lives. The Philosophical Lives of Richard Routley/Sylvan and Val Routley/Plumwood*, Cambridge, UK: The White Horse Press, 2014.

Routley, Richard, 'Is There a Need for a New, an Environmental Ethic?', *Proceedings of the XVth World Congress in Philosophy*, Varna, i, 205–10, 1973.

NICHOLAS GRIFFIN

PATRIARCH BARTHOLOMEW 1940–

Climate change is a profoundly moral and spiritual problem.
(Amazon Symposium, July 2006)

His All-Holiness Ecumenical Patriarch Bartholomew is the spiritual leader of 300 million Orthodox Christians worldwide. Born in Imvros Turkey (1940), he is 270th Archbishop of the 2000-year-old Church founded by St. Andrew, serving as Archbishop of Constantinople–New Rome and Ecumenical Patriarch since 1991.

He studied at the Theological School of Halki in Istanbul (1961), the Pontifical Oriental Institute of the Gregorian University (1966), the Ecumenical Institute of Bossey in Geneva (1967), and the University of Munich (1968). He served as Assistant Dean at the Theological School of Halki (1968–71), Personal Secretary to the late Ecumenical Patriarch Demetrios (1972–91), Metropolitan of Philadelphia (1973–91), and Metropolitan of Chalcedon (1990–1).

Ecumenical Patriarch Bartholomew has worked for unity among Christian Churches, affirmed religious tolerance through interfaith dialogue, and supported Orthodox countries emerging from religious oppression behind the Iron Curtain. His pioneering work for international peace and environmental protection placed him at the forefront of global visionaries as an apostle of love and reconciliation. In 1997, he was awarded the Gold Medal of the United States Congress. In 2008, he was named one of *Time* magazine's 100 Most Influential People in the World for "defining environmentalism as a spiritual responsibility."

Exceptional and pioneering leadership

His All-Holiness Ecumenical Patriarch Bartholomew has repeatedly emphasized that climate change is a moral and spiritual problem during his ministry. Indeed, over the past twenty-five years, no other worldwide religious leader has placed the ecological crisis at the forefront of his service and sermons as has Ecumenical Patriarch Bartholomew, spiritual leader of the Orthodox Church, which numbers over 300 million adherents worldwide. Yet, during the same period, the world has witnessed alarming rise in ecological degradation. Deforestation has removed about half of the world's forest cover, ocean acidification is causing a degradation of life in the seas; an excess use of fossil fuels is causing global climate change which is raising sea levels, increasing temperatures and causing greater storm intensities; and toxic wastes in food chain are culpable in a variety of diseases. Additionally, there is an increasing failure to implement environmental policies and an ever-widening gap between rich and poor.

For Patriarch Bartholomew, then, "eco-justice" is not an exercise in public relations; nor is it merely a fashionable advocacy. It is, rather, a deep theological, liturgical, and spiritual conviction; it is the understanding that Christianity is essentially and profoundly maximalist, materialist, and environmentalist. This is why Patriarch Bartholomew was prominently and exceptionally highlighted in Pope Francis' "green encyclical" *Laudato Si'* published in June 2015; this marked the first time ever that a pope included the name, cited the words, and featured the ministry of someone outside the Roman Catholic Church.

For the Ecumenical Patriarch, then, the environment is not only a political or a technological issue; it is, as we have come to appreciate, primarily a religious and spiritual issue. The profound disfiguration and degradation of creation, the disharmony and environmental destruction that are now "everywhere present" become a form of sacrilegious "graffiti" on the holy temple of the earth that can only be cleansed and eliminated by the cleansing of our personal temples and by eliminating the sinful causes of ecological destruction in our minds and souls. The ecological crisis at heart is a moral, ethical, and spiritual issue. This is why religion has a key role to play. A spirituality that remains uninvolved with outward creation is ultimately uninvolved with the inward mystery too.

The distinctive features of the Patriarch's vision are humble simplicity (*ascesis*), liturgical communion (*koinonia*), and the upholding of the patristic tradition that discerns the divine presence of the Holy Trinity in all things and all creation. In all that he says or does, Patriarch Bartholomew is aware that everyone without exception – irrespective of confessional or religious conviction – must be included. Likewise, every science and discipline should contribute; every culture and age should concur. Moreover, the Ecumenical Patriarch is also aware that environmental issues are intimately related to numerous social issues: war and peace, social justice and human rights, poverty and unemployment. We have become increasingly aware of the effects of environmental degradation on people, especially the poor. "Our mission," declares His All-Holiness,

> is to make it well known throughout the world that we are all the caretakers and not the owners of our common "oikos," [earthly home] ... Churches indeed have, through sermons, catechism and continuous education, a serious responsibility to inform our faithful about these matters.
>
> *(September 16, 2015)*

As Patriarch Bartholomew introduces the task of caring for creation, he also points to the further dimensions of our mission as caretakers of creation. We are to commit ourselves to making our bodies and souls fit temples of the Holy Spirit, so that our thoughts, feelings, and actions become life-enhancing, earth-healing, and death-destroying according to the divine pattern established once and for all by the life, death and resurrection of our Lord Jesus Christ Himself.

Theology of a worldview

Orthodox Christianity retains – in its theology, liturgy and spirituality – a profoundly sacred view of the natural environment, proposing a world richly imbued by God and proclaiming a God intimately involved with creation. What is at stake, His All-Holiness has declared, is not just respect for biodiversity, but our very survival:

> Scientists calculate that those most harmed by global warming will be the most vulnerable and marginalized ... And there is a bitter injustice about the fact that those suffering its worst ravages have done least to contribute to it. The ecological crisis is directly related to the ethical challenge of eliminating poverty and advocating human rights. This means that global warming is a moral crisis and a moral challenge. The dignity and rights of human beings are intimately and integrally related to the poetry and – we would dare to say – the rights of the earth itself.
>
> *(Philippines, February 26, 2015)*

The underlying and all-embracing conviction is that, by disconnecting creation from the Creator, we have in fact desacralized both. For the way we relate to the world around us directly reflects the way we pray to "our Father in heaven." So we respond to the natural world with the same sensitivity that we address God. All of us understand that we must care for other human beings, as created "in the image of God" (Genesis 1:26); it is time to appreciate the need to care for everything, as containing "the trace of God" (Tertullian, 2nd century). Indeed, Orthodox Christians perceive the notion of sin as the stubborn refusal of humanity to regard the created world as a gift of communion – as nothing less than a sacrament.

In the seventh century, two mystics of the Eastern Christian Church eloquently described this intimate relationship between nature, humanity, and God. St. Maximus the Confessor spoke of the world as a "cosmic liturgy," a magnificent altar upon which human beings worship in thanksgiving and glory. The entire world comprises an integral part of this sacred song; God is praised by the sun and moon, worshipped by the trees and birds (Psalm 18:2). And Abba Isaac the Syrian invited his spiritual disciples to "acquire a merciful heart, burning with love for all of creation: for humans, birds, and beasts." If today we are guilty of relentless waste in our world, it may be because we have lost the spirit of worship and the spirituality of compassion.

History of a movement

While its ecological initiatives date back to the mid-1980s, since 1989, the Ecumenical Patriarchate – located in the ancient Christian See of Constantinople (modern day Istanbul, Turkey) – has invited Orthodox Christians throughout the world to reserve September 1, the official opening of the Church Calendar, as a day of prayer for environmental preservation; numerous Christian communions have followed suit, encouraged by the World Council of Churches and most recently the Evangelical Environmental Network. After his election in 1991, Ecumenical Patriarch Bartholomew launched a series of environmental activities, including a Pan-Orthodox conference in Crete, an unprecedented meeting of Orthodox Patriarchs, inviting them to endorse his ecological vision, and a series of five ecological summer seminars on education, ethics, communications, justice, and poverty.

In 1995, the Ecumenical Patriarch established the Religious and Scientific Committee, which to date has organized seven international, interfaith, and inter-disciplinary symposia in the Aegean Sea (1995) and the Black Sea (1997), along the Danube River (1999) and in the Adriatic Sea (2002), in the Baltic Sea (2003) and on the Amazon River (2006), as well as in the Arctic (2007) and on the Mississippi River (2009). Since that time, he has embarked on a series of more focused, more impactful "Halki Seminars," inviting diverse representatives of the scholarly and spiritual fields, as well as the institutional and corporate sectors, to conversations about how to transform society. He firmly and fervently believes that "any chance of reversing climate change and the depletion of the earth's resources requires first and foremost a radical change in values and beliefs in order that people can include the ethical and spiritual dimension of environmental sustainability in their lives and practices."

In 2002, Ecumenical Patriarch Bartholomew co-signed the "Venice Declaration" with Pope John Paul II, the first joint statement by the two world leaders. The recipient of several environmental awards (including the Sophie Prize, the Binding Foundation Award, and the United Nations Environmental Protection Award) and the U.S. Gold Medal of Congress, His All-Holiness has been labeled the "Green Patriarch" by Al Gore and the media, and honored by *Time* magazine (USA) and *The Guardian* (UK) as one of the world's most influential environmental authorities.

Yet the hallmark of the Ecumenical Patriarch's initiatives is not success, but humility and service. In beholding the larger picture, the Ecumenical Patriarch recognizes that he stands before something greater than himself, indeed something greater than his (or any) Church. For Patriarch Bartholomew, healing a broken environment is a matter of truthfulness to God, humanity, and the created order. He was the first to dare broaden the traditional concept of sin – beyond individual and social implications – to include environmental damage! Thus, while the Ecumenical Patriarch has been widely recognized for his pioneering work in confronting the theological and ethical imperative of environmental protection, he has courageously declared the abuse of the natural environment as sinful:

> To commit a crime against the natural world is a sin. To cause species to become extinct and destroy the biological diversity of God's creation; to degrade the earth's integrity by causing climate change, stripping the earth of natural forests or destroying wetlands; to injure other human beings with disease and contaminate the earth's waters, land, air, and life … these are sins.
>
> *(Santa Barbara, USA, 1997)*

A new vision of heaven and earth

The sacramental image of the world is represented in color through the well-known icon depicting the hospitality of Abraham and Sarah as they welcome three strangers in the desert of Palestine. It is an icon of the communion between the three persons of the Trinity as they relate to creation: "The Lord appeared to Abraham by the oaks of Mamre, as he sat at the entrance of his tent in the heat of the day" (Genesis 18:1).

If we pay close attention, not only do "the oaks of Mamre" provide refreshing shade for the Patriarch of Israel, but they are the occasion for divine revelation. By analogy, then, not only do the trees of the world provide sustenance for humankind in diverse ways, but they reflect the very presence of God. Cutting them down implies eliminating the divine presence from our lives. Indeed, the Hebrew interpretation of this text implies that the oak trees themselves – just as the visitors who appeared at the same time – somehow actually reveal God. For it is not until Abraham recognized the presence of God in the trees (i.e., in creation) that he was also able to recognize God in his visitors (in human beings).

The Patriarch recognizes that the crisis before us is not primarily ecological. It is a crisis about the way we envisage or imagine the world. We treat our planet in a godless manner because we fail to see it as a divine gift, which it is our obligation in turn to transmit to future generations. So before we can effectively deal with problems of our environment, we must change the way we perceive the world. Otherwise, we are simply dealing with symptoms, not with their causes. We require a new worldview if we desire "a new heaven and a new earth" (Revelation 21:1). This is our calling; and this is God's command. As His All-Holiness Ecumenical Patriarch Bartholomew declared jointly with the late Pope John Paul II: "It is not too late. God's world has incredible healing powers. Within a single generation, we could steer the earth toward our children's future. Let that generation start now" (Venice, 2002).

See also in this book
St. Francis of Assisi, Nasr, White

Bartholomew's major writings
Ecumenical Patriarch Bartholomew, *Encountering the Mystery: Understanding Orthodox Christianity Today*, New York: Doubleday, 2008.
John Chryssavgis (ed.), *On Earth as in Heaven: Ecological Vision and Initiatives of Ecumenical Patriarch Bartholomew*, Bronx, NY: Fordham University Press, 2011.

Further reading
John Chryssavgis, *Bartholomew: Apostle and Visionary*, New York: Harper Collins, 2016.
John Chryssavgis, *Beyond the Shattered Image: Insights into an Orthodox Ecological Worldview*, Minneapolis, MN: Light and Life, 1999.
Frederick Krueger, *Greening the Orthodox Parish: A Handbook for Christian Ecological Practice*, Santa Rosa, CA: Orthodox Fellowship of the Transfiguration, 2012.
Elizabeth Theokritoff, *Living in God's Creation: Orthodox Perspectives on Ecology*, Yonkers, NY: St. Vladimir's Seminary Press, 2009.

JOHN CHRYSSAVGIS and FREDERICK W. KRUEGER

WANGARI MAATHAI 1940–2011

What I have learned over the years is that we must be patient, persistent, and committed. When we are planting trees sometimes people will say to me, "I don't want to plant this tree, because it will not grow fast enough." I have to keep reminding them that the trees they are cutting today were not planted by them, but by those who came before. So they must plant the trees that will benefit communities in the future. I remind them that like a seedling, with sun, good soil, and abundant rain, the roots of our future will bury themselves in the ground and a canopy of hope will reach into the sky.

(Unbowed: A Memoir, *p. 289)*

Born on April 1, 1940, in the village of Ihithe in the Central Highlands of British Kenya, Wangari Muta Maathai was deeply rooted in this landscape dense with forests and wildlife. Her father was a mechanic and driver for a British settler, and her mother farmed. When she was very young, her mother gave her a small piece of land to tend, on which she planted sweet potatoes and beans. Maathai watched, amazed, as her vegetables grew and birds and insects came to feast, but still left plenty for the harvest.

One day, Maathai's elder brother, asked his mother "Why doesn't Wangari go to school like the rest of us?" Her mother paused and then replied: "There's no reason why not," even though schooling the daughters of rural Kenyans was extremely unusual (*Unbowed: A Memoir*, p. 39). Her mother's answer provided the genesis for the boundary-breaking "firsts" Maathai would go on to achieve as a woman in academic and public life, including, in 2004, becoming the first African woman and first environmentalist awarded the Nobel Peace Prize.

In the local primary school, Maathai excelled and developed an aptitude for biology and formed strong mentoring relationships with many of the Irish and Italian nuns who taught her. They encouraged Maathai's love of science and she admired them for their selfless service. A similar commitment to the greater good, joined to a deep love for the environment and a strong sense of justice inherited from her Kikuyu forebears would animate Maathai's adult life.

In 1959, Maathai received a scholarship to attend university in the US, awarded to promising young East Africans whose nations were poised to achieve independence. She graduated in 1964 from Mount Saint Scholastica (now Benedictine College) in Atchison, Kansas with a degree in biological sciences, and then studied for a Master's degree at the University of Pittsburgh.

During her years in the US, Maathai observed and absorbed the tumult and triumphs of the civil rights movement and the conduct of the movement's leaders; their tactics, informed Maathai's later struggles to secure democratic space, human and civil rights, and environmental and social justice in Kenya.

In 1966, Maathai returned home, exultant to live in a now-liberated nation filled with possibility and promise. In 1971, she completed a Ph.D. in veterinary anatomy at the University College of Nairobi. In doing so, Maathai became the first woman in Eastern and Central Africa to obtain a doctoral degree. At the University of Nairobi, she became chair of the Department of Veterinary Anatomy and an associate professor in 1976 and 1977 respectively – the first woman in each position.

An elite woman by virtue of her education and professional and social status, Maathai was expected to join the boards of civic organizations like the Red Cross and the Kenya Association of University Women that were seeking to "Africanize" their leadership, and she did so enthusiastically.

In the early 1970s Maathai became directly acquainted with the emerging movements for environmental protection and women's equality that were propelled by two seminal United Nations conferences: on the human environment and on women. The environment conference led to the founding of the UN Environment Programme (UNEP) in Nairobi, and the women's conference launched the first international women's year. Maathai participated in processes linked to each that strongly influenced her thinking on the relationships binding the environment, development, gender, poverty, economics, democracy, rights and responsibilities – and grassroots mobilization.

She was asked to join the local board of the Environmental Liaison Centre (ELC), established to ensure civil society's participation in UNEP's work. Through the Centre, Maathai began exchanging ideas and forming friendships with environmentalists from the US, Europe, and Asia. The issues weren't entirely unfamiliar. Maathai was a biologist who'd grown up in rural Kenya where people's lives depended on the health of the environment. But she found the holistic perspective being brought to the natural sciences particularly intriguing. "A whole different world opened up to me," Maathai writes in her autobiography, *Unbowed*, and recalls that her ELC work became "my second full-time career." (*Unbowed: A Memoir*, p. 120).

By the early 1970s, Maathai had joined the National Council of Women of Kenya (NCWK), which later she would chair. Before the 1975 UN women's conference in Mexico City, Maathai and other NCWK members convened Kenyan women to learn about their concerns to inform Kenya's government delegation to the conference. Some of the women were from the Central Highlands. They testified about rivers drying or silting up, making drinking water scarce; they had to walk further to collect firewood because forests were dwindling; and how hunger was a daily reality for themselves and their children.

Maathai had witnessed changes in the environment when she'd visited her family and conducted academic fieldwork. She saw that the forests were mostly gone, cleared for commercial timber plantations. Farms like her mother's that grew a diverse array of food crops were planted edge-to-edge with "cash" crops of coffee and tea. Wild animals were rarely seen. Maathai asked herself how this environmental degradation could be addressed. "Now, it is one thing to understand the issues. It is quite another to do something about them," Maathai writes in *Unbowed*. "But I have always been interested in finding solutions." And the solution, she recalls, just came to her: "Why not plant trees?" (*Unbowed: A Memoir*, p. 125).

Trees would provide wood that would enable women to cook nutritious foods, serve as fencing and fodder for cattle and goats, bind the soil, and provide food, too. The trees would also help regenerate the ecosystem. Thus germinated the seed that would grow into the Green Belt Movement (GBM), the organization Maathai founded that has since mobilized tens of thousands of rural citizens in Kenya – principally women – to grow, plant, and protect tree seedlings on private and public lands; to expand democratic space; and to stand up to violations of ordinary people's environmental and human rights. Maathai formally launched GBM on World Environment Day in 1977 by planting seven trees in downtown Nairobi.

Maathai imbued those trees with a symbolism that magnified their meaning. She dedicated them to women and men, whose service to the nation had been forgotten by most Kenyans and ignored by the post-independence government. Those seven trees became the first of tens of thousands of "green belts." By 2016, GBM had planted more than 50 million trees in Kenya alone and encouraged numerous similar efforts around the world, at scales both smaller and much larger (e.g. UNEP's Billion Tree Campaign, which Maathai co-chaired).

From the outset, Maathai encouraged local women to use their knowledge of and expertise at planting to become "foresters without diplomas" – a rebuke to the notion that environmental conservation could only be handled by educated (male) government employees. Use your "woman-sense" she'd tell them. She also sought to instill a belief in self-empowerment, which became a core GBM value.

Maathai sought to extend the impact of the women's efforts, and of GBM itself, through ongoing mobilization. Once the women had planted the seedlings, she encouraged them to visit close-by areas and convince others to plant trees, too. As she writes in *Unbowed*: "This was a breakthrough, because it was now communities empowering each other for their own needs and benefit" (*Unbowed: A Memoir*, p. 137).

It wasn't a coincidence that GBM was born and expanded rapidly during the UN Decade for Women (1976–85), or that it was midwifed by the UN Voluntary Fund for Women. In 1982, the Fund gave the movement its first large grant, and its first director, Margaret Snyder, became a strong supporter and close personal friend of Maathai. (The Voluntary Fund eventually became UN Women).

As Maathai began running GBM full-time, she became known in global environmental circles. In 1984, she received the Right Livelihood

Award, often called the "alternative Nobel"; in 1987, Maathai was included on UNEP's "Global 500" list.

Maathai increasingly perceived environmental degradation in Kenya as being exacerbated by "land-grabbing" by political and economic elites, cronyism, corruption, autocratic rule, as well as a populace disengaged from politics and over trusting of a demagogic leadership that exploited ethnicity. A return to values and heightened critical consciousness would, as Maathai saw it, enable communities to solve problems using their own knowhow, and ensure political leadership that was accountable, acted in the best interests of *all* communities in Kenya, and protected the natural resources of the country.

Maathai began to conduct "civic and environmental education" seminars in which she challenged GBM communities to question how they'd become poor, disenfranchized, and powerless—particularly by analyzing their own actions or inaction and what was within their control to change. This conscientization became a central pillar of GBM's work, transporting it beyond "mere" environmentalism. It also made her more of a threat to the government's hold power.

In 1989, Maathai embarked on her first high-profile environmental campaign, which catapulted her onto the government's enemies list, as well as the global stage. She waged a struggle, virtually singled-handed, against the government's plan to build a skyscraper and statue of then-president Daniel arap Moi in Uhuru Park in Nairobi, which provided much-needed green space for millions of people. Maathai also wielded her pen to great effect. She wrote many letters in clear, pungent prose protesting the "park monster." Moi and his cronies were not amused. Maathai was vilified personally in parliament, harangued as a woman out of step with her defined (subservient) place, and even an enemy of the state. Still, she persisted until the project was quietly shelved, and Maathai called upon GBM members to join her in the park to for a victory dance, declaring that the project was as dead as a dodo.

The tenacity of Maathai's campaign gave her an even higher global profile. In 1991, Maathai helped found the Women's Environment and Development Organization (WEDO); in 1992 at the "Earth Summit" in Rio, she spoke on behalf of civil society; and in 1994, she became a commissioner of the Earth Charter, an effort to instantiate the ethic of environmental protection within international policy and *praxis*.

To Maathai and Kenyan political analysts, the Uhuru Park victory was the beginning of the end of the one-party state; it was also the beginning of Maathai and GBM's immersion in Kenya's movement for democratization. Maathai became a household name, admired for her refusal to be silenced.

"I don't tend to invite challenges," she wrote, "but I meet them ... That, perhaps, has been my strong point" (*Unbowed: A Memoir*, p. 194).

The 1990s weren't easy. Violence and fear were rife as the government beat, jailed, or killed its opponents. In 1992, Maathai was assaulted and hospitalized when police violently broke up protests by the mothers of young men whom the government had imprisoned and tortured for political activity. Security forces physically attacked Maathai on many occasions; she was once barricaded inside her house before being jailed, and had to travel incognito to safe houses.

In 1998, she learned that developers were about to gazette parts of Karura Forest on the outskirts of Nairobi to build luxury housing for government cronies. Rallying GBM activists, students, and members of the opposition, Wangari faced down hired thugs in Karura and was beaten and bloodied for her pains.

Finally, in 2002, the first genuinely free-and-fair elections in 24 years ushered a new coalition into power. Maathai ran for the parliamentary seat of Tetu, near her home village, and won with an overwhelming majority. She was appointed deputy minister for the environment, and worked to ensure that environmental protection and a respect for cultural biodiversity were woven into Kenya's new constitution. In her constituency, Maathai created systems to ensure citizens participated in development decisions, challenging a culture of dependency and encouraging innovation and solidarity. She refused to promise what she couldn't deliver and required her constituents to "rise up and walk!" with her.

In 2004, she was awarded the Nobel Peace Prize for her contribution to sustainable development, democracy, and peace. In the years before her death in September 2011 from ovarian cancer, Wangari travelled the world promoting sustainable development, the opening of democratic space, and protection and restoration of forests, to combat the effects of climate change. In 2010, in partnership with the University of Nairobi, she founded the Wangari Maathai Institute for Peace and Environmental Studies (WMI), to bring the wisdom and outside-the-box thinking of GBM's work to the academy and vice versa.

Maathai's life reveals her multivalent identities. A highly educated academic, she never forgot her roots and was entirely comfortable planting trees with the rural women in GBM networks. She was political (she ran for office several times, including once for the presidency), yet her greatest effect on the Kenyan *polis* lay outside parliament.

She welcomed genuine assistance from Western donors, yet worried about the poor in not using that aid to develop self-reliance. She was warm and engaging, yet relentless and fearless when facing down thugs or politicians who vilified her. Although a world celebrity, she remained deeply committed to Kenya, even being tear-gassed in April 2008 at a Nairobi demonstration. She was always optimistic, but rueful, too. She'd refer to the date of her birth and exclaim that perhaps she was the greatest fool there'd ever been.

Perhaps Maathai's most important legacy is her integration of grassroots-based, indigenous African environmental tradition into democratic and development norms, and her conviction that all of us, in whichever part of the economic or social spectrum we find ourselves, can do our part in sustaining the environment for future generations of humans and other species.

See also in this book
Mendes, Shiva

Maathai's major works
The Green Belt Movement: Sharing the Approach and the Experience (New York: Lantern Books, 2003).
Unbowed: A Memoir (New York: Anchor Books, 2007).
The Challenge for Africa (New York: Pantheon, 2009).
Replenishing the Earth: Spiritual Values for Healing Ourselves and the World (New York: Doubleday Image, 2010).

Further reading
Florence, Namulundah, *Wangari Maathai: Visionary, Environmental Leader, Political Activist* (New York: Lantern Books, 2014).
Kennedy, Kerry, *Speak Truth to Power: Human Rights Defenders Who Are Changing Our World* (New York: Umbrage Editions, 2003), pp. 172–5.
Lappé, Anna, and Frances Moore, *Hope's Edge: The Next Diet for a Small Planet* (New York: Tarcher, 2002), pp. 167–95.
Ndegwa, Stephen N., *The Two Faces of Civil Society: NGOs and Politics in Africa* (West Hartford, CT: Kumarian Press, 1996).

MIA MACDONALD

J. BAIRD CALLICOTT 1941–

There is no survival value in pessimism. A desperate optimism is the only attitude that a practical environmental philosopher can assume.[1]

For more than four decades, environmental philosopher John Baird Callicott has argued that philosophy and ethics lie at the root of our global environmental problems. He has steadfastly clung to a 'desperate optimism' that philosophy and ethics can both elucidate and help resolve these problems:

> Although an ethic, whether environmental or social, is never perfectly realized in practice, it nonetheless exerts a very real force on practice. Ideals do measurably influence behaviour. In envisioning, inculcating, and striving to attain moral ideals, we make some progress both individually and collectively, and gain some ground.[2]

Baird Callicott was born in Memphis, Tennessee, on 9 May 1941. He was educated at Rhodes College and Syracuse University, receiving his PhD in philosophy (focused on the philosophy of Plato) from Syracuse in 1971. He has taught and lectured at a vast number of universities in the United States and abroad, and is currently University Distinguished Research Professor and Regents Professor of Philosophy (ret.) at the University of North Texas. His contribution to the field of environmental ethics has been immeasurable. He was there at the beginning: teaching the very first university course in the world in environmental ethics in 1971 at the University of Wisconsin – Stevens Point, publishing in the very first issue of the original journal in the field in 1979, and establishing himself as one of the founders of the field.[3] He has been referred to as 'the man who practically invented environmental ethics'.[4]

Although Baird Callicott's publishing career did not begin until age 38, his reputation for insightful and creative argument, lucid and engaging prose, and provocative thought have earned him the highest recognition from many contemporary environmental thinkers. One can hardly pick up an issue of *Environmental Ethics*, *Environmental Values* or any other journal in the field (and many related fields) without encountering numerous references to and comments upon Callicott's work. As the editor of *Environmental Values* once wrote: 'Sustained critical interest in the work of J. Baird Callicott … just won't lie down.'[5] Renowned environmental activist Dave Foreman even acknowledged that 'in scholarship, sincerity, and openness … Callicott stands head and shoulders above his academic colleagues'.[6] An introduction at a wilderness conference in Montana once invoked the words of Henry Miller to comment on the status of Callicott's contribution: '"Only a very few souls, at any time in man's history have been privileged to battle with the great problems, the problems of man." Baird Callicott is just such a soul.' And always, Callicott's efforts have included a progressive attempt to bring philosophy out of the ivory tower of academia and apply it not only to real-world environmental problems but to other disciplines as well. He has written for many non-philosophical journals, various encyclopedias, textbooks in conservation biology, and has served on natural resource advisory boards.

Callicott's interest in environmental ethics grew out of his serious commitment to the discipline of philosophy. It has remained philosophically grounded ever since: 'My work has always been connected to philosophy; I see environmental ethics both as philosophy and as something that is challenging and transforming philosophy.'[7] His sense is that in the years to come the progress made by environmental philosophers on this front will be positively acknowledged:

> I've bet my life on the belief that environmental philosophy will be regarded by future historians as the bellwether of a twenty-first-century intellectual effort to think through the philosophical implications of the profound paradigm shifts that occurred in the sciences during the twentieth century.[8]

If so, Baird Callicott will be one of the philosophers most deserving of the credit.

Callicott is most notably recognized as the leading interpreter of the philosophical legacy of Aldo Leopold. Leopold's recognition that evolution and ecology altered our fundamental assumptions about ourselves and the world around us marks him as an early environmental philosopher. However, Leopold was not a philosopher in the formal sense. His ideas required unpacking. Baird Callicott provided the conceptual and philosophical foundations upon which to ground Leopold's metaphysical and ethical assumptions. Just as it is difficult to see where the ideas of Socrates leave off and those of Plato (his student and scribe) begin, it is difficult at times to tell where Leopold's thoughts end and Callicott's emerge. However, both Leopold and Callicott view the evolutionary/ecological world-view as a dismissal of the modern mechanistic paradigm which, until quite recently, has been taken as a given. Denying the sharp divisions between self and nature and forcing a re-thinking of an atomistic and mechanical world in terms of an organic and systematically related world, Leopold and Callicott assert that such scientific paradigm shifts cannot be viewed in isolation – they have profound metaphysical and ethical implications, they challenge and change both. Building upon the work of biologist Charles Darwin and philosophers David Hume and Adam Smith, Leopold and Callicott point out that one's sense of ethical inclusiveness corresponds with one's sense of a shared community. And, since evolution and ecology can be seen as portraying the 'soils and waters, plants, and animals, or collectively: the land'[9] and human beings as part and parcel of a shared social community, Leopold and Callicott have argued that the ethical duties we admittedly owe to one another can be, and indeed ought to be, prompted and extended to this land community as well. Leopold refers to this set of ethical obligations as the 'land ethic', and Callicott's most acknowledged role has been that of defender of the land ethic. The power of the work of Leopold and Callicott, then, is that they portray the world as significantly more morally fertile than previously perceived.

Within the larger debate surrounding the extension of moral obligations to encompass the land, Callicott has argued that the land possesses intrinsic value: said to be value beyond or in addition to merely instrumental or use value. Such a move designates Callicott as an ecocentrist – or one who attributes direct moral standing to such things as species, ecosystems, watersheds, biotic communities and even the biosphere as a whole, not to mention those individuals which constitute those biological collectives[10] – and places him in the company of other environmental philosophers such as Holmes Rolston III, Arne Naess and Val Plumwood. The debate surrounding the ascription of intrinsic value to environmental parts or wholes, and Callicott's contribution to this debate, has remained at the centre of environmental ethics from its inception.

Of course, anyone familiar with Baird Callicott's work knows that he has made deep contributions in a multiplicity of other areas as well. Such sundry topics as environmental education, aesthetics, the distinction between animal welfare ethics and environmental ethics, Judeo-Christian stewardship, conservation biology, ecological restoration, hunting ethics, agriculture, health and wellness, and environmental activism, have all garnered his attention.

Callicott's work has been uncannily provocative. If arguing that nature possessed intrinsic value and that we owe moral obligations to the land was not enough, a number of other topics he has taken up over the past thirty years have launched him into the centre of, sometimes, controversial debates.

Callicott became one of the earliest theorists on the environmental attitudes and ethics expressed by the overlapping world-views of North American Indian societies. He argues that an examination of the cosmology of American Indian tribes displays an environmental ethic worthy of notice – and one, interestingly, that shares strong affinities with Leopold's land ethic. As he once wrote:

> The implicit overall metaphysic of American Indian cultures locates human beings in a larger social, as well as physical, environment. People belong not only to a human community, but to a community

of all nature as well. Existence in this larger society, just as existence in a family and tribal context, places people in an environment in which reciprocal responsibilities and mutual obligations are taken for granted and assumed without question or reflection.[11]

This line of thought later developed into commentary on the environmental attitudes and values expressed in a wide range of world cultures and religious traditions – from Christian to Islamic, from Pagan to Australian Aboriginal – which was published as his critically acclaimed book *Earth's Insights*.

Callicott was also at the centre of the highly contentious debate over the concept of wilderness. Along with historian William Cronon, Callicott has argued that the concept of wilderness is a product of social construction; a product desperately in need of reconstruction. Callicott 'believes that the received wilderness idea has been mortally wounded by the withering critique to which it has been lately subjected'.[12] However, although this point is often misunderstood or ignored, he is no enemy of wilderness, but rather a friendly critic: 'I am as ardent an advocate of those patches of the planet called 'wilderness areas' as any other environmentalist. My discomfort is with an idea, the received concept of wilderness, not with the ecosystems so called.'[13] Callicott also emphasizes that through a conceptual re-thinking of wilderness those areas we refer to as wilderness will be better protected.

Most recently, Callicott has come full circle, returning to an early essay written by Aldo Leopold in 1923 (though unpublished until 1979) wherein Leopold speculates about an 'Earth Ethic': '[T]he "dead" earth is an organism possessing a certain kind and degree of life, which we intuitively respect as such.'[14] A postulation of a moral obligation to the earth as a whole, Callicott argues, is required for addressing environmental ethics dilemmas, such as global climate change, which operate at a global scale:

> [t]he moral philosophies we that we have inherited from the past are woefully inadequate … we need to think up a moral philosophy that is commensurate with the spatial and temporal scales of the wholly novel, utterly (and literally) unprecedented ethical issues with which we are not confronted.[15]

A journey through the writing and thoughts of J. Baird Callicott is always insightful, always challenging, always instructive, and always a lesson in the power of sound reasoning and good writing. And at all times in his work there is a sense of an empowering optimism, an affirmation that a successful ethical relationship between humans and the non-human world can and will be forged.

Notes

1 From 'Benevolent Symbiosis: The Philosophy of Conservation Reconstructed', in J. Baird Callicott and Fernando J.R. da Rocha (eds), *Earth Summit Ethics*, p. 157.
2 *Earth's Insights*, p. 3.
3 He also established one of the first environmental studies programmes in the United States at the University of Wisconsin – Stevens Point.
4 Arthur Herman, *Community, Violence, and Peace: Aldo Leopold, Mohandas K. Gandhi, Martin Luther King, Jr, and Gautama the Buddha in the Twenty-first Century*, Albany, NY: State University of New York Press, p. 234, 1999.
5 Alan Holland, 'Editorial', *Environmental Values*, 9/1, p. 1, 2000.
6 Callicott, 'The Ever-robust Wilderness Idea and Ernie Dickerman', *Wild Earth*, 8/31, p. 1, 1998.
7 Personal communication, March 1999.
8 'Introduction: Compass Points in Environmental Philosophy', in *Beyond the Land Ethic*, p. 4.
9 Leopold, *A Sand County Almanac*, p. 204.
10 Callicott defines an ecocentric environmental ethic as 'An environmental ethic that takes into account the direct impact of human actions on non-human natural entities and nature as a whole', *Earth's Insights*, p. 10.
11 'Traditional American Indian and Western European Attitudes Toward Nature: An Overview', in *In Defense of the Land Ethic*, pp. 189–90.
12 'Introduction', in J. Baird Callicott and Michael P. Nelson (eds), *The Great New Wilderness Debate*, p. 12.
13 'The Wilderness Idea Revisited', in J. Baird Callicott and Michael P. Nelson (eds), op. cit., p. 339.
14 'Some Fundamentals of Conservation in the Southwest', in J. Baird Callicott and Susan Flader (eds), *The River of the Mother of God and other essays by Aldo Leopold*, p. 95.
15 *Thinking Like a Planet*, p. 268.

See also in this book
Black Elk, Darwin, Leopold, Naess, Rolston

Callicott's major writings
In Defense of the Land Ethic: Essays in Environmental Philosophy, Albany, NY: State University of New York Press, 1989.
Earth's Insights: A Multicultural Survey of Ecological Ethics from the Mediterranean Basin to the Australian Outback, Berkeley, CA: University of California Press, 1994.

Beyond the Land Ethic: More Essays in Environmental Philosophy, Albany, NY: State University of New York Press, 1999.

Thinking Like a Planet: The Land Ethic and the Earth Ethic, Oxford, UK: Oxford University Press, 2013.

Further reading

Callicott, J. Baird (ed.), *Companion to 'A Sand County Almanac': Interpretive and Critical Essays*, Madison, WI: University of Wisconsin Press, 1987.

Callicott, J. Baird and Ames, Roger T. (eds), *Nature in Asian Traditions of Thought: Essays in Environmental Philosophy*, Albany, NY: State University of New York Press, 1989.

Callicott, J. Baird and Flader, Susan L. (eds), *The River of the Mother of God and Other Essays by Aldo Leopold*, Madison, WI: The University of Wisconsin Press, 1991.

Callicott, J. Baird and Nelson, Michael P. (eds), *The Great New Wilderness Debate*, Athens, GA: The University of Georgia Press, 1998.

Callicott, J. Baird and Nelson, Michael P., *American Indian Environmental Ethics: An Ojibwa Case Study*, Upper Saddle River, NJ: Pearson Prentice-Hall, 2004.

Callicott, J. Baird and Rocha, Fernando J.R. da (eds), *Earth Summit Ethics: Toward a Reconstructive Postmodern Philosophy of Environmental Education*, Albany, NY: State University of New York Press, 1996.

Freyfogle, Eric T., *Bounded People, Unbounded Land: Envisioning a New Land Ethic*, Washington, DC: Island Press/Shearwater Books, 1998.

Hargrove, Eugene C., *Foundations of Environmental Ethics*, Englewood Cliffs, NJ: Prentice-Hall, 1989.

Leopold, Aldo, *A Sand County Almanac: And Sketches Here and There*, New York: Oxford University Press, 1949.

Nelson, Michael P. and Callicott, J. Baird, *The Wilderness Debate Rages On: Continuing the Great New Wilderness Debate*, Athens, GA: University of Georgia Press, 2008.

Rolston III, Holmes, *Environmental Ethics: Duties To and Values in the Natural World*, Philadelphia, PA: Temple University Press, 1988.

MICHAEL PAUL NELSON

BOB HUNTER 1941–2005

Fantastic points were being made about how one life-form lives in relation to another, how this weird little creature couldn't possibly survive if this thing on that plant weren't around to turn this into that, and so on, ad infinitum. To begin to understand the world, it seemed you could start anywhere, and the interwoven chains and bracelets of life would inevitably take you everywhere there was to go. There were no real divisions. Everything was One.[1]

Robert (Bob) L. Hunter was born in the Canadian prairie city of Winnipeg, Manitoba. After high school, he decided not to follow the normal path and go to university. Instead, Hunter decided to burn his university acceptance letter on the steps of his high school and walk to Vancouver. He could not have known then that the series of events that followed would give him a central role in changing the world.

From his work as an activist/journalist for the rights of mental health patients in Vancouver, to his role at the Vancouver Sun as the counter-culture reporter, to his joining Quakers Irving Stowe and Jim Bohlen on the "Don't make a Wave" committee, to his embarking with the Phyllis Cormack to protest nuclear weapons testing in the Aleutian islands, to being a co-founder of Greenpeace and seeing it through campaigns against the wholesale slaughter of whales off the west coast of the US, to the south pacific to Australia, and his championing, as Greenpeace's first president, of the international movement to eliminate the Canadian seal hunt, his life and work are characterized by a frenetic sense of urgency to care for the environment and all its creatures and he did indeed, change the world. But he remained, to the end of his days in 2005 when he died of cancer, a humble and self-effacing advocate for the planet. He was fond of pointing out that "I" was never the most important thing in the room. Hunter had a strong sense of the necessity for collective thought and action and he worked tirelessly to bring attention, not just to events in the world, but perhaps more importantly to the critique of western thought as characterized by a self-defeating sense of "operationalism" and fragmentation.

While working at *The Vancouver Sun* newspaper, he'd been asked to cover a protest at the Peace Arch on the Canadian–US border. That event would see the only closure of the 'longest undefended border' in the world in protest against American nuclear weapons testing. He had been asked by the organizers to make a speech at the protest, which his editor had forbidden: he was a reporter, so report. The organizers hadn't been sent that memo and he took the stage and made his first public protest speech. Later, after the formation of the "Don't Make a Wave" committee, the decision was made to extend the protest into dramatic action. Members of the group decided that they would hire a boat and sail into the waters around Amchitka, the underground test site. What started as a concern for the possibility that such testing would cause tectonic shock waves, which could spark tsunamis that would swamp the whole north western coast, became a movement. At the end of a meeting of the group, "Irving Stowe made his usual V sign and said, 'Peace'. The youngest member of the committee, a twenty-three-year-old Canadian named Bill Darnell, said: 'Make it a *green* peace,'"[2] and a worldwide environmental movement was born. Up to that point none of the members of the committee had made the move from talk to action, and after that, they would all place the emphasis on the latter, since the former was proving so ineffectual.

That first "run" deepened Hunter's sense of purpose. Between noticing the total lack of life, marine or terrestrial, on islands in the area around Amchitka, and the realization that "a flower is my brother," Hunter and the crew of the *Phyllis Cormack* found themselves at the heart of something quite new. Across Canada and the US, people had heard of the action in back-page news reports, but it was only after they had failed to achieve their goal, and the decision made to return to port that Hunter, dejected and demoralized, heard over the shortwave that on October 6, 1971, 10,000 high school students had converged on the US consulate in Vancouver. The group realized in that moment that their defeat had become a victory. "I found myself out at the bow, crying as I had not cried in years, with joy, the kind of crying people did during the war when news came through of a great victory. We all felt the kind of awe and love that mass movements inspire in those who take part in them, a feeling of unaloneness."[3] The concept of 'unaloneness' runs throughout Hunter's work, in print and in action. It was an ideal that he formulated in his early writing.

His first book, *Erebus* (1968), was described as "distasteful, disrespectful, ugly and angry, all those things which upset the average, middle-class, puritan sensibility … This is the industrial, polluted environment in which young people are forced to grow up—it is sheer madness propagating further madness."[4] The novel explores the anaesthetizing effect of working and living in a world that seems utterly indifferent to life as such. At the time, Hunter's focus was on the increasing encroachments of industry on humanity. The novel was followed in 1970 by Hunter's first work of popular philosophy on the subject, *The Enemies of Anarchy* in which he explores the question of the sustainability of "civilization." The first 100 pages of the book are intended "to show that we cannot hope to survive by making minor adjustments here and there … A reversal of our *basic premises*, behavior, and methodology, is necessary at this stage if we are to survive."[5] The following year Hunter published *The Storming of the Mind*, which pushes the thesis of *Enemies* to the next stage by looking at the question of "how" we think those "basic premises." His conclusions in the book point to the necessity of what he calls "a revolution in consciousness," shifting from the operationalism of post-industrial society, the view that all of nature is but a resource produced for human exploitation, to engagement with nature as an integrated whole. Operationalism, he argues, is patently self-defeating. A civilization that depends on the resources of nature cannot hope to continue to flourish without care and attention to the preservation and protection of that same nature. The complexity of the issue though, is certainly not lost on Hunter, and in the years after these books, Hunter took his ideas, and his words, and put them into action.

After the embryonic movement had become Greenpeace, Hunter found himself in a position to garner attention to environmental issues on a scale that hadn't yet been imagined. Following the campaign at Amchitka, Greenpeace's next objective was the protection of whales and the elimination of factory ship whaling. Fraught with even greater logistical problems than the relatively simple protest at Amchitka, the organization had to "beg, borrow or steal" just about everything it needed. As with the first expedition, they had a boat, renamed *The Greenpeace*, a captain and crew, but not much else. Attributed to a planetary awakening, Hunter remembered the time as "freaky," as resources, funds and equipment seemed to fall from the sky. He says that his experiences with Greenpeace in those early days bore all the hallmarks of fate. If one coincidence occurs, that's a coincidence, but once five or more happen at the same time, "you have to move into the realm of small 'm' miracle."[6] In his chronicle of Greenpeace, *Warriors of the Rainbow*, Hunter remembers that "During some of the Greenpeace campaign, miracles were not only commonplace—we got a few of them on film."[7]

"Getting them on film" became a key feature of Greenpeace activism. Going back to the influence of Irving Stowe, Jim Bohlen, and some of their Quaker beliefs, Hunter knew that what they were involved in with Greenpeace was a "bearing witness" in the age of mass media. At the end of *The Storming of the Mind*, Hunter explores the implications. "In the past, the true revolutionary who wanted to make man more humane ... had only a few weapons in his arsenal ... The only medium through which a revolution could communicate itself was armed struggle."[8] But the advance of technology has produced a new host of weapons and "a mass communication system exists." What this mass communication system provides is a "delivery system" for "mindbombs." And by that term he meant literally to bomb the minds of people such that they too "bear witness."[9] A lifelong advocate of non-violence, Hunter first used a mindbomb during Greenpeace's first anti-whaling campaign over the Mendocino Ridge.

After a long pursuit of the Russian whaling fleet, the Greenpeace crew found the fleet of pursuit and killer boats heading back to their processing factory ship. Recognizing that the time they would have to make any impact was short, and that an inflatable zodiac with two "eco-freaks" on it could do little to stop one of these behemoths, Hunter had his speed boat positioned between a Russian explosive harpoon and an exhausted whale. Initially believing that the harpoonists wouldn't dare risk human life by firing, they had a camera man on another boat film the event. As luck would have it, the harpoonist held no such qualms and, as he fired at the whale, the harpoon flew a few short feet above their heads. Crucially, this was caught on film, which was brought to port, developed and sent to as many media outlets as possible. Now, everyone within sight of a television set saw it happen and bore witness to the callousness of an industry that could easily be replaced by other things. "The idea was to take cameras out there and make everybody bear witness, and presumably that would have the same effect on other people as it was for the person who was just seeing for themselves."[10] Thus modern environmental activism was born and public opinion began to change.

This same tactic was employed on numerous occasions to chronicle, among other things, the inhumanity of the Canadian seal hunt, the attack on a Greenpeace boat near Mururoa and the subsequent beating of the Greenpeace crew by French soldiers. The simple philosophy behind it is that when transparency is brought to bear, it becomes uncomfortable if not impossible to continue "business as usual."

Ecology, Hunter has said, consists in the idea that nature is by necessity and design diverse, interdependent and finite. Self-interest on an individual and collective human level is pursued at the expense of everything we come into contact with, from the largest biosphere to the smallest cell, and rationalizes destruction and violence to a degree that could not be imagined by any one individual. After leaving Greenpeace, Hunter returned to the world of journalism, moving from Vancouver to Toronto, where he worked as an 'ecology specialist' for a number of media outlets. Over the course of his years in Toronto, Hunter became deeply involved in the developing story of climate change, or, as he preferred to call it, climate collapse. One of the reasons for using that term was that he saw a relative of the operationalism described in *The Storming of the Mind* now at work in the environmental movement itself. The obfuscating debate about the causes of collapse had served to soften the language used by climate scientists in an effort to appease oil industry (what he called the 'Juice-cans') pseudo-science in the reports of organizations like the Conference of Parties (COP) and the Intergovernmental Panel on Climate Change (IPCC). Named in 2000 by *Time* magazine as one of the twentieth century's "Eco-Heroes," he published his last book in 2002, titled *2030: Confronting Thermaggedon in Our Lifetime.* That work demonstrates the collected knowledge of a lifetime of activism and writing on the environment. Hunter brings to bear his considerable breadth on the issue of climate and environmental collapse, tracing his personal development from "A Passion for Wheels" as a member of a civilization addicted to oil, to the need to take a dramatic step back from our current trajectory. He ends the book with a pledge to his grandson worth quoting at length:

> The climate crisis isn't out there somewhere. Its causes are close at hand, literally, as close as the night-lamp switch. I hold them in my hand when I dig the keys out of my pocket to start the car. Don't judge me by my words, which are many, someone said, but by my actions, which are few. And that should be said about all of us here, now, in the belly of the beast, mainlining coal and oil. My vow to you, Dexter, is that I am coming off the stuff as fast as I can. It took me years to learn to stop smoking. It will take a while to learn to stop climate-wrecking.[11]

Notes

1 *Warriors of the Rainbow*, p. 87.
2 Ibid. p. 31.
3 Ibid. p. 109.
4 Smith, Ronald F. "A Dark World", *The Journal of Commonwealth Literature*, 1972, 7, 107–8.
5 *The Enemies of Anarchy*, p.3.
6 *Warriors of the Rainbow*, p. 21.
7 Ibid. p. 22.
8 *The Storming of the Mind*, p. 216.
9 Ibid.
10 "The 'Don't Make a Wave Committee' Were the Founders of Greenpeace," on Public Radio International's *Living on Earth*, available at http://loe.org/shows/shows.html?programID=96-P13-00023
11 *2030: Confronting Thermageddon in Our Lifetime*, p. 276.

See also in this book
Gore, McKibben, Mendes, Naess

Hunter's major writings

The Enemies of Anarchy, Toronto, Canada: McClelland and Stewart (1970).

The Storming of the Mind, New York: Doubleday (1971).

Warriors of the Rainbow (40th Anniversary Edition), Fremantle, Australia: Fremantle Press (2011).

The Greenpeace to Amchitka: An Environmental Oddessey (with Robert Keziere), Vancouver, Canada: Arsenal Pulp Press (2005).

2030: Confronting Thermageddon in Our Lifetime, Toronto, Canada: McClelland & Stewart (2002).

Occupied Canada (with Robert Calihoo), Toronto, Canada: McClelland & Stewart (1991).

Further reading

Hunter, Robert L. *Red Blood: One (Mostly) White Guy's Encounters with the Native World*, Toronto, Canada: McClelland & Stewart (2000).

Hunter, Robert L. and Watson, Paul *Cry Wolf!* Vancouver, Canada: Shepherds of the Earth Publications (1985).

Clarke, Arthur C. *Childhood's End*, New York: Ballantine (1946).

Toffler, A. *Future Shock*, New York: Random House (1970).

THOMAS E. HART

SUSAN GRIFFIN 1943–

> We know ourselves to be made from this earth. We know this earth is made from our bodies. For we see ourselves. And we are nature. We are nature seeing nature. We are nature with a concept of nature … Nature speaking of nature to nature.[1]

Contemporary feminist poet Susan Griffin, who began writing at the age of 14, has published more than fifteen books of poetry, drama, fiction and non-fiction on subjects ranging from rape and pornography to war, *eros* and illness. She has received many book awards and honours, including the Ina Coolbirth Prize for Poetry, an Emmy Award, a National Endowment for the Arts grant, the Malvina Reynolds Award for cultural achievement, a Schumacher Fellowship, a Commonwealth Medal, a Women's Foundation Award, a MacArthur Foundation grant, and nominations for the Pulitzer Prize and National Book Critics Circle Award. Although nature is a concern throughout Griffin's body of work, her long prose-poem *Woman and Nature: The Roaring Inside Her* (1978) stands out as a key text of environmental thought and a germinative work of ecofeminism, a movement that originated in the 1970s and that has become an influential voice in environmental discourse of the twenty-first century.

Griffin's background is pertinent to understanding her work, for, concurring with the feminist insight that 'the personal is political', Griffin at times interweaves autobiography with cultural critique, a literary form she calls 'social autobiography'. Born in California in 1943, Griffin came of age during the Cold War, years marked by nuclear testing, anti-communist propaganda and social conformity, resistance to which caused her to identify herself as a radical. Griffin's parents divorced when she was six, and she and her older sister were separated and sent to live with different relatives, their mother's alcoholism rendering her unable to care for either child. This early experience of abandonment and separation struck deeply into the psyche of Griffin, whose later writing would probe these wounds to provide insight into Western culture and whose years of therapy pre-disposed her to view the collective mind of Western civilization from a psychological perspective. Griffin was raised by conservative Republican grandparents near Hollywood, California, where she grew up loving movies and becoming a fan of Eisenstein, whose film montages and juxtaposition of images almost certainly inform the associative collage technique of much of her writing. As a teenager, Griffin lived with a close friend's Jewish family, where her consciousness was raised about the historical treatment of Jews; in later years the Holocaust became an important image and racism a recurrent theme in her work.

Attending the University of California, Berkeley, during the student unrest of the 1960s, Griffin became involved in the Free Speech movement, the Civil Rights movement and protest against the Vietnam War. She transferred to what is now San Francisco State University, where she graduated *cum laude* in English in 1965, received her MA in 1973 and worked for the radical magazine *Ramparts* as an editorial assistant, becoming troubled by the sexist attitudes of the staff. During the late 1960s and 1970s, Griffin married, became a feminist, gave birth to a daughter, divorced and became a lesbian. Simultaneously, she taught writing and developed her own writing career with several volumes of poetry, short stories and an award-winning play, *Voices* (1975), which were published by feminist presses and reflected her experience as a woman in society. Early exposure to the diverse worlds of gentile and Jew, conservatism and radicalism, heterosexuality and homosexuality, marriage, motherhood and divorce, allowed Griffin to become what she calls a 'bridge figure', someone who straddles boundaries rather than reinforcing them.[2] 'As a writer', Griffin says, 'I have always felt myself to be a kind of crucible, my mind a medium in which the many voices, spoken and unspoken, belonging to our age, are melted, mixed and transformed.'[3]

As Griffin reflects on her classic work *Woman and Nature* she notes that two voices (each set in a different typeface) engage in an extended dialogue, 'one the chorus of women and nature, an emotional, animal, embodied voice, and the other a solo part, cool, professorial, pretending to objectivity, carrying the weight of cultural authority'.[4] The book opens with a stunning, heavily researched and annotated, chronologically arranged compendium of statements, or, rather, parodies of statements from Plato through Einstein – magisterial voices of science, philosophy, and religion from Western civilization – proclaiming parallel 'truths' about nature and women. For example, referring to Aristotle, Griffin writes, 'It is decided that matter is passive and inert, and that all motion originates from outside matter ... It is decided that the nature of woman is passive, that she is a vessel waiting to be filled.'[5] Later, paraphrasing Lamarck on evolution and the Marquis de Sade on women, Griffin writes: 'It is declared … [t]hat "the stronger and the better equipped … eat the weaker and … the larger species devour the smaller". And it is stated that if women were not meant to be dominated by men, they would not have been created weaker.'[6] Only occasionally, in this first section, do the voices of women and nature speak, anguished cries such as *'Our voices diminish … We become less … And they say that muteness is natural in us'*.[7]

The next section of the book focuses on the tandem mistreatment women and nature have received at the hands of a patriarchal culture dominated by the mind-set chronicled above. In 'Timber', for example, drawing from her reading of forestry manuals and office management textbooks, Griffin juxtaposes the economics of timber harvest (where trees are referred to as so many 'board feet') with the efficient supervision of stenographers. Other chapters compare factory farming with modern childbirth, horse training and dressage with facelifts and breast implant surgery, nuclear waste disposal with hiding the body of a murdered woman, and strip mining with rape. Needless to say, these pages are deeply disturbing. Casting her extensive research into a poetic form, Griffin's goal is to evoke feeling even as she awakens consciousness. These pages convince the reader that the comparisons Griffin makes point beyond metaphorical similarities to systemic unity; namely, that these various cruelties are all part of the same system, founded on, in Griffin's words, 'a philosophy that is also a submerged psychology'.[8]

In *Woman and Nature* and throughout her later work, Griffin develops a diagnosis of the illness of the Western mind; we are suffering from a form of insanity that lies at the heart of our destruction of the environment. According to Griffin in essays such as 'Split Culture' and 'Ideologies of Madness' (the latter collected in *The Eros of Everyday Life*), we live in a culture of fear. We are afraid of physical pain, illness, change and death, and we are likewise terrified by the power of nature over our lives. In the face of such terror, we resort to denial and domination. We deny our physical natures, imagining instead that we are our minds, that we possess an immortal soul. We attempt to control nature, to master it, subdue it, shape it to our desires. However, argues Griffin, the repressed always returns to haunt us in our dreams. So, we project onto an 'other' the parts of ourselves that we wish to disown. In our culture, white men have been in power; thus, men have defined themselves as above matter, and they have construed women as closer to nature. Man's domination of women, of racial 'others', and of nature can be understood as part of his ongoing efforts to be in control of himself, to triumph over the body and to deny death. 'In a culture of delusion', Griffin writes, 'women symbolize a denied self who experiences what it is to be human, to be in and of nature. This self knows that we die, this self feels, suffers pain, knows love without boundary, grieves loss, knows the world through sensation, through the body, accepts that we are sometimes powerless before the powerful circumstances of this earth.'[9]

Why have women come to be allied so closely with nature in men's psyche? Griffin, paralleling the work of Dorothy Dinnerstein, reasons that mothers are a child's first experience of nature: she has the power to feed and to comfort; likewise, she has the awful power to withhold food and to abandon the child. It is this early feeling of helpless dependence on mother/nature that causes the grown man to strive for independence, creating a culture built upon fear of connection and alienation from nature. Paradoxically, though, in dominating nature, we threaten the very grounds of our continued existence. Griffin writes: '[W]e belong to a civilization which is bent upon suicide, which is secretly committed to destroying Nature and destroying the self that is Nature.'[10]

What is the way out of this madness? In psychotherapy, the first step in healing is naming. We must become conscious of the problem. Griffin conceives of herself as a witness, someone who is 'able to speak the unspeakable, to break the silence'.[11] She explains, '[B]ecause the assumptions that belong to a culture are often invisible in their fullest dimensions and consequences, one must make them visible before discerning change. The very process of seeing the structure of thought *is* itself a crucial kind of change and genesis.'[12] She contends that the root of the problem is our culture's construction of masculinity. Griffin notes differences between socially masculine and feminine values: 'The roles society [has] given to men and women [have] produced different thinking and different ways of being in us … [M]en, valuing power, produce nations, conflict and wars, and … women, valuing life, produce relationship, continuity and peace.'[13] '[T]here are lots of reasons why males are violent', she observes, 'and they have more to do with tradition than testosterone.'[14] What is needed, according to Griffin, is 'a deep transformation of consciousness'.[15] Hallmarks of this shift will be the celebration of sensual knowledge, respect for a multiplicity of views rather than a single perspective, a view of the earth as being imbued with intelligence and intrinsic meaning, and, most important, the reunification of body and spirit and nature and culture in our conception of humanity. Griffin hopes, 'If human consciousness can be rejoined not only with the human body but with the body of earth, what seems incipient in the reunion is the recovery of meaning within existence that will infuse every kind of meeting between self and the universe, even in the most daily acts, with an eros, a palpable love, that is also sacred.'[16]

In her analysis of patriarchy and articulation of a healthier cultural alternative, Griffin, along with mutually influential writers Carolyn Merchant and Adrienne Rich, is widely regarded as a leading ecofeminist thinker. *Woman and Nature* has been called 'fundamental to an ecofeminist library', a 'cultural feminist classic', and a 'touchstone text' for 'virtually all ecofeminists'.[17] Ecofeminism joins feminist thought with ecological thought, insisting that one cannot fully understand the oppression of women without understanding how Western civilization has regarded nature, and, conversely, one cannot adequately understand our civilization's abuse of nature without taking into account how our culture conceptualizes women. As Griffin explains: '[T]he social construction (exploitation, destruction) of nature is implicit in and inseparable from the social construction of gender.'[18] Although scholars have noted that there are different varieties of ecofeminism, all versions seem to agree that patriarchy rests upon a conceptual foundation of hierarchical dualism in which reality is categorized by oppositional pairs (such as spirit/matter, intellect/emotion, mind/body, man/woman, culture/nature), in which the first term of the pair is accorded greater worth, privilege and power than the second. In this system, man is allied with culture, spirit and intellect, while woman is identified with nature, the body and emotion. While some feminists seek to liberate women from the sphere of the natural and some separatist ecofeminists wish to bar men from that sphere, Griffin and the majority of ecofeminists celebrate the woman–nature bond and urge that men likewise cultivate a closer relationship with nature and their own material bodies. In general, ecofeminists aspire to move beyond dualistic thinking and to establish relationships based not on hierarchy and domination, but on caring, respect and awareness of interconnection.

Griffin's writing, which since 1976 has been published by major trade presses, reveals an exceptionally broad understanding of interconnection. Her studies of rape and pornography reveal motivations and mechanisms of domination that also explain our relationship to nature. Her poetry is intimately related to her prose, which itself is highly poetic, reflecting her conviction that poetry is 'a powerful way of knowledge' that arises out of bodily experience and 'teaches political theory imagination'.[19] Griffin's *A Chorus of Stones* draws connections between the private psyche formed in childhood and public acts of violence in war, showing conversely how war creates violence in private life. Her recent *What Her Body Thought* connects the story of the flamboyant nineteenth-century courtesan featured in the movie *Camille* with Griffin's own illness from Chronic Fatigue Immune Dysfunction Syndrome, which in turn, 'like canaries in a mine', becomes 'a signal of the sickness of the planet'.[20] Revealing the economics of illness, Griffin indicts society for failing to support those in need. As one supporter has noted, 'By refusing to respect the "commonsense" distinctions among historical, social and personal issues, Griffin creates a kind of network of meaning in which everything illuminates everything else.'[21] In the context of environmental thought, Griffin's profound insight that gender issues and ecological issues are interconnected has been responsible for transforming both feminism and environmental thought.

Notes

1　*Woman and Nature*, p. 226.
2　Griffin, quoted in 'Susan Griffin', *Utne Visionaries: People Who Could Change Your Life, 1995 Profiles*, www.utne.com/visionaries/95profiles2.html.
3　*Made from this Earth*, p. 3.
4　Ibid., p. 82.
5　*Woman and Nature*, p. 5.
6　Ibid., p. 27.
7　Ibid., p. 26, original emphasis.
8　'Ecofeminism and Meaning', p. 216.
9　*Made from this Earth*, p. 18.
10　'Split Culture', p. 199.
11　*The Eros of Everyday Life*, p. 12.
12　Ibid., p. 6.
13　*Made from this Earth*, pp. 14–15.
14　Griffin, quoted in 'Susan Griffin', *Utne Visionaries*.
15　*The Eros of Everyday Life*, p. 20.
16　Ibid., p. 9.
17　Judith Plant (ed.), *Healing the Wounds: The Promise of Ecofeminism*, Philadelphia, PA: New Society Publishers, p. 255, 1989; Ynestra King, 'Healing the Wounds: Feminism, Ecology, and Nature/Culture Dualism', in *Gender/ Body/Knowledge: Feminist Reconstructions of Being and Knowing*, ed. Alison M. Jaggar and Susan R. Bordo, New Brunswick, NJ: Rutgers University Press, p. 124, 1989 ; Patrick D. Murphy, *Literature, Nature, and Other: Ecofeminist Critiques*, Albany, NY: State University of New York Press, p. 40, 1995.
18　'Ecofeminism and Meaning', pp. 219–20.
19　*Made from this Earth*, pp. 16, 242.
20　Griffin, 'The Internal Athlete', Ms. v. 2.6, p. 38, 1992.
21　'Susan Griffin', *Utne Visionaries*.

See also in this book
Aristotle, Lovelock, Plumwood, Schumacher, Wright, J.

Griffin's major writings
Woman and Nature: The Roaring Inside Her, New York: Harper & Row, 1978; new edn, Sierra Club Books, San Francisco, CA, 2000.

Pornography and Silence: Culture's Revenge Against Nature, New York: Harper & Row, 1981.

Made from this Earth: An Anthology of Writings, London: Women's Press, 1982; New York: Harper & Row, 1983.

'Split Culture', *The Schumacher Lectures*, vol. 2, ed. Satish Kumar, London: Blond & Briggs, pp. 175–200, 1984.

The Eros of Everyday Life: Essays on Ecology, Gender and Society, New York: Doubleday, 1995.

'Ecofeminism and Meaning', *Ecofeminism: Women, Culture, Nature*, ed. Karen J. Warren, Bloomington, IN: Indiana University Press, pp. 213–26, 1997.

Bending Home: Selected & New Poems, 1967–1998, Port Townsend, Washington, DC: Copper Canyon Press, 1998.

Further reading

Adams, B., 'Susan Griffin', *Contemporary Lesbian Writers of the United States: A Bio-Bibliographical Critical Sourcebook*, ed. Sandra Pollack and Denise D. Knight, Westport, CT: Greenwood Press, pp. 244–51, 1993.

Macauley, D., 'On Women, Animals and Nature: An Interview with Ecofeminist Susan Griffin', *American Philosophical Association Newsletter on Feminism and Philosophy*, 90, 3, 116–27, 1991.

Merchant, C., 'Earthcare: Women and the Environmental Movement', *Environment*, 23, 5, 6–13, 38–40, 1981.

'Susan Griffin', *Contemporary Authors, New Revision Series*, vol. 50, ed. Pamela S. Dear, Detroit, MI: Gale Research, pp. 169–72, 1996.

'Susan Griffin', dialogue with Nannerl Keohane in 1980, collected in *Women Writers of the West Coast Speaking of Their Lives and Careers*, ed. Marilyn Yalom, Santa Barbara, CA: Capra Press, pp. 40–55, 1983.

CHERYLL GLOTFELTY

CHICO MENDES 1944–1988

My dream is to see this entire forest conserved because we know that it can guarantee the future of all the people who live in it ... If a messenger from heaven came down and guaranteed me that my death would help to strengthen our struggle it would even be worth it. But experience teaches us the opposite ... I want to live.[1]

Francisco 'Chico' Alves Mendes Filho, man of courage, words and deeds, hero of the rubber tappers of the Amazon, played a major role in the transformation of the landscape of the Brazilian rainforest. Chico Mendes was born on 15 December 1944 on a rubber estate in Xapuri, Acre, in northwest Brazil. Forty-four years later, on 22 December 1988, he was brutally assassinated, leaving wife Ilzamar G. Bezerra Mendes and their two children, Helenira aged 4 and Sandino aged 2. Mendes' short life was devoted to leading the rubber tappers' fight to defend the Amazon forest and its fragile eco-system against exploitation by powerful and wealthy land speculators and ranchers.

Mendes was born into poverty. His parents had come from the northeast during the Second World War, having been sent to cut rubber for the Allied war cause. He received no formal education and became a *seringueiro*, a rubber tapper, at the age of 9. He learned to read and write around the age of 20.

> My life began just like that of all rubber tappers as a virtual slave bound to do the bidding of the master. I started work at nine years old, and like my father before me, instead of learning my ABC I learned how to extract latex from a rubber tree ... schools were forbidden on any rubber estate in the Amazon. The estate owners wouldn't allow it ... If a rubber tapper's children went to school they would learn to read, and write, and add up, and would discover to what extent they were being exploited.[2]

Ruthless exploitation from a variety of sources was to become the dominating force in the rubber tappers' existence, and resistance to this the focus of Mendes' life. Traditionally rubber tappers were at the mercy of a system of debt bondage, but during the 1960s and 1970s this system faced collapse in Xapuri. Ranchers from southern Brazil began to buy up rubber estates and clear vast areas of the forest for cattle grazing. Many tappers were forcefully, often brutally, evicted. Others retreated deeper into the forest to continue their work, only to be exploited by local merchants.

Chico Mendes knew that the future of the forests and of the rubber tappers were inextricably linked; that in order to secure a future for the people, the forests had to be protected and managed by those who understood the eco-system and how to live in it sustainably. From his endeavours emerged the concept of 'extractive reserves', which are legally protected forest areas that are held in trust for people who live and work on the land in a sustainable manner.

Early in the 1970s, the Xapuri Rural Workers' Union was founded, and Mendes was elected its president. As exploitation and conflict intensified, the union developed the technique of the '*empate*' or 'stand-off'. During the dry season ranchers hire labourers to clear the forest for pasture. Just before the rains come in September the cleared areas are fired. Faced with eviction the rubber tappers assembled at sites about to be cleared, preventing the clearing and persuading the labourers to lay down their chainsaws. During the months of June, July and August in the 1970s and 1980s the forests of the upper Acre valley were the scene of numerous *empates*.[3] In 1985, Mendes and other leaders founded the National Council of Rubber Tappers (CNS) and gained increasing international support for their cause and passive resistance demonstrations. The movement was recognized as a force not only for social justice, but also against environmental destruction. The rubber tappers were able to propose a socially equitable and environmentally sustainable development policy for the region based on securing and improving their way of life, rather than official investments in ranching and colonization projects that would lead to disaster both for them and for the forest.[4]

Chico Mendes played a crucial role in negotiating with governments, with the World Bank and the Inter-American Development Bank. For example, in 1987 he visited the USA at the invitation of the Environmental Defense Fund and the National Wildlife Federation in order to discuss an Inter-American Development Bank-funded road paving project in Acre. Chico's message of caution was that the project would be disastrous if environmental conditions in the loan were not fulfilled. The loan was later suspended.

In addition to a great deal of respect and support, Mendes won two international prizes for his efforts. He was awarded the Ted Turner's Better World Society Prize and the United Nations Global 500 Environmental Prize. In 1988, responding to ever-increasing international pressure and support for the cause, the Brazilian government established the first ever extractive reserve. Yet as rewards and support increased, so too did risk to the rubber tappers, and, as their leader, to Chico Mendes in particular. Despite the creation of the CNS and the increasing level of organization of the tappers, the political power of the landowners was formidable. Their movement, the União Democrática Ruralista (UDR), was enormously influential throughout the country and in Congress. It had successfully defeated land reform proposals in the Constituent Assembly.

> Here in Xapuri, the UDR is beginning to make its presence felt. Since April 1988, when it formally set itself up in Acre, the number of hired gunmen in Xapuri has increased, as have the number of assassinations and attempted assassinations of workers. These gunmen are in effect the armed wing of the UDR and we are the targets.[5]

Chico Mendes was well aware of the threat to his own life; perhaps he foresaw his death. The quotation at the opening of this account is taken from a letter he had written shortly before his assassination by the son of local cattle rancher Darli Alves da Silva.[6]

Perhaps the most significant element of the legacy of Mendes is the enhanced power and voice of the organizations associated with him and the rubber tappers' cause – the National Council of Rubber Tappers and the Amazon Work Group from whose membership emerged a new generation of environmental leaders and activists. In Acre, Mendes' co-campaigners won important elective offices. For example, Marina Silva, co-founder with Mendes of the union movement and the Workers' Party in Acre, was elected to the Federal Senate in 1994; colleague Jorge Viana was elected mayor of Acre state capital in Rio Branco in 1992 and governor in 1999; and environmentalist João Alberto Capiberibe was re-elected in 1999 as governor of Amapa. Such political successes for the Mendes cause have transformed national debate in Brazil on the Amazon region. The new generation of environmentalists have a major task ahead – the environmental, ecological and social crisis of Amazonia remains critically serious. Yet the political conditions for potential change have never been better as state and federal policies which promote and support sustainability are framed.

At the time of writing, a total of twenty-one extractive reserves and extractive settlements have been established in the seven states in Brazil, covering an area of 3.3 million hectares, together with a number of state reserves. By law, residents of the reserves must prepare a management plan for their area in order to obtain long-term rights of use. Both local communities and government have rights and responsibilities which encompass principles of ecological sustainability. Beyond Brazil, international agreements enforce protection of rainforest ecosystems. Yet federal extractive reserves account for some 1.5 per cent of the Amazon area. Deforestation rates are as high as ever in many regions, land degradation becomes an increasingly significant issue as time goes by, fires are more frequent and harder to control, and illegal logging practices continue to strip hardwoods from within protected areas. Furthermore, rubber prices have fallen so low that the extractive reserves are not producing the income to support even the basic needs of some communities. The poverty, degradation and destruction of Amazonia are amongst the greatest socio-environmental challenges of the present day; challenges brought onto the world's political stage and the agendas of NGOs as a result of various significant influences. The charismatic and courageous leader of the Brazilian rubber tappers' union must surely be one of the most significant of all.

In early 1989, in the aftermath of Mendes' death, which made a great impact not only in Brazil but worldwide, the Second National Congress of Rubber Tappers was held in Rio Branco. Rubber tappers of Brazil were joined there in force by tappers from Bolivia, by indigenous communities from Acre and elsewhere, and by representatives of government, human rights groups, the Church and political organizations. The meeting published twenty-seven demands concerning environmental protection, social development and human rights protection. It also published the *Declaration of the Peoples of the Forest* in memory of Chico Mendes and in the hope of the fulfilment of his vision for the future of the Amazon:

> The traditional peoples who today trace on the Amazonian sky the rainbow of the Alliance of the Peoples of the Forest declare their wish to see their regions preserved. They know that the development of the potential of their people and of the regions they inhabit is to be found in the future economy of their communities, and must be preserved for the whole Brazilian nation as part of its identity and self-esteem. This alliance of the Peoples of the Forest, bringing together Indians, rubber tappers and riverbank communities, and founded here in Acre, embraces all efforts to protect and preserve this immense, but fragile life-system that involves our forests, lakes, rivers and springs, the source of our wealth and the basis of our cultures and traditions.[7]

Notes
1   *Fight for the Forest, Chico Mendes in His Own Words*, p. 6.
2   Ibid., p. 15.
3   T. Gross in *Fight for the Forest*, p. 2.
4   Ibid.
5   *Fight for the Forest*, p. 80.
6   Ibid., p. 6.
7   National Council of Rubber Tappers, Union of Indigenous Nations, Rio Branco, Acre, March 1989. In *Fight for the Forest*, p. 85.

See also in this book
Gore, Hunter, Maathai

Mendes' major writings
Mendes, C., *Fight for the Forest, Chico Mendes in His Own Words*, London: Latin American Bureau (research and action), 1989.

Further reading

BBC London, *Living with Chico Mendes*, 2008, available at www.bbc.co.uk/worldservice/documentaries/2008/05/080507_living_with_chicomendes.shtml (documentary marking the 20th anniversary of the assassination).

Branford, S. and Glock, O., *The Last Frontier: Fighting Over Land in the Amazon*, London: Zed Books, 1985.

Caulfield, C., *In the Rainforest*, London: Heinemann, 1985.

Hyman, R., 'Rise of the Rubber Tappers', *International Wildlife*, 18 (5), 24–8, 1988.

Miranda Productions, *Voice of the Amazon*, 1989, available at www.mirandaproductions.com/voice/ (documentary about Chico Mendes' life and death).

Revkin, A., *The Burning Season: The Murder of Chico Mendes and the Fight for the Amazon Rain Forest*, New York: Plume, 1994.

Rodrigues, G. and Rabbin, L., *Walking the Forest with Chico Mendes: Struggle for Justice in the Amazon*, Austin, TX: University of Texas Press, 2007.

JOY A. PALMER COOPER

BARRY LOPEZ 1945–

Historically, humanity has more often benefited from the genius of the community than the genius of the individual. And people with no faith in their own wisdom in hard times have perished waiting for a genius to appear and lead them.[1]

Barry Holstun Lopez was born January 6, 1945 in Port Chester, New York, a town near where his parents, Mary Frances Holstun Brennan and John Edward Brennan, resided in Mamaroneck, about 25 miles northeast of Manhattan. The couple had married in 1942; it was a second marriage for both. Prior to this marriage, Mary had worked in journalism and written a column for the *Birmingham News*. John Brennan continued his work in advertising and commuted to New York City. In 1948, after their second son Dennis was born, the family moved to California's San Fernando Valley. In 1950, Brennan, whose previous marriage had never been officially terminated with divorce proceedings, abandoned his new family and returned to his first wife and son. He had no contact with Barry and Dennis from that time on. After she divorced John Brennan, Mary taught home economics in local schools and also used her dressmaking skills to support her family.[2]

Lopez spent these years of his childhood in a semi-rural, agricultural area, closely observing the plants, birds, and other animals in the San

Fernando Valley farming communities. His mother frequently took her young sons on trips to the western deserts, beaches, and mountains, all within driving range of the home she had purchased in Reseda, a San Fernando Valley town. Lopez should have experienced a completely idyllic life in a safe community with a loving mother, but his childhood was traumatized by the repeated sexual abuse of a serial pedophile.[3] In 1955 Mary Brennan, married Adrian Lopez, a magazine publisher who lived in Manhattan. Adrian Lopez adopted the two young boys, and they both took his name.

This marriage and the move to New York offered Lopez a release from the abuse. Additionally, the move introduced him, then 11 years old, to a fascinating and inspiring world of museums, concerts, and theater. He attended Loyola School, a Jesuit prep school, graduating in 1962. He then enrolled in University of Notre Dame where he majored first in aeronautical engineering, because, following the recent changes made in American and world imaginations by Sputnik and the new possibilities of space travel, he found himself "enthralled with ... the metaphor of flight."[4] This metaphor could not sustain him, and he quickly changed his major to communication arts, an interdisciplinary program of American studies, writing, and theater. Lopez graduated with honors from Notre Dame in 1966. After graduation he drove to Kentucky where he stayed at the Trappist Abbey of Gethsemane, home to Thomas Merton, as he considered, and then rejected, the vocational calling of monastic life.

Lopez married Sandra Landers in 1967. Following his completion of an MA in teaching at Notre Dame in 1968, the couple moved to Eugene, Oregon, where Lopez enrolled in the University of Oregon's Master of Fine Arts program for a semester, then matriculated at the university for another year before leaving to write fulltime in 1969. Lopez's first marriage ended in divorce in 1999. He married the writer and editor Debra Gwartney in 2007 at a small ceremony in Santa Fe, New Mexico. They live on the McKenzie River in rural western Oregon.

Lopez has been writing and publishing for over fifty years. His career as a magazine writer began in 1966, when he started writing for several Catholic magazines on the political left (*Ave Maria, a.d.*) and for automotive magazines (*Popular Imported Cars, Autodriver*) published by Adrian Lopez. By the early seventies he was writing largely for environmental publications (*National Wildlife, Audubon*) and, then and later, for general-interest publications (*Harper's, Popular Science, Outside*), focusing on a variety of subjects, including travel, the practical use of tools and machinery, natural history, and outdoor life. His essays, short stories, and interviews began appearing in literary journals (*Skywriting,*

Chouteau Review, Dalmo'ma) in 1972. By 1976, the year his first book, *Desert Notes*, was published, his short fiction and essays were beginning to appear regularly in *North American Review* and *Harper's*.[5]

Lopez received a National Book Award for *Arctic Dreams* in 1986 and a John Burroughs Award for *Of Wolves and Men* in 1979. Other honors and awards include: an Award in Literature from the American Academy of Arts and Letters, a Lannan Literary Award, a Guggenheim Fellowship, the John Hay Medal, five Antarctic Fellowships from the National Science Foundation, and a major award from the Association of American Geographers. The Barry Lopez Visiting Writer in Ethics and Community, a writing residency in Hawaii, was established in 2015 in honor of Lopez's seventieth birthday: "in recognition of his long career as an author whose work addresses the fundamental crisis of our times: the disintegration of dignified and ethical relationships among human beings, and between humanity and the natural world."[6] Ann Pancake, author of *Strange as This Weather Has Been*, was the award's first recipient. In 2003 Lopez was appointed Texas Tech University's first Visiting Distinguished Scholar; he visits the campus twice a year, working closely with students and faculty. His papers are included in the James Sowell Family Collection in Literature, Community and the Natural World in Texas Tech University's special collections library.

Often described as a "writer who travels," Lopez has visited both the Arctic and Antarctic polar regions, bustling cities, refugee and concentration camps, muggy jungles, and remote archipelagos. These locations, with their special combinations of people, cultures, attitudes, and problems of historical importance, are thoughtfully and intelligently presented in his writing. Readers find themselves captivated by his unique and compassionate vision and his ability to weave a complicated web of global life and issues into a powerful narrative. His non-fiction pieces combine reportage with a keen awareness of ethical and moral considerations. His work is widely translated and anthologized, including in publications such as annual editions of *Best American Essays* and *Best American Spiritual Writing*.

Lopez's work continues to illuminate the environmental and cultural issues challenging the contemporary world and its inhabitants. Yet Lopez has always been uneasy with the "nature writing" label. He suggests that

> the term of "nature writing" is an inadequate and often inappropriate term for a kind of literature that addresses injustice and hierarchy. The term marginalizes the work. For me, a certain strain of feminist writing, a certain strain of civil rights writing, and a certain strain of writing about the relationship between nature and culture are all embedded in the same impulse, a sense of outrage about injustice.[7]

Indeed, Lopez has become more than a "writer who travels"; he has become noted for writing and speaking about the directions taken by consumerist world cultures and oppressive political organizations. He makes frequent appearances and visits to campuses and other gatherings, sharing his thoughts and ideas with a younger generation.[8]

Of Wolves and Men and *Arctic Dreams: Desire and Imagination in a Northern Landscape*, Lopez's first two major non-fiction books, are milestones in literary history as they fundamentally changed both the expectations of readers and the goals of writers. *Of Wolves and Men* describes humans' interactions with wolves using fairy tale and myth, as well as biological and cultural constructs. Lopez's layered and nuanced approach to writing about wolves and humans stems from his belief that there is a multiplicity of ways of knowing. *Of Wolves and Men* reframed the "wild animal narrative" that had previously been so popular with both readers and writers, making a routine chronicle of a wild and exotic animal's daily activities cycling from spring to winter an insufficient and superficial endeavor. Anthropomorphizing, or applying the cuteness of a pathetic fallacy to animals, would no longer pass muster.

In an early interview, Lopez said, "My passion is language and landscape, and those two are inseparable for me. That is where the focus of my life is."[9] This passion is clearly articulated in *Arctic Dreams*, a narrative in which Lopez's precise and elegant language made landscape and all that that term encompasses—visually, sonically, historically; from the sweep of the wind to the call of the gulls to the iceberg's hues floating in freezing water, from shadowy wolverine tracks to bright patches of orange lichen—one of the narrative's most important characters.

In *The Rediscovery of North America*, the inaugural Thomas D. Clark lecture at the University of Kentucky which Lopez delivered in 1990, he describes the waves of European incursions into and across the New World, beginning with Columbus and continuing to modern times, culminating in acid rain and desiccated farmlands. The situation facing world populations, in 1990 as now, Lopez argues, was sanctioned by an assumption that wealth and power were the completely justifiable goals of individual endeavor. He concludes by encouraging his audience to "rediscover" the continent, to make cultural and personal connections to the landscape and its myriad inhabitants. In the face of despair, people can still take action, he suggests, as simple, metaphorically, as planting trees. "They will grow. They will hold the soil, provide shelter for birds, warm someone's home after we are gone."[10] *The Rediscovery of North America* was recently translated into Arabic because its discussions of land and land ownership, of the right of people to occupy their own traditional landscape, resonates with many in occupied Palestine.

Since the appearance of *Desert Notes*, Lopez has published 14 more books. In addition to the titles previously discussed his body of work includes essay collections *Crossing Open Ground* (1988) and *About This Life* (1998); *Crow and Weasel*, the iconic coming of age story wonderfully illustrated by Tom Pohrt; and *Giving Birth To Thunder, Sleeping with His Daughter* (1978), Lopez's retelling of Native American Trickster stories. Lopez's early fiction includes *Desert Notes* (1976), *River Notes* (1979), and *Field Notes* (1994). Since then he has written *Light Action in the Caribbean* (2000) and *Resistance* (2004), both short story collections; *Outside*, a new compilation of six previously published short stories (2014); and *Home Ground: Language for an American Landscape* (2006), a dictionary of landscape terms that Lopez co-edited with Debra Gwartney. He continues to publish short pieces in major journals. Recent work clearly articulates long-held concerns, notably "¡Nunca Más!" (*Manoa*, 2008) which records Lopez's response to visiting Auschwitz and Birkenau and his thoughts about government complicity in genocide, and "Sliver of Sky," which is a strong assertion of a child's ability to survive and surmount traumatic experiences. Lopez is currently working on a non-fiction manuscript, *Horizon: The Autobiography of a Journey*.

In "A Voice," an essay published in 1998 and in an earlier interview with E.O. Wilson, moderated by Edward Lueders, Lopez talks about writing the literature of hope, "a literature that can bring hope to bear on the things that confound us."[11] More recently, perhaps with a sense of urgency in the face of global conflicts and increasing concerns about environmental degradation, Lopez has offered a change of focus. "What I would say today," he writes, "is that I want my work to be part of a Literature of Resistance."[12] Hope without action will not sustain people and communities. Still, after fifty years, Lopez's work remains modulated with a powerful humanitarian ethic and an unwavering commitment to the principles of responsibility, reverence, and reciprocity. These are the foundation of successful communities and civilizations. They are, according to Lopez, traits worthy of remembrance and practice.[13]

Notes

1 Tydeman, William E. *Conversations with Barry Lopez: Walking the Path of Imagination.* p. 13.
2 For a more complete biography, see pp. 3–9, *Conversations with Barry Lopez.*
3 Lopez wrote about this experience in the essay "Sliver of Sky."
4 *Conversations*, p. 72.
5 This portion is an adaptation of the headnote to Warner's Lopez bibliography in *Conversations with Barry Lopez*. For a more complete bibliography of Lopez publications, see *Conversations*, pp. 151–94.
6 www.kickstarter.com/projects/1275873096/the-barry-lopez-visiting-writer-in-ethics-and-comm/posts/1152254

7 *Conversations*, p. 121.
8 *Conversations*, p. 47.
9 "Ecology and the Human Imagination: Barry Lopez and Edward O. Wilson," in *Writing Natural History: Dialogues with Authors*, ed. by Edward Lueders. Salt Lake City, UT: University of Utah Press, 1989. p. 32.
10 Lopez, Barry. *The Rediscovery of North America*. p. 51.
11 "Ecology and the Human Imagination," p. 16.
12 *Conversations*, p. 124.
13 *Conversations*, p. 14.

See also in this book
Muir, Nash, Tuan, Wilson

Lopez's major writings
About This Life: Journeys on the Threshold of Memory. New York: Knopf, 1988.
Arctic Dreams: Desire and Imagination in a Northern Landscape. New York: Scribner, 1986.
Crow and Weasel. Berkeley, CA: North Point, 1990.
"Sliver of Sky." In *Best American Essays 2014*, edited by John Sullivan Boston, MA: Houghton Mifflin, 2014. pp. 122–39.
Of Wolves and Men. New York: Scribner, 1978.
Rediscovery of North America. Lexington, KY: University of Kentucky Press, 1990.

Further Reading
Barry Lopez website: Barrylopez.com
Newell, Mike. *No Bottom: In Conversation with Barry Lopez*. Gambier, OH: XOXOX Press, 2008.
Tydeman, William E. *Conversations with Barry Lopez: Walking the Path of Imagination*. Norman, OK: University of Oklahoma Press, 2013.
Warren, James Perrin. *Other Country: Barry Lopez and the Community of Artists*. Tucson, AZ: University of Arizona Press, 2015.

DIANE WARNER

PETER SINGER 1946–

If it is in our power to prevent something very bad from happening, without thereby sacrificing anything of comparable moral significance, we ought to do it.

(Practical Ethics, p. 229)

Peter Singer has been described as having more positive influence on the world than any other living philosopher,[1] and in 2005 *Time* magazine listed him among the 100 most influential people worldwide. His book *Animal Liberation*, published in twenty-three languages and described as the Bible of the animal movement was listed by *Time* in 2011 as one of the 100 'All-TIME' best nonfiction books in English since *Time* started in 1923. *Practical Ethics*, available in eighteen languages, was named one of the world's 100 most significant philosophical texts.[2] Singer is currently De Camp Professor of Bioethics at Princeton, and Laureate Professor of Bioethics at Melbourne University, having had appointments at Oxford, New York, La Trobe and Monash. He stood for the Australian Senate as a candidate for the Australian Greens in 1996, and became Australian Humanist of the Year in 2004, and was made a Companion of the Order of Australia in 2012. He is the founder, and a board member, of *The Life You Can Save*, and on the Advisory Board of *Academics Stand Against Poverty* and *Incentives for Global Health*.

Singer was born on 6 July 1946, in Melbourne, eight years after his parents arrived in Australia, escaping the Nazi persecution of the Jews in Austria. His maternal grandfather, David Oppenheim, who co-authored research with Freud, died in Theresienstadt Concentration Camp. His paternal grandparents were taken by the Nazis to Lódz, and nobody heard from them again.[3]

Singer studied law, history and philosophy at the University of Melbourne, where he met his wife Renata – they have three daughters and four grandchildren – and participated in the movement against the Vietnam War. This experience inspired his first book *Democracy and Disobedience*, based on his 1971 Oxford BPhil supervised by R.M. Hare.

Utilitarianism

Richard Hare was a leading advocate of utilitarianism, an impartial form of consequentialism Singer was also attracted to. Utilitarians hold that we must consider all interests without bias and aim to achieve the best consequences, either directly, through individual actions, or indirectly, perhaps by following the right rules. Since utilitarians are also *welfarists* and *aggregationist*, the best outcome is for them the one containing the greatest *sum* of happiness or *utility*, regardless of its distribution. Utility can refer to the balance of pleasure over pain, or of preference satisfaction over frustration. Singer defended preference-utilitarianism until he co-authored *The Point of View of the Universe*. The book modifies some of his earlier views, defending the greater importance of positive experiences over preference satisfaction, vindicating act utilitarianism, and admitting the existence of objective moral truths and values.

Utilitarianism's concern to promote certain mental states naturally leads to the inclusion of all sentient creatures capable of such states, and of all the future individuals that will enjoy positive states or suffer, as a result of our actions. Both convictions can support environmental protection, particularly when combined, as the interests of all future animals can outweigh those of current humans. On the other hand, the greatest aggregate utility may involve the greatest sum of people with barely worth-living lives, which is not what environmentalists advocate. Utilitarians, however, need not commit to pro-natalism for greater utility may be achieved with smaller human populations or larger non-human populations, providing animal lives are not lives of net suffering.[4]

Singer believes that welfarist premises are hardly controversial, and provide a sound basis for an environmental ethic. It is therefore unnecessary to appeal to assumptions about the 'inherent worth of all life', or 'the intrinsic value of species and ecosystems'. As Singer explains, it is difficult to see how 'a species or an ecosystem can be considered as the sort of individual that can have interests, or a "self" to be realised', let alone that 'the survival or realisation of that kind of self has moral value, independently of the value it has because of its importance in sustaining conscious life'.[5] By contrast

> an ethic based on the interests of sentient creatures is on familiar ground. Sentient creatures have wants and desires. The question: 'what is it like to be a possum drowning?' at least makes sense, even if it is impossible for us to give a more precise answer than 'it must be horrible' ... But there is *nothing* that corresponds to what it is like to be a tree dying because its roots have been flooded.[6]

Seeking convergence towards positive change

A moral theory is stronger if it relies on weak premises like 'pain is bad', and 'it is wrong to cause it unnecessarily' than on controversial metaphysical assumptions. And Singer tries to appeal to people from different positions. Consider, for instance, his famous illustration.

> The path from the library at my university to the humanities lecture theatre passes a shallow ornamental pond. Suppose that on my way to give a lecture I notice that a small child has fallen in and is in danger of drowning. Would anyone deny that I ought to wade in and pull the child out? This will mean getting my clothes muddy and

either cancelling my lecture or delaying it until I can find something dry to change into; but compared with the avoidable death of a child this is insignificant.[7]

Now, one need not believe in maximizing utility to rescue the child. The *prioritarian* requirement to give priority to those who are worse off in absolute terms, the *sufficientarian* requirement to bring individuals above a minimum threshold, and the *egalitarian* requirement to help those who are comparatively worse off than ourselves, also dictate rescuing the child.[8] And so the example can strike a chord with people of different moral outlooks. Singer argues that a plausible principle requiring the rescue is this: 'If it is in our power to prevent something very bad from happening, without thereby sacrificing anything of comparable moral significance, we ought to do it.'[9]

This principle – which again is not necessarily utilitarian, rather than prioritarian, for example – maintains wide appeal, referring to a general phenomenon in moral reasoning. But it can be employed to defend drastic international redistribution and a radical change in our treatment of animals. For we can prevent human starvation and factory-farm suffering without a comparable sacrifice. He concludes we should become vegetarians and donate around 10 per cent of our income.[10]

Some find the case for vegetarianism more convincing because we actively cause animal suffering, while we merely allow humans to starve. Others find starvation worse because it concerns humans. Singer challenges the relevance of *both* distinctions. Other consequentialists have challenged the act/omission distinction.[11] Singer is known for opposing the species-based distinction.

Anti-speciesism

Singer argues that granting exclusive attention, or priority, to the alleviation of human suffering relies on *speciesism*, a discriminatory prejudice comparable to racism, that also focuses on group membership, disregarding the merits of each case.

Singer's anti-speciesism is often misinterpreted. First, anti-speciesists can accept that humans and animals are, in fact, different. Similarly anti-racists, and feminists, may accept the existence of racial or sexual differences since they need only deny that any such differences justify giving less importance to the interests of women or racial minorities. Second, anti-speciesists need not claim that killing an animal is as bad as killing a person. They may think that the badness of death depends on how much we lose by losing our life, and deem ape or cetacean long and culturally and emotionally rich lives more valuable than those of worms or shrimps. Anti-speciesists may also grant that death is worse for creatures with psychological contiguity, memories, plans, a sense of self, and a deep connection to their future. Singer's views on our unequal interest in continued existence – discussed in *Practical Ethics* and subsequent works – contrasts with *Animal Liberation*'s view that pain of the same type, intensity and duration is no more important because if it is part of a human life. His views on death and suffering are consistent, however, because the interests in suffering-avoidance and continued existence differ. For example, when doctors cannot spare both a mother and her fetus a certain amount of pain, mothers typically prefer greater pain for themselves than for the fetus. But when both lives cannot continue, it is the mother's that is generally saved. This difference provides the plot of a good number of war movies. During the siege, all the sedatives, or the cognac, are given to the most gravely maimed soldier, who is enduring the greatest mental and physical pain. But when at the end of the movie they cannot all be rescued – for example, because someone must remain to detonate the explosives – the volunteer is always the maimed soldier who, having lost so much, now has less to lose than others. While an individual's interest in life-extension depends on what sort of life it is going to be, the interest in suffering-avoidance is universal, and when its character, duration and intensity are identical, it matters equally, whether or not it is part of a certain kind of life.

This idea changed the life of Singer's student, Henry Spira. Having participated in the struggle for Civil Rights in the American South and for Trade Union reform in the U.S. Labour Movement, Spira devoted his last two decades to animal rights. Singer documented Spira's biography because his life expressed perfectly what the philosopher tried to say: that there is a natural progression from human liberation to animal liberation. The same compassion, the same sense of justice, the same opposition to cruelty and exploitation which made us reject slavery and, later, so many other forms of oppression and abuse, have to make us react against the systematic and prolonged torture of millions of sentient creatures crammed in laboratory cages and factory farms.

Seeking convergence again

Besides emphasizing the continuity between liberation movements, Singer stresses that by 'ceasing to rear and kill animals for food, we can make so much extra food available for humans that, properly distributed, it would eliminate starvation and malnutrition from this planet. Animal liberation is human liberation too.'[12] Furthermore, Singer argues that the meat industry is so environmentally damaging that it cannot be part of a sustainable lifestyle, and must also be rejected on the basis of intergenerational justice.[13] This convergence of values and policies is not universally subscribed. Awareness of climate change's harmfulness to the global poor has greened many human rights advocates and caused a shift towards climate issues in leading cosmopolitans like Simon Caney and Henry Shue. However, animals and the environment have also been incorporated as elements of some nationalist discourses.[14] Moreover, while the animal movement had sociologically flourished largely within the green movement, some of its branches are now vehemently anti-environmentalist. Some blame greens for "pest" controls and other interventions to protect native species and deny the value of nature or the natural (non-artificial). They see no point in biodiversity or ancient, pristine ecosystems, and so refuse to balance such considerations against large-scale interventions to help animals.[15]

Against this background, Singer remains a central, unifying figure. Regarding global justice, he remains strongly cosmopolitan stressing not only our duty to give generously, but also wisely and effectively.[16] He has rejected anti-immigration arguments,[17] publicly renounced his right to 'return' to Israel[18] and thinks of racism and nationalism as ideological construction with disastrous consequences by contrast with valuable forms of permissible partiality such as our natural and instrumentally beneficial concern with our own children.[19]

Regarding environmentalism, though Singer now admits the existence of objective value, his emphasis remains the suffering environmental destruction causes, arguing for conservation from the point of view of humans and animals, rather than 'the universe', and discouraging intervention on the basis of our track record, rather than because of a difference between harming and failing to aid humans and other sentient creatures.

In 'Animal Liberation is an Environmental Ethic'[20] Dale Jamieson argues that environmental and animal protection goals can sometimes come into conflict but so can the goals of preserving natural processes or biodiversity, or those of ending animal slavery or suffering. And Singer's emphasis on focusing on the global picture and doing the most good advises against sectarianism and favours seeking convergence to achieve positive change.

In an academic and political context favouring novel, complex, theories that ultimately lead to the same reformist conclusions, with authors sometimes appearing to care less about the issues than about having distinctive views about them, it is important that somebody renders these complex views accessible and returns to basic principles, with clear arguments and common sense, to inspire people to radically re-examine their lives, reach into their pockets, and take to the streets.

Notes

1 *The Philosophers Magazine*, 4, 1989, and R. Posner, 'Reply to Critics', *Harvard Law Review*, 111.7, p. 1816, 1998.

2 K. Worsley, 'Heartless Animal or Rational Beast?', *Times Higher Education Supplement*, 29 May 1998, p. 17.

3 See his *Pushing Time Away*, part VI.

4 Cf. Yew Kwang Ng 'Towards Welfare Biology', *Biology and Philosophy*, 10.3: 255–85 and Catia Faria, *Animal Ethics Goes Wild. The Problem of Wild Animal Suffering and Intervention in Nature*, PhD, Pompeu Fabra University, 2016.

5 *Practical Ethics*, p. 283.

6 Ibid., p. 277.

7 Ibid., p. 229. See also, 'Famine, Affluence and Morality', *Philosophy and Public Affairs*, 1: 229–43, 1972; 'Reconsidering the Famine Relief Argument', in *Food Policy: US Responsibility in the Life and Death Choices*, ed. P. Brown and H. Shue, New York: The Free Press, 1977, pp. 36–53; *Practical Ethics*, Chapters 8 and 9; *The New York Times*, 5 September 1999.

8 See, e.g. Shelly Kagan, *The Limits of Morality*, Oxford: Oxford University Press, 1989, pp. 3–4 and 16.

9 *Practical Ethics*, p. 229.

10 The 'Singer Solution to World Poverty', *New York Times Magazine*, September 5, 1999.

11 But see, e.g., *Practical Ethics*, pp. 206–13, 218, 222–9, 309 and D. Jamieson, *Singer and His Critics*, pp. 311ff.

12 End of the 1975 Prologue to *Animal Liberation*.

13 See *Practical Ethics*, pp. 287–8 and *How Are We to Live?*, pp. 44ff.

14 Some examples include the Welsh burning of holiday homes, and the Catalan bull-fighting ban while preserving *correbous*.

15 Catia Faria, ibid.

16 See *The Life You Can Save* and *The Most Good you Can Do*.

17 See 'The Ethics of Refugee policy' with Renata Singer in *Open Borders, Closed Societies*, ed. M Gibney New York: Greenwood Press, 1988, reprinted in *Practical Ethics*.

18 www.huffingtonpost.com/antony-loewenstein/israels-dubai-hit-continu_b_498255.html

19 See *One World Now*, pp. 185–8.

20 *Environmental Values* 7.1: 41–57, 1998. I thank Peter Singer for revising this entry.

See also in this book
Goodall, Midgley

Singer's major writings

Animal Liberation, New York: New York Review/Random House, 1975, 2nd edn 1990.

Practical Ethics, Cambridge, UK: Cambridge University Press, 1979, 3rd edn 2011.

The Expanding Circle, Oxford, UK: Oxford University Press, 1981.

The Reproduction Revolution. New Ways of Making Babies, (with Deane Wells) Oxford, UK: Oxford University Press, 1984.

Should the Baby Live? The Problem of Handicapped Infants, with Helga Kuhse, Oxford, UK: Oxford University Press, 1985.

How Are We to Live?, Melbourne, Australia: Text Publishing, 1993.

The Great Ape Project, ed. with Paola Cavalieri, London: Fourth Estate, 1993.

Rethinking Life and Death. The Collapse of our Traditional Ethics, Melbourne, Australia: Text Publishing, 1994.

Ethics into Action: Henry Spira and the Animal Rights Movement, Lanham, MD: Rowman & Littlefield, 1998.

A Darwinian Left, London: Weidenfield and Nicholson, 1999.

Writings on an Ethical Life. New York: Ecco, 2000.

One World. The Ethics of Globalization. New Haven, CT: Yale University Press, 2002; 3rd edn, *One World Now*, 2016.

Pushing Time Away. My Grandfather and the Tragedy of Jewish Vienna, New York: Ecco Press, 2003.

The Ethics of What We Eat, with Jim Mason, New York: Rodale, 2006.

The Life You Can Save. Acting Now to End World Poverty. New York: Random House, 2009.

The Most Good You Can Do: How Effective Altruism is Changing Ideas About Living Ethically. New Haven, CT: Yale University Press, 2015.

Famine, Affluence and Morality, Oxford, UK: Oxford University Press, 2016.

Ethics in the Real World: 82 Brief Essays on Things That Matter, Princeton, NJ: Princeton University Press, 2016.

Further reading

Jamieson, D., (ed.) *Singer and His Critics*, Oxford, UK: Blackwell, 1999.

Schaler, J. A., (ed.) *Peter Singer Under Fire: the Moral Iconoclast Faces His Critics*, Chicago, IL: Open Court Publishers, 2009.

PAULA CASAL

AL GORE 1948–

For civilization as a whole, the faith that is so essential to restore the balance now missing in our relationship to the earth is the faith that we do have a future. We can believe in that future and work to achieve it and preserve it, or we can whirl blindly on, behaving as if one day there will be no children to inherit our legacy. The choice is ours; the earth is in the balance.[1]

Albert Arnold (Al) Gore, like his father, would make his first foray into national politics on behalf of the American state of Tennessee. He was born in Washington DC on March 31, 1948. He would serve in both the House of Representatives (1976–84) and in the Senate (1984–8), before eventually becoming the Vice-President of the United States. He has written a dozen books on topics ranging from family, faith, economics, governance, and the environment. Gore's convictions about the abuse of the environment, he says, go back to his days on a Tennessee farm.

My earliest lessons on environmental protection were about the prevention of soil erosion on our family farm ... Unfortunately little has changed: even now, about eight acres' worth of prime topsoil floats past Memphis every hour. The Mississippi River carries away millions of tons of topsoil from farms in the middle of America, soil that is now gone for good.[2]

It is this notion of "gone for good" that has motivated Gore's activism.

His environmental conscience met his political one when during his House years the intersection between technology and the climate crisis was slowly taking hold, and it matured in the period between a failed bid for the Democratic nomination for president and his eventual acceptance as Bill Clinton's Vice-Presidential running-mate. And while the environment continued to fuel his politics, his faith, and his drive for change, his political career demanded that it be stifled. Politically, Gore is often seen as a "cardboard" cut out, especially in the application of his "Vice-Presidential stare", but there is good reason to believe that this is due to the trade-off of quieting his environmental voice for the potential good to be had in the exercise of political power on its behalf in the future. It was required, Gore would come to believe, because of what was broken in the American political apparatus: the desire to create and watch stultifying spectacle, rather than develop social and moral leadership for the common good. Gore was instrumental in the early 1980s in the creation of the first congressional hearings on climate change, drawing to political attention the scientific findings of James Hansen, the leading climate scientist from NASA's Goddard Institute.

Gore's first book on the environment, *Earth in the Balance: Ecology and the Human Spirit*, was published in 1992 and was the result of a "search for truths" which accompanied his attendance to his son's recuperation after a life-threatening car accident. Taken out of the halls and contexts of political struggle, Gore found himself in a crisis that was as "mid-life" as it was environmental. In order to deal with the former he studied existentialism and phenomenology at Vanderbilt, and this ultimately informed the introspection which produced a coherent, if shallow-ecological, and circumspect position on the environment. In *Earth in the Balance*, Gore makes an attempt to convince an audience of his peers and the wider public about the integrated reality of nature and humanity. He believes that this relationship had profoundly changed in the preceding 20–40 years due to the explosion of scientific and technological discovery, which had amplified humanity's ability to exploit the natural world. This led to what he called a "relatively new way of thinking [of humanity], as somehow separate from nature, isolated individuals entitled to exploit nature as much as we want."[3] While perhaps not new for those outside of the halls of power, it certainly was new for those in government. And in spite of Gore's repeated and credible calls for urgent attention to environmental issues, the other half of the relationship between his politics and his convictions about the environment would bring those concerns to heel for the period of his public service. It was the environment more than any other issue that Gore was passionate about, and he would not be able to demonstrate his commitment to environmental reform until he was "let off the leash" after conceding the presidency to George W. Bush in one of the most bizarre elections in US history. Once released from "handlers," spin doctors, and the Democratic Party's machinery, Gore's evangelical championing of environmentalism, global climate crisis, and activism would come to the fore.

To characterize Gore as an environmental activist may be seen as a bit of a stretch by those with a deep-ecological outlook, but there is no gainsaying the importance and impact of his awareness raising book and film, *An Inconvenient Truth* (2006). The message he brought to the mass public was awarded the Nobel Peace Prize for 2007 jointly with the Intergovernmental Panel on Climate Change (IPCC). "Al Gore has for a long time been one of the world's leading environmentalist politicians … His strong commitment, reflected in political activity, lectures, films and books, has strengthened the struggle against climate change. He is probably the single individual who has done most to create greater worldwide understanding of the measures that need to be adopted."[4] One of the things that gives Gore this ability is a style which allows him to incorporate an enormous amount of information in a way that engages the reader and demonstrates the extent of the research and understanding that goes into his work.

The concept of the "inconvenient" truth is one that Gore employs to great effect in his 2007 book *The Assault on Reason*, and later in *The Future: Six Drivers of Global Change* (2013). In both of these books, as well as the earlier *Earth in the Balance*, Gore's message and interest are clear: global climate crisis and our willful disregard for our connectedness to nature will bring about an end to human civilization in its current form. On the other hand, in paying conscious, rational, and moral attention to these issues, he sees the real possibility of the emergence of a "Global Mind" which will struggle against the current mentality of ignorance, obfuscation, and elite manipulation of resources and people. The Global Mind he speaks of is, while still capitalist and growth oriented, miles away from the current status quo. He is harshly critical of the operation of the American system and feels that there is little about it, political, economic, social, or environmental, that is not in urgent need of radical reform. Clearly not the words of someone who may hold hopes of holding high office again, Gore is "all in" at this point: changing to a committed vegan diet in 2012 and using the sway and dollars of his investment portfolio to promote green alternative research and development, vegan restaurant and grocery alternatives, and funding internet and crowd sourced initiatives to increase the speed and regularity of data gathering on environmental conditions. He has traveled extensively in pursuit of an understanding of the state of the environment and to be able to offer first-hand accounts of the situation. And apart from the institutional attitudes and political spinelessness on environmental issues, Gore feels that the main thing lacking is a moral compass based on the obvious intersection of human civilization and environmental issues.

> This is not a political issue. This is a moral issue ... It is not a question of *Left vs. Right*; it is a question of *right vs. wrong*. Put simply, it is wrong to destroy the habitability of our planet and ruin the prospects of *every* generation that follows us.[5]

In the concluding section of *The Future: Six Drivers of Global Change*, Gore delivers a rapid-fire list of the social, political, and economic institutions that he feels should be the focus of a more publicly minded and environmentally responsible America. The role of reason takes center stage in his reforms, but he cautions that

> Arming ourselves with the "weapons of reason" is necessary, but insufficient. The emergence of the Global Mind presents us with an opportunity to strengthen reason-based decision making, but the economic and political system within which we implement even the wisest decisions are badly in need of repair.[6]

He sees the root cause of the lack of moral compass as the fact that "Democracy and capitalism have been hacked. The results are palpably obvious in the suffocating control of policy decisions by elites."[7]

In *The Assault on Reason*, he quotes from Upton Sinclair, who, although writing a century ago, puts the problem in its clearest terms for Gore: "It is difficult to get a man to understand something when his salary depends upon his not understanding it."[8] And this may be the most salient point in the analysis. Movement on global climate and environmental issues will not be brought about without a rethinking of individual interest and political will. It will come about through collective action and individual initiative. "Ultimately, it is about who we are as human beings and whether or not we have the capacity to transcend our own limitations and rise to this new occasion."[9] He cites Abraham Lincoln's address to a joint session of Congress in 1863 "The occasion is piled high with difficulty, and we must rise to the occasion. As our case is new, we must think anew, and act anew."[10] As one reads through the material that Gore has written on the environment, it becomes clear that this is a very well-read, introspective thinker who, perhaps because he was never part of the academy, is able to use information, analysis, and explication from the best of climate science and his own activity to great effect. He demonstrates a skillful use of the material to clarify, at some length, that there is no debate about the climate crisis,

> an opportunity to experience something that few generations ever have the privilege of knowing: a common moral purpose compelling enough to lift us above our limitations and motivate us to set aside some of the bickering to which we as human beings are naturally vulnerable.[11]

We can choose to continue to exploit everything within our insatiable grasp and hasten the end of civilization (Gore is under no illusions about nature's ability to survive *us*), or we can rise to the occasion and build a more sensitive, symbiotic and sustainable version of our relationship with nature.

What will be required though, is the restoration of the ability to "communicate clearly and candidly with one another in a broadly accessible forum about the difficult choices we have to make."[12] This he believes is the only way to counteract the damage that has been done as a result of the delivery of policy and planning into the hands of the elites, which has precipitated the loss of any moral authority vested in public officials. America may no longer be "government of the people, by the people," but Gore maintains that the American constitution is still a model document for governance and understanding human nature.

> [The] enduring genius of the U.S. Constitution stemmed from its authors' clear-eyed, dead-on understanding of human nature—even

though it was limited to white males—and their design of structural safeguards that discouraged the impulse to egotistical power-seeking and incentives that rewarded the impulse to resolve their differences through collective reasoning that maximized the likelihood of creative compromises based on the pursuit of the greater good.[13]

In the final analysis Gore feels that it is only through, among other things, clear and untrammeled communication, a transition of democratic institutions to the internet, movement away from ad-driven corporate news media, integration of externalities into economic analysis and benchmarks, the better to understand the real environmental and human cost of our comfort, a tax on CO_2 emissions used to offset the cost of educational and scientific developments towards a more sustainable future and the restoration of public goods and commons that we will be able to begin to navigate the world that we have created through the abuse of nature. "Human civilization has reached a fork in the road we have long traveled. One of two paths must be chosen. Both lead to the unknown."[14]

Notes
1 Gore, *Earth in the Balance*, p. 368.
2 Ibid. p. 3.
3 Interview with Al Gore on *Earth in the Balance*, C-Span, November 27, 2006.
4 Nobel Prize Press Release, October 12, 2007.
5 Gore, *The Assault on Reason*, p. 213.
6 Gore, *The Future*, p. 368.
7 Ibid. p. 365.
8 Upton Sinclair quoted in Gore, *The Assault on Reason*, p. 211.
9 Gore, ibid., p. 212.
10 Gore, *The Future*, p.357.
11 Gore, *The Assault on Reason*, p. 214.
12 Gore, *The Future*, p. 369.
13 Ibid. p. 363.
14 Ibid. p. 374.

See also in this book
Ehrlich, Heidegger, Hunter, McKibben, Naess

Gore's major writings
Earth in the Balance: Ecology and the human spirit, Boston, MA: Houghton Mifflin (1992).

An Inconvenient Truth: The planetary emergency of global warming and what we can do about it, Emmaus, PA: Rodale. Gore & Melcher Media (2006).
The Assault on Reason, London: Penguin Press (2007).
The Future: Six drivers of global change, New York: Random House (2013).

Further reading
Turque, B. (2000). *Inventing Al Gore: A biography*, Boston, MA: Houghton Mifflin.

THOMAS E. HART

VANDANA SHIVA 1952–

Biotechnology, as the hand-maiden of capital in the post-industrial era, makes it possible to colonise and control that which is autonomous, free and self-regenerative. Through reductionist science, capital goes where it has never been before. The fragmentation of reductionism opens up areas for exploitation and invasion … It is in this sense that the seed and women's bodies as sites of regenerative power are, in the eyes of capitalist patriarchy, among the last colonies.[1]

Born in the lap of the Himalayas on November 5, 1952, Vandana Shiva inherited her environmental values from her mother, a farmer with a deep love of nature, and her father, a Conservator of Forests. Initially educated at St. Mary's School in Nainital, and the Convent of Jesus and Mary in Dehradun, she then trained as a physicist, and was awarded a Doctorate in Philosophy at the University of Western Ontario for her quantum theory thesis.

Conducting inter-disciplinary research in science, technology and environmental policy at the Indian Institute of Science and the Indian Institute of Management in Bangalore, she became a leading theoretical physicist in the ecology movement, but in 1981 set aside her professional career to devote her next decade to environmental activism. She began by founding the Research Foundation for Science Technology and Ecology (RFSTE) to independently address emerging ecological and social issues in partnership with local communities and social movements. Here she drew the parallels between "poverty and underdevelopment" which were worryingly integral to the world's emerging third industrial revolution in its endeavors to engineer biological process.

It can be said that when Shiva writes, the world reads. To this day, she is a prolific author, with each of her numerous volumes on biodiversity, biopiracy, biopolitics, biotechnology, ecofeminism, globalization and food security reflecting a profound multidisciplinary scholarship. Her 1988 debut volume, *Staying Alive*, created common ground for feminists and environmentalists, providing an exemplary insight into the plight of women throughout developing regions. Drawing on historical evidence of the feminized poverty resulting from colonial rule, she identified "modern development" as a product of Western patriarchy which further eroded women's productivity by removing land, water and forests from their management, while simultaneously impairing ecological productivity and sustainability via the destruction of soil, rivers and vegetation. Central to her argument was the theft of the natural biodiversity and food security which women had safeguarded over centuries by a Eurocentric science and economics which reshaped the earth and its seed to fit with the latest in patriarchal delusions. Shiva deemed the postcolonial development paradigm to be *maldevelopment* – a process subjugating women and nature while creating twinned social and environmental injustice – and called for the *sanctity of life* to replace the *sanctified development concept rooted in patriarchy*.

Shiva's 1991 volume, *The Violence of the Green Revolution*, challenged the accepted gospel that Norman Bourlag's hybridized semi-dwarf high-yielding wheat seeds had transformed the region's austerity into prosperity. While simultaneously correcting the widely held perception that the contemporary bloodshed which left 15,000 Punjabis dead in the 1980s was due to religious fundamentalism, Shiva exposed that Bourlag's Nobel-Prize-winning miracle had turned the soil into waterlogged expanses or salinated deserts; and with the environmental destruction came community violence which was hardest felt by women and children. In effect, with control over both the environment and people essential to the tactics of the Green Revolution integrating Third World farmers into the global market of fertilizers, pesticides and seeds, the ecological collapse, together with the political disruption of society, were predictable outcomes of a paradigm which had disconnected nature from society. Shiva foresaw that seed was being colonized through a political process which transferred control over biological diversity from peasant farmers to corporate interests. Paraphrasing Shiva, the corporately produced seed had divested peasants, robbed them of their livelihoods, and was the very instrument of their poverty. Whether farmers owned or leased their land, biotechnology's genetically-programmed seed was corporate property, with its lifespan regulated by corporations, rather than by farmers guided by generations of traditional wisdom.

It can also be said that when Vandana Shiva speaks, the world listens. She is a prized speaker on the global conference circuit, captivating audiences with her eloquence, her passion, and her unquestionable logic. From the beginning, her research synchronized with that of feminist activists and multidisciplinary academics representing every world region. Included were biologists, sociologists, engineers, political scientists and a consortium of experts in bioethics, development, economics, environment, law, medicine, nuclear hazards and science and technology.

In 1991, aware that the world's land, forests, rivers, oceans and atmosphere were either colonized, eroded or polluted, and that global capitalism was seeking new territories – plants, animals and women's bodies – to invade and exploit in the quest for further wealth, Shiva convened a seminar on "Women, Health and the Environment" in the Indian city of Bangalore. Gathering feminists from an international circle, each committed to reconstructing the links with nature which a patriarchal and technocratic environmental science had incrementally destroyed, this audience saw *"environment,"* as opposed to any hypothetical perspective, as the place where they lived, translating into *everything* which affected their lives. Their faith in the earth-body and the human-body continuum meant that environmental hazards were health hazards. Their human rights and health ethics were at odds with the population control, racism and misogyny advocated by neo-Malthusians to salvage the earth's resources, and for them the South was not the source of every environmental problem, any more than the North, for all its technology and capital, but rather the source of every environmental solution. Shiva and her allies therein launched ecofeminism into a political movement which was destined to become the strongest opponent of environmental degradation, economic exploitation, cultural globalization and institutionalized gender and indigenous discrimination that the world had ever witnessed; and which by the end of the twentieth century had grown to influence global policy on the numerous interlinked paradigms of these issues.[2] Two years later, in 1993, she was awarded the alternative Nobel Prize, known as the Right Livelihood Award, the first of many awards acclaiming her research into the environmental and social injustice which underpins corporate solutions to the earth's declining renewable resources.

Shiva saw that the second Green Revolution paved the way for human rights, including the right to a livelihood, to be exchanged for property rights protecting the processes of biotechnology. She laid bare the loopholes in the General Agreement on Tariffs and Trade (GATT) which allowed transnational corporations (TNCs) to market agricultural commodities without restriction, regulation or responsibility. Encouraging the free trade of agricultural components, and aided by World Bank and IMF Structural Adjustment programs, GATT destroyed local food markets, converting subsistence Third World food production into a lucrative emporium for corporations. Small producers, most of whom are women, were destined for displacement by GATT.

By 1998, Intellectual Property Rights shaped by the World Trade Organization (WTO) were ordained to deny the world's poorest farmers both free access to their own seed, and the liberty to exchange their own seeds between themselves. Shiva gave the opening keynote address at the First Grass Roots Gathering on Biodevastation: Genetic Engineering in the US city of St. Louis. Interviewed afterwards, she was asked to further explain her "Third World perspective."[3] She answered that following European colonization, the Third World was left with only its biodiversity, and a solitary renewable resource, the seed, to meet health and nutritional needs, and retain a semblance of agricultural viability. Consequently, in the Third World, where the majority are totally dependent on agriculture for survival,

> You can't have a consumer society with poor people and therefore what you will have is deprivation, destitution, disease, hunger, epidemics, hunger, malnutrition, famine and civil war. What is being sown is the greed of the corporations in stealing the last resources of the poor.

While the Green Revolution invaded the seed to become a source of ecological disruption, biotechnology went further, colonizing the seed by destroying its fertility and self-regenerating capacity; and via GATT patent protection (Trade Related Intellectual Property Rights or TRIPs) transferring the ownership of laboratory-spliced and/or relocated genes to the seed's "genetic tailors," most of whom were U.S.-based TNCs and institutions. Shiva argued that TRIPs denied Third World farmers both their intellect and their rights, and calculated that the resulting transfer of funds from poor to rich countries had the potential to exacerbate Third World debt ten times over. She also emphasized that biotechnology, in addition to devaluing the seed "from a living renewable resource into a mere raw material," demeaned women in the same fashion. Patriarchy's construction of nature, and its politics of separation and fragmentation, failed to wash with Shiva, and she instead embraced the partnership which women shaped with nature in their everyday lives as the sustainable paradigm for "dynamic and diverse" regeneration.

Readers of the print media are privy to her research on a regular basis. In 1997, writing in *The Guardian*, she drew European attention to the double standards of TNCs seeking to abduct global food security, and exposed the folly of biotechnology's answer to famine: "The introduction of herbicide-resistant crops destroys biodiversity and rural livelihoods, which are supported by the full variety of nature. Herbicide use in societies where people collect 'weeds' for vegetables and fodder can destroy nutrition and women's work. In India women gather more than 130 species of greens, or weeds – the most important source of vitamin A in rural areas. The irresponsible spread of herbicides through herbicide-resistant crops will aggravate malnutrition in poor communities."[4]

In 1999, via *The Hindu*, she warned of Monsanto's impending agenda to monopolize global water supplies.[5] To Shiva, it was clear that Monsanto's water initiative, like its seed and aquaculture trade, risked expanding the company's monopolies over the basic ingredients of life; with the privatization and commodification of water undermining the right to life. For Shiva, water was a commons and had to be managed as such, rather than "controlled and sold by a life sciences corporation (i.e. Monsanto) that peddles in death."[6] Her 1999 fears of Monsanto's water agenda proved prophetic. Ten months later, just as Medha Patkar and Arundhati Roy took the Narmada Bachao Andolan struggle against displacement and human rights violations by the Sardar Sarovar large dam project to the Second World Water Conference in The Hague, the World Water Commission for the 21st Century put forward its report proposing the global privatization of water supply and sanitation services. Within hours, the report was condemned by various nongovernment organizations, women's groups and individual ecofeminists, all of whom shared Shiva's view that corporate control over water would end any concept of a universal right to water and sanitation, thereby replacing yet another human right with a free market concept commodifying water.

While the popular catchphrase of the 1990s called on people to "think global, act local," Shiva thought and acted at every level. At home in India, she successfully filed Public Interest writs in the Supreme Court on a variety of environmental and trade-related issues; and in 1991, she founded Navdanya (meaning nine seeds), a national movement designed to protect the diversity and integrity of living resources from corporate appropriation. She met with farmers to demystify GATT, and explain TRIPs and the Agreement of Agriculture. And, with extraordinary stamina, she made significant contributions to numerous grassroots campaigns, including the mobilization of 500,000 farmers against GATT in 1993. Similarly, she played a pioneering role in linking TRIPs to Biodiversity and Indigenous Knowledge and to the Convention on Biological Diversity; launched the idea of collective rights to defend indigenous knowledge; and was the first to suggest that the *sui generis* option in TRIPs should be based on community and farmers' rights.

Less locally, in mid-1998, Shiva openly reminded Professor Mohammad Yunus in neighboring Bangladesh that the Grameen Bank's impending partnership with Monsanto was a betrayal of the very women to whom his microcredit scheme promised self-reliance. Yunus listened, and a month later abandoned the agreement brokered between the Grameen Bank and Monsanto.

Internationally, Dr. Shiva initiated the women's movement on food, agriculture, patents and biotechnology. Launched in Bratislava, Slovakia in May 1998, as "Diverse Women for Diversity", she led the movement to the WTO Ministerial meeting in Seattle in 1999 to protest against international trade regulations which discriminated against the environment, women and the Third World. Paraphrasing her description at the time, the rebellion on the streets and within the WTO negotiations indicated the start of a new democracy movement; one where citizens from around the world and governments of the South refused to be bullied and excluded from decisions in which they had due voice.[7] In 2016, having already overstayed her promised term as an activist by more than two decades, she also assists movements in Africa, Asia, Latin America, Ireland, Switzerland, Austria and Australasia with their campaigns against genetic engineering to the present day.

Vandana Shiva is certainly in 2017 one of today's key environmental voices on the global stage. Her leadership and commitment to the preservation of the planet and the world's expanding underclass of poverty-stricken farmers, found mostly in developing regions, and the majority women, is unprecedented, and she was deservedly rewarded with the 2010 Sydney Peace Prize. But to say that Shiva is an environmentalist, or an ecofeminist, is to sell her short. Vandana Shiva is a fearless intellectual, tireless in her efforts to overturn existing or impending injustice which prevails within the institutional halls of government and academy.

Notes
1   Shiva. "The seed and the earth." In *Minding Our Lives: Women from the South and North Reconnect Ecology and Health.* Kali for Women, 1993.
2   Maria Mies and Vandana Shiva. *Ecofeminism.* Zed Books, 1993.
3   Nic Paget-Clarke. An interview with Vandana Shiva. *In Motion Magazine,* August 14, 1998. www.inmotionmagazine.com/shiva.html
4   Shiva. "Genetic seeds of hope and despair." *The Guardian,* December 17, 1997.
5   Shiva. "Monsanto's expanding monopolies." *The Hindu,* May 1, 1999.
6   Ibid.
7   Shiva. "The Historic Significance of Seattle." RHR Press, December 12, 1999. https://ratical.org/co-globalize/HSoS.html

Shiva's major writings

Biopiracy: The Plunder of Nature and Knowledge. Cambridge, MA: South End Press, 1997.

Stolen Harvest: The Hijacking of the Global Food Supply. Cambridge, MA: South End Press, 2000.

[These and other important volumes are referenced within *The Vandana Shiva Reader* (Foreword by Wendell Berry). Lexington, KY: University Press of Kentucky, 2014.]

Seed Sovereignty, Food Security – Women in the Vanguard of the Fight against GMOs and Corporate Agriculture [Editor]. Berkeley, CA: North Atlantic Books; Melbourne, Australia: Spinifex Press, 2016.

Further reading

Bandarage, Asoka. *Women, Population and Global Crisis.* London: Zed Books, 1997.

Hartmann, Betsy. *Reproductive Rights & Wrongs: The Global Politics of Population Control.* Cambridge, MA: South End Press, 1995 [Revised Edition].

Roy, Arundhati. *The Cost of Living.* London: Flamingo, 1999.

Salleh, Ariel. *Eco-Sufficiency and Global Justice.* New York: Pluto Press; Melbourne, Australia: Spinifex Press, 2009.

LYNETTE J. DUMBLE

BILL McKIBBEN 1960–

> Global warming … is a negotiation between human beings on the one hand and physics and chemistry on the other. Which is a tough negotiation, because physics and chemistry don't compromise. They've already laid out their nonnegotiable bottom line: above 350 ppm (atmospheric carbon) the planet doesn't work. In this case, the good and the essential and the perfect and the adequate are all about the same.[1]

William (Bill) Ernest McKibben wrote this in 2010. The grass-roots movement he started with a handful of graduate students, and which to date has precipitated, organized or participated in climate change protests around the world takes its name from nature's bottom line of 350 ppm atmospheric carbon. As of late 2016 the number of parts per million of atmospheric CO_2 is over 400. On the face of it, the planet is broken. McKibben accepts this, but does not see it as a reason to roll over, but rather as a reason to adapt and re-think our lifestyle and make the changes necessary to make human life and civilization better fit the "tough new world" we have created. Life, human or otherwise, has always struggled, this is fundamental. McKibben believes that the most important struggle that humanity faces is not one with nature: nature will not change according to our desires, our pleading or our arguments. The struggle that we face and the one to which he feels we must focus our attention, ingenuity and effort, is with ourselves. The only way McKibben sees human civilization surviving the coming changes in environmental circumstances is to stop trying to change nature and instead start working, aggressively, to change *our* nature. And in many respects, this may be the more difficult of the two to overcome.

McKibben began his career as a staff writer for *The New Yorker* magazine where, from 1982 to 1987, he wrote "Talk of the Town," a listing of cultural and entertainment events in New York until he left in 1987. The internal politics of the magazine's new ownership led him to recognize that a withdrawal from that type of work was in order. Global warming, climate collapse and the science behind the two figured more and more prominently in his thoughts as he communed with his environment in the state of Vermont, his adopted home. "What mattered most to me was the inference I drew from [early climate] science: that for the first time human beings had become so large that they altered everything around us."[2] Initially, he says, he believed that he would write a book, people would read it and then change. It didn't work out that way. As a result, he became what he calls an accidental activist, never having considered stepping out of his comfort zone, let alone into a prison cell, as part of his nature. The realization that everyone will need to make hard, uncomfortable changes in the coming age drove his transformation. His first book on the environment, *The End of Nature* (1989), is to many people their *Silent Spring*[3] and is one of the first books to bring global warming and climate collapse to popular attention in America. The title of his first book is meant to imply that humanity has inadvertently usurped nature, that "Hurricanes and thunderstorms and tornadoes [have] become not acts of God, but acts of man."[4] Bill McKibben is one of the most influential American activists on environmental issues of the modern age, having written a number of popular books on the subject, appeared on television and radio broadcasts, and been appointed to the American Democratic Party's National Committee. He was a leader in the People's Climate March in New York and Toronto; started Step It Up '07, a nationwide environmental campaign to demand action on climate change, which became 1Sky; and he eventually founded 350.org. He was one of the chief protestors against the Canada–US Keystone XL pipeline, which proposed to bring the world's filthiest forms of oil from the tar sands in

Alberta, Canada to refineries on the Gulf of Mexico and was finally rejected by President Barrack Obama in 2015.

McKibben's main argument is that humanity in general, and America in particular, has created a model of growth which is entirely unsustainable and totally absurd. This has precipitated the now cliché notion of "too big to fail", used to excuse the irresponsible abuses of the American financial sector in the early 2000s. McKibben's response to this notion is quite simply that "Anything too big to fail is by definition too big."[5] The idea that things have become too big is central to McKibben's objectives as an environmental writer and activist. His 2010 book, *Eaarth: Making a Life on a Tough New Planet*, takes an Emersonian vision for the future of American society. Nature as a set of ideas that helps explain humanity to itself. The book outlines the chief problems: Global warming and its negative consequences are already here, not part of some future result of current practices. "Given all that we know about topics ranging from the molecular structure of carbon dioxide to the psychology of human satisfaction, we need to move decisively to rebuild our local economies."[6] The modern economic growth paradigm must be changed if not reversed and pared down. "The pain of the recession—a word that, after all, literally means getting smaller—has been entirely real, because our economy is geared to work only with growth."[7] But most of all, the colossal, global, top-down models of production and distribution—be it of energy, of food, of consumer products of every sort—need to be downsized and localized. And to this jeremiad list the conclusion seems somewhat counter intuitive: McKibben believes there is still hope. That hope resides in our "willing[ness] to embrace reality, to understand that we live on the world we live on, not the one we might wish for."[8] The one we wish for, McKibben maintains, is the American golden age of growth, from the end of the Second World War to the mid 1960s, when the myth of infinite growth was created. McKibben feels that that world simply no longer exists, and that "It is the contrast between the pace at which the physical world is changing and the pace at which the human society is reacting that constitutes the key environmental [issue] of our time."[9]

There is a palpable sense of, if not capitulation, realist frustration with what might have been possible solutions to the problems we face, and his answers at times feel to some people somewhat fanciful. McKibben has argued for one child per family, for a return to village rather than city as the primary unit of social organization coupled with the access to information that is afforded by the internet to bridge the knowledge gap of "doing for oneself and one's own." But is that being realistic? At this stage in history, for the first time, more people on the planet live in urban centres than in the countryside, and it is hard to imagine a practical way for everyone to develop relationships with their local farmers and food producers. McKibben has been criticized on a number of fronts and arguably not without justification. We must ask ourselves if giving up on what is the case and looking to what may be is either realistic or the right move. Part of the problem here is a veiled kind of despair. "Forget the grandkids; it turns out this [global warming] was a problem for our parents."[10] Knowing what we now do about accelerating environmental change, the last time that something sufficient might have been done to halt that change was likely the 1970s. "We are no longer at the point of trying to stop global warming, it's too late for that. We're at the point of trying to keep it from becoming a complete and utter calamity."[11] No action was taken, even after the warnings had been tabled, in part because the science wasn't yet there and so the will, political and otherwise, wasn't either. Unfortunately, even though the science is there now, the will still hasn't become general enough for it to matter politically. A good part of the activism in McKibben's work has been to generate and foster that will.

In the film "Do The Math," which stems from an article McKibben wrote for *Rolling Stone Magazine* called "Global Warming's Terrifying New Math" he tries to make the point in as dramatic and as simple a way as possible: The Copenhagen Accord of 2015 identified 2 degrees Celsius as the allowable limit of global temperature increase. McKibben says that "in political terms, it is the only thing that anybody's agreed to."[12] That limit has been calculated to be possible if the total amount of carbon put into the atmosphere remains below 565 billion tons. McKibben points out that this is indeed a very large number, but knowing that we currently produce about 30 billion tons a year, and that number increases by roughly 3 percent per year, we will surpass the 565 billion ton mark by 2030. He ends the analysis of "the new math" with a number calculated by the Carbon Tracker Initiative in London, which is a financial rather than environmental group, who concluded that the total amount of carbon awaiting removal to the fossil fuel industry is 2795 billion tons, which is about five times the accepted limit.[13]

Although Bill McKibben has been enormously successful at motivating people to act and at organizing activities to protest the current situation, there is one area that is more likely than any other to have an effect on the fossil fuel industry. McKibben points out that the success of the top five most profitable petrochemical companies comes from a number of factors that define them as outlaws; "Not outlaws against the laws of the state, they get to write those for the most part, but they are outlaws against the laws of physics."[14] As such the call to arms on behalf of the environment is not a call to action in the home, at work or in the lab, but a call to arms in the stock market. These companies simply will not heed any warning other than the financial one; they make too much money. If their business becomes unprofitable, it will change its direction. "The logic of divestment couldn't be simpler: if it's wrong to wreck the climate, it's wrong to profit from that wreckage ... The hope is that divestment is one way to weaken those companies—financially, but even more politically. If institutions like colleges and churches turn them into pariahs, their two-decade old chokehold on politics in DC and other capitals will start to slip."[15] The principle of divestment as a way to create and drive political change was a feature of the apartheid regime of South Africa. That, of course, requires a certain degree of reason to be at play, but dealing with corporations is something McKibben likens to misbehaving children.

> In fact, corporations are the infants of our society—they know very little except how to grow (though they're very good at that), and they howl when you set limits. Socializing them is the work of politics. It's about time we took it up again.[16]

The movement is making strides, major universities, the Catholic Church and even the Rockefeller Family, whose forebear founded Standard Oil, have divested from the fossil fuel industry, but perhaps the greatest success this movement has made comes not from the financial and oil sector at all. For McKibben, nothing will change if minds do not. And given how resistant to change society seems, it will take a new society, fueled by a new sense of place for the necessary changes to take hold. Divestment has made great inroads into the way current university students around the world relate to the administrations of their institutions. Proving to these institutions that it is in their 'corporate' best interest to move away from fossil fuel is a great step in the right direction. The steps that are becoming necessary will depend as much on that action, but "we also need a new mental model of the possible."[17]

Notes
1 *Eaarth: Making a Life on a Tough New Planet*, p. 81.
2 *The End of Nature*, p. xviii.
3 *Silent Spring*, written by Rachel Carson in 1962, is often regarded by Americans as the work that began the environmental movement.
4 *The End of Nature*, p. xviii.
5 *Eaarth*, p. 102.
6 *Deep Economy: The Wealth of Communities and the Durable Future*, p. 2.
7 *Eaarth*, p. 102.
8 Ibid. p. 9.

9   *The End of Nature*, p. xv.
10 *Eaarth*, p. 10.
11 Ibid. p. 28.
12 Nyks, Kelly & Scott, Jared P. PF Pictures (2013). *Do The Math*. Retrieved from https://youtu.be/IsIfokifwSo
13 Ibid.
14 McKibben, B. (July 19, 2012) "Global Warming's Terrifying New Math" *Rolling Stone Magazine*.
15 Nyks and Scott, op.cit.
16 McKibben, B. (Feb. 22, 2012) "The Case for Fossil-Fuel Divestment" *Rolling Stone Magazine*.
17 *The Bill McKibben Reader*, p. 171.

See also in this book
Carson, Emerson, Gore, Hunter, Sukhdev

McKibben's major writings

The End of Nature, New York: Random House, 1989.

Maybe One: A Personal and Environmental Argument for Single-child Families, New York: Simon & Schuster, 1998.

Deep Economy: The Wealth of Communities and the Durable Future, New York: Holt, 2007.

Eaarth: Making a Life on a Tough New Planet, New York: Holt, 2010.

Oil and Honey: The Education of an Unlikely Activist, New York: Holt, 2013.

Further reading

Carson, R. (1987). *Silent Spring* (25th anniversary ed.), Boston, MA: Houghton Mifflin.

Emerson, R. W. (1961). *Emerson's Essays*, New York: Crowell, 1844.

Grady-Benson, J., & Sarathy, B. (2016). "Fossil Fuel Divestment in US Higher Education: Student-led Organising for Climate Justice". Local Environment, 21(6), 661–81.

Kirk, A. G. (2007). *Counterculture Green: The Whole Earth Eatalog and American Environmentalism*, Lawrence, KS: University Press of Kansas.

McKibben, B. (2008). *The Bill McKibben Reader: Pieces from an Active Life*, New York: Holt.

THOMAS E. HART

PAVAN SUKHDEV 1960–

By assigning economic values to the services flowing from nature to people, policy makers and the global economy can start to account for the costs of biodiversity loss, as well as reward responsible custodians for the benefits that natural ecosystems provide. This will also help conserve what is left of natural capital as a societal asset for future generations, rather than burn it up in a frenzy of GDP growth-fixated policies.[1]

Pavan Sukhdev, born in Delhi, India, in 1960, built a career in the financial sector, holding several leading positions at Deutsche Bank, including head of their global emerging markets division in London. But by 2003 Sukhdev had developed an interest in green national accounting, and helped establish the Green India States Trust (GIST), which produced a series of influential reports that adjusted national and state gross domestic product (GDP) accounts to reflect the economically 'invisible' benefits of nature and the hidden costs of its degradation. In 2007, Germany and the European Union launched the 'TEEB' initiative (The Economics of Ecosystems and Biodiversity) to analyse the global economic benefit of biodiversity, the costs of its loss, and the failure to take protective measures versus the costs of effective conservation. Sukhdev was appointed to lead the initiative and lead authored its synthesis report in 2010. This, together with his role as head of the United Nations Environment Programme's Green Economy Initiative and his leading of its report 'Towards a Green Economy', and his numerous popular articles, has made Sukhdev one of the most significant defenders of the value of an economic approach to nature conservation. While his career in international banking gives him an atypical profile for a key environmental thinker, it has arguably made business and policy leaders more receptive to his arguments.

There exists an extensive debate on environmental problems within mainstream economics which has led to the establishment of an important sub-discipline, namely environmental economics. The theoretical foundations for it were laid down by some of the main twentieth-century economists, such as Pigou, Hicks, Kaldor, Hotelling, Dasguptha, Coase, Solow and Arrow.[2] While Pavan Sukhdev is, therefore, by no means the intellectual father of the idea of the economic valuation of nature as such, he, and the projects and initiatives he has led, have played a substantial role with regard to the spreading of the idea in non-academic circles, particularly among business and policy leaders. Sukhdev's contribution may be compared in certain respects to the Brundtland report, which similarly was not the first to propose environmentally friendly growth but gave this idea a much wider audience through introducing the notion of sustainable development. Another comparison may be made to what Nicholas Stern's 2006 report did for the issue of climate change. Like Stern, Sukhdev brought his issue of biodiversity loss and ecosystem degradation to new prominence by rigorously advancing the case for the economic costs of a business-as-usual approach.

Sukhdev was once posed the question by a friend, 'Why are some things worth money and other things not?'[3] As Sukhdev recalls, 'I understood her question, but had no answer! That bothered me, so I started off on a quest for an answer.'[4] In the course of reflecting on fundamental economic theory concerning externalities, he discovered one of the pioneering works in environmental economics, *The Blueprint for a Green Economy* by David Pearce, who he cites as one of his inspirations.[5] Sukhdev arrived at the standard answer as to why many ecosystems and much biodiversity are 'not worth money': their status as public goods. A public good has the following two features: (i) they are non-excludable, i.e. it is difficult to exclude people from using them; and (ii) they are non-rival, i.e. one person using it does not reduce the amount available for another. If it is impossible to exclude users from a good, then it cannot be traded in markets and therefore will have no price and will be treated as available for free. In other words, 'natural capital' – as Sukhdev, in common with other environmental economists, refers to biodiversity and ecosystems – is economically invisible. Therefore, since the problem lies in the fact that zero value is attached to natural capital, the solution offered by environmental economists is to place a monetary value on it. The TEEB initiative, to which Sukhdev was appointed head in 2008, involves hundreds of researchers in a global study to put a monetary value on the societal benefits of ecosystems and biodiversity and the social consequences of their loss and degradation, and so make them economically visible. It calculated that the annual loss of land-based natural capital is between US\$2 trillion and US\$4.5 trillion,[6] which constitutes a significant proportion of global GDP, estimated as US\$63 trillion in 2008.[7]

Sukhdev proposes a range of answers to address the economic invisibility of nature's value. First, he advocates the 'greening' of national GDP accounts, and his GIST project did exactly this for India. It calculated, for example, that the losses to just the nation's forest ecosystem services, such as erosion and flood prevention, over the period 2001–2003 amounted to a loss of 1.1 per cent of national GDP, with up to 6 per cent loss in some regions.[8] Taking such negative growth into account presents a very different image of India's economic performance. What economies are doing is consuming nature's capital, rather than living on the interest it generates. It is these potential economic costs that are relevant when assessing the opportunity costs of conservation. Sukhdev cites Balmford et al. (2000), who claim that the annual US$45 billion cost of protecting 15 per cent of the land and 30 per cent of the sea would be a hundred times less than the estimated annual value of the goods and services such protection would deliver, namely $4.5–$5.2 trillion.[9]

Second, Sukhdev proposes that part of the solution for capturing nature's economic value is through designing market-based instruments with incentives and price signals. Here, much current debate is about payments for ecosystem services (PES). A famous example of PES is to be found in Costa Rica: transfers funded by transportation taxes are paid to farmers who preserve forest patches in their land, thereby benefitting all those who enjoy the public benefits of these forests. Sukhdev is aware that such PES schemes are no panacea. For instance, payments may go to areas where the risks of deforestation were low already. Nonetheless, he sees such economic instruments as a prominent way to tackle environmental degradation. For instance, Sukhdev holds up the Reducing Emissions from Deforestation and Forest Degradation (REDD+) programme – whereby payments from the developed world are made to replant forests as carbon stocks in less developed countries – as a promising initiative. As well as these market-based instruments Sukhdev also argues for a broad range of non-market measures, including environmental regulation, public investment in green infrastructure and community-based conservation.[10]

There are two ways in which Sukhdev deviates to a certain extent from the mainstream environmental-economic storyline. First, he maintains a focus on the crucial importance of natural capital for the poor throughout his work. Nature-based production is crucial for the livelihoods of poorer households, which are almost always hit hardest by the mismanagement of environmental resources because they so heavily rely on them. Sukhdev's GIST reports showed that 'although the value of forest services, such as fresh water, soil nutrients and non-timber forest products only was around 7% of national GDP, it amounted to some 57% of the income of India's rural poor people.'[11] Secondly, he has more recently focussed on the importance of the corporate sector for achieving environmental objectives. For dealing with global commons problems, such as climate change and biodiversity loss, we tend to look at intergovernmental organisations and treaties, such as the United Nations Framework Convention on Climate Change and the Convention on Biological Diversity. However, so far their results have been limited. Therefore, Sukhdev argues, in order to realise effective climate change or biodiversity solutions one must recognise the crucial role of the private sector. The private sector accounts for 60% of the global GDP. However, it also creates substantial externalities: 'in 2008, annual corporate "externalities" – the costs to society of the leading 3,000 public companies globally, in the form of emissions, freshwater use, pollution, waste and land-use change – added up to US$2.15 trillion.'[12] This amounts to 3.6% of global GDP,[13] or a third of these companies' combined profits.[14] But their worth is only calculated in relation to their shareholder-owned financial capital, thereby ignoring all other groups of stakeholders. Sukhdev argues for redefining corporate success, for example through regulations which require they disclose the cost of their externalities.

Although Sukhdev proposes government regulation and community-based solutions for managing many environmental problems, insofar as we are examining his arguments as representative of an approach to environmental problems that centrally involves monetary valuation techniques and the creation of new markets, we will also outline three criticisms of this broadly economic approach. Before this, however, it is worth mentioning two more general sources of unease about the economic approach. First, there is general scepticism among sections of the environmental movement that, after decades of treating capitalism as one of the central sources of environmental problems, the answer is to expand it further into the environmental sphere and become, as Sukhdev characterises it, 'total capitalists'.[15] Secondly, while many environmental economic decision tools do not involve the creation of actual markets, there is a 'slippery slope' concern that the expansion of market concepts and norms – utilitarianism, value commensurability, monetisation for the purposes of valuation and decision making – into the environmental sphere makes the further and potentially problematic steps of commodification, marketisation and financialisation more likely.[16]

There appear to be at least four more substantive criticisms. First, if environmental goods are commodified, there might be a fairness problem since poor people have less means to acquire these goods.[17] Second, many have argued that it is impossible to value different environmental goods using a single scale. Comparing goods is always done with regard to a specific comparative value (such as economic benefit, beauty or health). Since there is no overall comparative value (being simultaneously better on all dimensions), converting all goods to a single scale can only be done by favouring one comparative value (reductionism). This idea is heavily discussed in the debate on value (in)commensurability.[18] In addition, if this single scale is operationalised as a monetary scale, this implies favouring an economic perspective, excluding valuations, based on a plurality of reasons, that are either hard or inappropriate to express in monetary terms.[19] The third line of criticism states that even if monetary valuation could provide a good estimate of individuals' valuations of nature, it does not necessarily provide the relevant criterion for environmental policy design. Policies should be based on what people deem in the public interest (citizen perspective, reasons) rather than (or next to) the sum of costs and private interests (consumer perspective, preferences).[20] Alternative, non-monetary valuation methods such as participatory and deliberative techniques are seen as performing better in enabling the articulation of the plurality of environmental values and reasons to preserve them.[21] Fourthly, there is some evidence that the introduction of market-based instruments such as payments for ecosystem services schemes can have a transformative effect on environmental motivations. Such price incentives might push aside – 'crowd out' – intrinsic motivation for environmental action and decrease the desired pro-environmental behaviour.[22] Together these criticisms have led to the rise of a new, more interdisciplinary field, namely ecological economics, as a reaction to the – in their view – narrow neoclassical basis of environmental economics.[23]

Sukhdev's motivation to campaign for the recognition of nature's value has similar sources to many of the thinkers in this book: long treks in nature with his family in his childhood, a life-changing holiday in India's Kaziranga National Park with his newly wed wife and birdwatching with his eldest daughter.[24] However, it is clear that he believes focussing on the economic arguments has the potential to effect behavioural and policy change where previous arguments have failed. Indeed, through GIST, TEEB, the Green Economy Initiative and now Corporation 2020, Sukhdev and his colleagues have shown that the economic arguments for nature conservation are very strong, and he has been very successful in making the economic valuation of nature argument persuasive and accessible. Because of this the arguments for the cost-effectiveness of conservation have reached corporate, financial and political circles that had previously been sceptical of or resistant to conservation. It might be that it is here that the most substantial benefits of Sukhdev's economic arguments is to be found, namely in persuading those who have seen nature as a resource to be (unsustainably) exploited, and have tended to put a zero value on conserved nature.

Notes
1 Sukhdev, Pavan 2011. 'Putting a Price on Nature: The Economic of Ecosystems and Biodiversity', *Solutions*, 1: 34–43.
2 Pearce, David 2002. 'An Intellectual History of Environmental Economics', *Annual Review of Energy and the Environment*, 27: 57–81.

3 Sahgal, Bittu 2008. 'Meet Pavan Sukhdev', in *Sanctuary Asia*, Vol. 27(1), www.sanctuaryasia.com/component/content/article/136-interviews/1584-meet-pavan-sukhdev.html
4 Ibid.
5 Ibid.
6 Sukhdev, Pavan 2011. 'Putting a Price on Nature: The Economic of Ecosystems and Biodiversity', *Solutions*, 1: 34–43.
7 World Bank data on GDP, http://data.worldbank.org/indicator/NY.GDP.MKTP.CD
8 Sukhdev, Pavan & Justine Leigh-Bell 2012. 'Importance of Green Accounting', The Hindu Business Line, June 22, www.thehindubusinessline.com/opinion/importance-of-green-accounting/article3559301.ece
9 Balmford, Andrew et al. 2002. 'Economic reasons for conserving wild nature', *Science* 297: 950–953, cited in Sukhdev, Pavan 2011. 'Putting a Price on Nature: The Economics of Ecosystems and Biodiversity', *Solutions*, 1: 34–43.
10 Sukhdev, Pavan 2011. 'Putting a Price on Nature: The Economics of Ecosystems and Biodiversity', *Solutions*, 1: 34–43.
11 Sukhdev, Pavan 2009. 'Costing the Earth', *Nature*, 462: 277.
12 Sukhdev, Pavan 2012. 'The Corporate Climate Overhaul', *Nature*, 486: 27–28.
13 Ibid.
14 Karunakaran, Naren 2010. 'Nature In Numbers'. *The Economic Times*, 30: 11.
15 Sukhdev, Pavan 2011. 'Three-dimensional capitalism'. *The Guardian*, www.theguardian.com/sustainable-business/blog/three-dimensional-capitalism-market-economy
16 Kill, Jutta 2015. 'Economic Valuation and Payment for Environmental Services: Recognizing Nature's Value or Pricing Nature's Destruction?'. Heinrich Böll Foundation.
17 Martinez-Alier, Joan 2002. *The Environmentalism of the Poor: A Study of Ecological Conflicts and Valuation*. Cheltenham, UK: Edward Elgar.
18 Anderson, Elizabeth 1993. *Value in Ethics and Economics*. Cambridge, MA: Harvard University Press.
19 O'Neill, John 1993. *Ecology, Policy and Politics*. London: Routledge.
20 Sagoff, Mark 2008. *The Economy of the Earth: Philosophy, Law, and the Environment*. 2nd edn; Cambridge, UK: Cambridge University Press; O'Neill, John 1993. *Ecology, Policy and Politics*. London: Routledge.
21 Smith, Graham 2003. Deliberative Democracy and the Environment. London: Routledge; Spash, C.L., 2007. 'Deliberative monetary valuation (DMV): issues in combining economic and political processes to value environmental change'. *Ecolological Economics*, 63: 690–699.
22 Frey, B.S., A. Stutzer 2008. 'Environmental morale and motivation' in Lewis, A. (ed.), *The Cambridge Handbook of Psychology and Economic Behaviour*. Cambridge, UK: Cambridge University Press, pp. 406–428; Rode, Julian, Erik Goméz-Baggethun and Torsten Krause 2015. 'Motivation crowding by economic incentives in conservation policy: A review of the empirical evidence', *Ecological Economics*, 117: 270–282.
23 Van den Bergh, Jeroen 2001. 'Ecological Economics: Themes, Approaches, and Differences with Environmental Economics', *Regional Environmental Change*, 2: 13–23.

24 Sahgal, Bittu 2008. 'Meet Pavan Sukhdev', in *Sancuary Asia*, 27(1). Available atwww.sanctuaryasia.com/component/content/article/136-interviews/1584-meet-pavan-sukhdev.html

See also in this book
Ehrlich, Schumacher

Sukhdev's major writings

UNEP (United Nations Environment Programme) 2011. *Towards a Green Economy: Pathways to Sustainable Development and Poverty Eradication, Synthesis for Policy Makers.* Nairobi: UNEP, http://web.unep.org/greeneconomy/resources/green-economy-report

TEEB 2010. *The Economics of Ecosystems and Biodiversity: Mainstreaming the Economics of Nature: A synthesis of the approach, conclusions and recommendations of TEEB. Corporation 2020. Transforming Business for Tomorrow's World.* London: Island Press, 2012.

Further reading

Daly, Herman E. & Farley, Joshua 2004. *Ecological Economics, Second Edition: Principles and Applications.* Washington, DC: Island Press.

Dasguptha, Partha, 2001. *Human Well-Being and the Natural Environment.* Oxford, UK: Oxford University Press.

Pearce, David, Markandya, A., & Barbier, E. 1989. *Blueprint for a green economy.* London: Earthscan.

Stern, Nicholas H. 2007. *The economics of climate change: The Stern review.* Cambridge, UK: Cambridge University Press.

PAUL KNIGHTS and STIJN NEUTELEERS